AF517714

Crop Production in Salt Affected Soils

The Authors

Dr. I.C. Gupta (b. 1941), was awarded Ph.D. on a subject thesis of environmental soil science by CAZRI/University of Jodhpur in 1968. He worked at Central Soil Salinity Research Institute from 1970 to 1986 (Principal Scientist). Dr. I.C. Gupta served Central Arid Zone Research Institute from 1986 to 2001 where he worked as Head of Regional Research Station, Bikaner and Head, Division of Natural Resources and Environment. A Fellow of Indian Water Works Association, United Writers Association of India & Indian Society of Salinity Research Scientists, Dr. Gupta was elected Vice-Chairman of Arid Zone Research Association of India. He has been working as Secretary of Indian Society of Salinity Research Scientists & Chief Editor Current Agriculture since 1977. Dr. Gupta had been conferred several awards notable being Glaxo Scientific Honours Club, Dr. Rajendra Prasad Puraskar, First Prize, Outstanding Book Award by ISAE, Dr. Gorakh Prasad Puraskar First Prize, S.P. Unvala Memorial Prize by IWWA and Life Achievement Award by IARS. He was nominated as member of the 'Environment Conservation Council of Rajasthan' and of 'Special Committee on Environment Impacts' of INCID, Ministry of Water Resources, Govt. of India. He has about 250 publications including 30 books and two inventions (Jaltripti and Jalshuddhi). Dr. Gupta was appointed Ph.D. examiner by ten universities. He was also nominated for one of the most INTRIGUING PEOPLE of 2003 by American Biographical Institute, North Carolina, USA, after close scrutiny of biographical archives of over one million individuals.

Dr. S K Gupta (b. 1949), obtained B. Tech. and M. Tech (Agril. Engg.) from the Punjab Agricultural University, Ludhiana, in 1970 and 1976 respectively and Ph.D. in Civil Engineering from JLN Technological University, Hyderabad, Telangana, India in 1984. After serving CSSRI for about 40 years, Dr. Gupta superannuated as Project Coordinator (AICRP on Management of Salt Affected Soils and Use of Saline Water in Agriculture). Dr. Gupta has been conducting researches in drainage for land reclamation, irrigation water management, hydrology, and use of saline water in agriculture. He has published more than 150 research papers besides publishing 15 books. Currently, he is an independent Researcher. Dr. Gupta is Fellow of the National Academy of Agricultural Sciences, Indian National Academy of Engineering and Indian Society of Agricultural Engineering. Dr. Gupta has been consultant to WAPCOS-Louis Berger, MoWR, MoRD, ActionAid International, Synergics Hydro India-Oromia Water Works Enterprise, Ethiopia and others. He has been bestowed with Rafi Ahmad Kidwai Award by ICAR. Besides he is recipient of several awards from NAAS, ISAE, IE (India), CBIP and MoWR among other organizations. He is the Chief Editor of Journal of Water Management of the Indian Society of Water Management.

Crop Production in Salt Affected Soils

Second Revised & Enlarged Edition

I.C. GUPTA, Ph.D.

Formerly Principal Scientist,
CSSRI, Karnal
Head Regional Research Station, Bikaner
Head, Division of Natural Resources and Environment
Central Arid Zone Research Institute, Jodhpur

S.K. GUPTA, Ph.D.

Formerly Principal Scientist and Head,
Division of Irrigation and Drainage Engineering,
Project Coordinator
INAE Distinguished Professor
Central Soil Salinity Research Institute, Karnal

Published by

SCIENTIFIC PUBLISHERS (INDIA)
5 A, New Pali Road
P.O. Box 91
Jodhpur - 342 001 INDIA
E-mail: info@scientificpub.com
Website: http://www.scientificpub.com

Print: 2019

ISBN: 978-93-88812-30-6 (Paperbound)
ISBN: 978-93-87991-58-3 (E-book)

Printed in India

PREFACE

Salt affected soils have been a part of the Indian land scape from time immemorial. Reference to *Usar* land has been made in the Epic Ramayana written by poet saint Tulsidas in *Vikram Samvat* 1631 (1574 CE). At one place he compares some of the issues that have little possibilities to happen with sowing of seeds in *usar* land having no possibility of germination. He says: ´ऊसर बीज बएँ फल जथा´ meaning ऊसर में बीज (बोने की भाँति यह सब) व्यर्थ जाता है. With the first reported incidence of water logging and soil salinity as a result of irrigation in the western Jamuna canal command in Haryana in 1855, we have travelled a long way. Currently, 6.73 million hectare (M ha) area is known to be affected by salts where nothing can be grown. On the other hand, these soils are potentially productive but for the excess accumulation of salts. If the salts are removed or one can perfect the art of living with the salts (manage), these soils can ensure livelihood to nearly 7 million farm families (assuming an average holding size of 1 ha) and contribute to the extent of 55 million tons to the national food grains kitty.

Systematic studies on reclamation and management of salt affected soils have been carried out for more than 150 years. Tremendous advances however, have only been made during the last five decades especially following the establishment of CSSRI, Karnal. Realizing the emerging challenges, land reclamation is a major thrust area of research in India. Moreover, to prepare the young graduates to learn the art of dealing with the daunting task of land reclamation, this subject is taught in all the state agricultural universities and deemed universities. The current publication titled ´Crop Production in Salt Affected Soils´ provides a comprehensive treatise of the fundamentals of soil salinization-desalinization and reclamation and management of salt affected soils. It also describes crop production technologies to green these white manmade deserts. Our endeavor has been to make it a comprehensive text book for the students of agriculture and agricultural engineering. Besides, our efforts are directed to make this publication a handy practical guide for the personnel of the line departments such as water resources, agriculture, irrigation, private enterprises and non-government organizations who are called upon to implement projects related to reclamation and management of salt affected soils.

The book has been divided into 16 chapters. Although we have not categorized the chapters in different sections, broadly they can be grouped as under:

First two chapters are the introductory chapters that discuss the categorization of salt affected soils, their extent and distribution and physical, chemical and biological properties of these soils.

Next eight chapters describe the basic principles and practices that go into any land reclamation program. These chapters include on-farm development, hydrology, irrigation practices, drainage methods, leaching and soil salinization, chemical amendments, new innovative techniques of land reclamation and agronomic and cultural practices. Since segregating the water quality problem from that of saline and alkali land reclamation is almost impossible, chapter eleven describes the characteristics of saline/alkali water, their distribution and management options.

Crop production practices for select cereal, oil seeds, sugar, fiber and forage including green manure crops are included in chapters twelve to fourteen. Cultivation of forage grasses and potential of forest plantations on salt affected soils is subject matter of discussion of chapter fifteen.

Chapter sixteen covers the economic evaluation and social issues involved in any land reclamation program. In all the chapters wherever relevant, illustrations and solved examples have been included for the benefit of the students. Glossary of commonly used terms has also been included.

The textbook is written in a very clear and easy-to-understand language by a team of two researchers, a soil scientist and an agricultural engineer. Both of them remained actively associated with land reclamation activities throughout their service career.

Thus, besides the basic concepts, field examples are included to explain the impacts of various interventions.

The authors place on records the contributions of CSSRI and several state agricultural universities for their contributions in developing technologies to reclaim salt affected soils and manage saline water for irrigation. Few of the SAUs contributing significantly to this field and having close working relationship in one form or the other with CSSRI are: PAU, Ludhiana, CCSHAU, Hisar, SK Bikaner Agricultural University, Bikaner, Rajmata Vijayaraje Scindia Krishi Vishwavidyalaya, Gwalior, Professor Jayashankar Telangana State Agricultural University, Hyderabad, Acharya NG Ranga Agricultural University, Lam, Guntur, Tamil Nadu Agricultural University, Coimbatore and University of Agricultural Sciences, Raichur. Authors are also grateful to many of their colleagues in the profession, who have given their inputs on many of the issues through suggestions, supply of their publications and discussions during preparation of the book. Special thanks are due to Dr. S.K. Chaudhary, ADG, ICAR, New Delhi, Dr. P.C. Sharma and Dr. D.K. Sharma, current and former directors respectively and Dr. D.S. Bundela, Head Division of Irrigation and Drainage Engineering, and Sh. Madan Singh of the Central Soil Salinity Research Institute, Karnal for active cooperation. Only because of this guidance and support, it has been possible to collect and synthesize the available information included in this book. The authors at the end express their heartfelt thanks to their respective family members. Without their full support, it might not have been possible to bring this project to its logical end.

The authors believe that the current publication will serve to meet the widely felt need of a comprehensive textbook and guide on land reclamation. We hope that it will find a place in the shelves of libraries of all the universities and personal collection of teachers, researchers and field functionaries. We will highly appreciate the feedback and contributions in the form of case studies or research results that might help to enrich the future editions of this textbook.

I.C. Gupta
S.K. Gupta

CONTENTS

CHAPTER 1

SALT AFFECTED SOILS: EXTENT AND DISTRIBUTION

HISTORICAL BACKGROUND

During the Vedic period (c 2000-600 B.C.), the peasants dug wells and constructed canals to supply water to the crops. Until this period, there are no records to show that there has been a rise in the salinity in these irrigated tracts. The large scale development of canal irrigation began in the second half of the 14th Century at the initiative of Sultan Feroz Shah Tughlak (1351-88). During his reign, extensive irrigation works were planned and executed to stabilize the agricultural economy. Five big canals were dug to distribute the water of the Jhelum, Yamuna and the Sutlej for irrigation, of which the most important was the Western Yamuna Canal, the oldest canal of north India. It carried the waters of the Yamuna to the orchards of Hansi (Haryana) and additionally provided irrigation to the limited cultivated areas in what are today known as Karnal, Rohtak, Hisar and Jind districts. The system as abandoned later but canal irrigation in the vicinities of Delhi got an impetus during the times of the Sher Shah Suri and later, during the reign of Akbar (1556-1605). Even at that time, the proportion of the areas irrigated by canals relative to that watered by wells was minimal. The salty patches in the soil, said to have developed due to canal irrigation, disappeared when the canals went into the disuse in the year 1750, after the fall of the Mughal Empire. In the primitive system of well irrigation, there was no appreciable increase in *reh* (salt affected) lands. The Western Yamnua Canal (WYC) was again commissioned in 1821, and re-opened for regular irrigation in 1839. The construction of the Eastern Yamnua Canal also began in 1820 and was opened in 1830. The Ganges Canal was commissioned in 1855. Soil salinization was reported within 10-15 years after opening of these system as well as some systems in Punjab.

The first reported incident of land degradation due to salts is from village Munak now in Haryana on the banks of WYC. Alarmed by the progressive deterioration of the land fed by the Western Yamuna Canal, the then Government of Punjab felt it necessary for the first time in 1858 to draw the attention of the then Governor General to the seriousness of the problem of the *reh* or *usar* (salt affected) tracts. The Chief Commissioner of Punjab reported, ``For some time past, it has been known that many villages on the banks of this canal in the Panipat, Delhi and Rohtak districts have been suffering from a destructive saline efflorescence. The accounts of poverty in some of these villages have been quite distressing. At present, about 60 villages, or one tenth of the total number of canal villages, are affected. The mischief is increasing yearly and it will soon attain a very considerable proportion and entail a fiscal loss to the government and great suffering to the people" (Whitcomb, 1971).

A petition was made in 1876, by David Roberts, an indigo planter in the Aligarh district of Uttar Pradesh, to the Board of Revenue for relief on the grounds that his lands, in many parts of Meerut, Aligarh and the *Kali Nadi* valley had deteriorated owing to the appearance of saline efflorescence, as a result of the introduction of the Ganges Canal irrigation (Agarwal and Gupta, 1968).

Voelcker (1897), who toured the country in 1891, found enormous tracts affected by salts especially in the plains of North India. According to him, in the North-Western Province (now Uttar Pradesh) alone, the salinity affected area was estimated to cover altogether 10,000-12,000 km^2. In this desolation, the patches of crops stood out, isolated in clumps, some in miserably poor condition like oases in the salt covered desert around them (Whitcomb, 1971).

Ever since then, the development of irrigated agriculture in the country has been taking place at a much faster rate, especially after 1950-51, the beginning of the era of planned development. As per the estimates, ultimately irrigation potential of India earlier assessed at 113 M ha, now stands revised to 140 M ha (CWC, 2013): 58.5 M ha through major and medium irrigation and 81.43 M ha through minor irrigation (Table 1.1). Irrigation potential created increased from 22.6 M ha in 1951 to 108.21 M ha by March 2010, i.e. at the end of the first three years of 11th plan. However, this achievement was not unblemished. The inadequate development of irrigation facilities at the project level and the poor water management at the farm level have been responsible for the high wastage of irrigation water, the low irrigation efficiency, low crop yield, water logging, salinity and alkalinity, and also various other environmental hazards. One of the inevitable consequence has been the rising water table in irrigation commands resulting in soil salinization since water logging and soil salinity are often called twins, the later following the former. It may not be out of place to mention that as per current assessment degraded and wastelands have been assessed to cover 114.01 M ha out of which salt affected soil constitute 6.73 M ha and waterlogged 6.41 M ha. Within water logging 1.66 M ha suffers from surface ponding while 4.75 M ha from subsurface water logging (ICAR-NAAS, 2010).

Table 1.1. Ultimate irrigation potential and water resource availability in India

Sr.No.	Sector	Irrigation Commission (2nd) (1972)	Current assessment
1.	Major and medium Irrigation	58.47	58.47
2.	Minor irrigation	55.00	81.43
a.	Surface water	15.00	17.38
b.	Groundwater	40.00	64.05
3	Total	113.47	139.90

ORIGIN AND FORMATION

Occurrence of saline and alkali soils in India can be traced back even to periods before the advent of canal irrigation, yet often it is linked with the introduction of canal irrigation. It is believed that introduction of canal irrigation exuberates the extent and degree of these soils in the landscape. However, it is now accepted that it is the one and not the only cause of the formation of such soils. Consequent upon the introduction of the canal irrigation, extensive seepage and leakage from the network, on-farm poor water management and impeded drainage resulted in the inevitable consequence of rise of the water table resulting in shallow water table causing water logging. Water logging is caused both by natural and manmade factors. Some of the factors responsible for water logging are as follows:

- Poor natural drainage as a consequence of unfavorable topography or unfavorable sub-soil geology like existence of hardpan at shallow depths
- Heavy storm rainfall coupled with poor natural drainage
- Spilling of rivers resulting in submergence of agricultural lands
- Introduction of irrigation without taking into account characteristics of soils and sub-soils for their irrigability
- Heavy losses of water due to seepage from canals, distributaries and watercourses
- Poor on-farm water management resulting in poor application efficiencies
- Development activities such as construction of roads, bridges, railway lines and buildings resulting in choking of natural drainage
- Poor maintenance of existing drainage system and outlets
- Deforestation and poor upkeep of watersheds
- Hydraulic pressure of water from upper irrigated areas resulting in seepage outcrop in low lying areas.

A combination of factors, which are mainly geological, climatic and hydrologic in nature, is usually involved in the formation of saline and alkali soils (Agarwal *et al.*, 1979).

(i) **Geological factors** - weathering of rocks and decomposition of soil minerals;

(ii) **Climatic factors** - arid and semi-arid climate, high evapotranspiration;

(iii) **Hydrologic factors -** shallow water table, surface runoff and evaporation of drainage water in depressions;

(iv) **Groundwater related factors**: mineralized groundwater associated with shallow water tables; and irrigation with saline waters.

EXTENT OF THE PROBLEM

The salt affected soils are found in all the continents of the globe, especially in arid and semi-arid zones and in coastal ecologies. The data on the extent of salt affected soils in the world as shown in the UNESCO/FAO world soil map and in the map of the salt affected soils of Europe (both in the scale 1 : 5,000,000) reveal that an area of 952 M ha has been affected by soil salinization (Szabolcs, 1977). Later Szabolcs (1989) described the global distribution of saline and sodic soils in different continents that sum up to 932 M ha (Table 1.2). According to a more recent report published by FAO in 2000, the total global area of salt affected soils has been pegged at 831 M ha of which saline soils constitute 397 M ha and sodic soils 434 M ha (Martinez-Beltran and Manzur, 2005). It also states that of the current 230 M ha of irrigated land, 45 M ha is salt affected (19.5%), and of the 1500 M ha under dry land agriculture, 32 M ha is salt affected to varying degrees (2.1%).

Table 1.2. Extent of salt affected soils in the world

Continent	Area (M ha)		
	Saline	**Sodic**	**Total**
North America	6.2	9.6	15.8
Central America	2.0	–	2.0
South America	69.4	59.6	129.0
Africa	53.5	27.0	80.5
South Asia	83.3	1.8	85.1
North and Central Asia	91.6	120.1	211.7
Southeast Asia	20.0	–	20.0
Europe	7.8	22.9	30.7
Australasia	17.4	340.0	357.4
Total	351.5	581.0	932.2

The distribution of salt affected soils in India is quite extensive occurring in almost all climatic zones. These soils occur principally in the following six areas:

(i) Throughout the Indus valley

(ii) The valleys and basins of western India

(iii) The Ganga valley west of about 80ºE longitude

(iv) The uplands of the Deccan plateau especially between the Tapti, the Godavari and the Bhima rivers

(v) The saline marshes of the sea coast and of the deltas of the Ganga, the Cauvery and the Mahanadi; and

(vi) The coastal salt flats along the Rann of Kutch.

It is estimated that in India, about 6.73 M ha area is affected by salts. The state wise distribution of the salt affected soils is reported in Table 1.3 (Mondal, 2011). The extent of the area is highest in Gujarat followed by Uttar Pradesh.

NOMENCLATURE AND CLASSIFICATION

No uniform nomenclature is followed in different parts of the world. It is because of the wide variations in characteristics of salt affected soils from one region to another. As such, different approaches are being followed in different countries to classify the salt affected soils. India opted to follow the United States Department of Agriculture (USDA) classification and therefore, the discussions are restricted to the USDA classification followed by the classification criteria adopted in India.

USDA Classification

The United States Salinity Laboratory Staff (Richards *et al.*, 1954) divided the salt affected soils into three main classes, viz. (i) saline soils (ii) saline alkali soils, and (iii) non-saline alkali soils, primarily on the basis of EC_e, ESP and pH_s. EC_e (Electric conductivity of the saturation extract of the soil) is defined as the reciprocal of the electrical resistance of the extract of the soil which is one centimeter long and has a cross sectional area of 1 cm^2. It is

Table 1.3. Extent of salt affected soils in India (ha)

State	Saline	Alkali/Sodic	Coastal saline*	Total
Andhra Pradesh	0	196609	77598	274207
Andaman & Nicobar Island	0	0	77000	77000
Bihar	47301	105852	0	153153
Gujarat	1218255	541430	462315	2222000
Haryana	49157	183399	0	232556
Karnataka	1307	148136	586	150029
Kerala	0	0	20000	20000
Madhya Pradesh	0	139720	0	139720
Maharashtra	177093	422670	6996	606759
Odisha	0	0	147138	147138
Punjab	0	151717	0	151717
Rajasthan	195571	179371	0	374942
Tamil Nadu	0	354784	13231	368015
Uttar Pradesh	21989	1346971	0	1368960
West Bengal	0	0	441272	441272
Total (ha)	1710673	3770659	1246136	6727468
Total (M ha)	1.71	3.77	1.25	6.73

*There is mismatch between the figures quoted since the extent as high as 3.1 M ha for coastal saline soils are being reported.

generally expressed as dS/m (deci-siemens per meter) at 25ºC and is used to express the salinity of the soil. ESP (exchangeable sodium percentage) is defined as the degree of saturation of the soil exchange complex with sodium. Mathematically, it is expressed as:

$$ESP = \frac{\text{Exchangeable sodium (me/100g)}}{\text{Cation exchange capacity (me/100g)}} \times 100 \qquad (1.1)$$

The pH_s (pH of saturated soil paste), is defined as the negative logarithm of the hydrogen ion activity of the soil paste and is used to distinguish the relative degree of acidity or alkalinity of the soil. The classification is reproduced in Table 1.4.

Table 1.4. Classification of salt affected soils

Class of the soil	Limiting values		
	EC_e (dS/m)	ESP	pH_s
Saline	> 4	< 15	<8.5
Saline alkali	>4	> 15	> 8.5
Non-saline alkali	< 4	> 15	>8.5

The saline soils were sub-divided into four classes, viz. (a) salt free (EC_e < 4 dS/m, total soluble salts (TSS) < 0.15%), the salinity effect being negligible except for the most sensitive plants; (b) slightly saline (EC_e 4-8 dS/m, TSS 0.15 - 0.35%), the yields of many crops restricted; (c) moderately saline (EC_e 8-15 dS/m, TSS 0.35 - 0.65%, only tolerant crops grow; (d) strongly saline (EC_e > 15 dS/m, TSS > 0.65%); only very tolerant crops grow.

In terms of ESP, Branson (1960) listed the critical limits as < 10, 10-20, 20-40, 40-60 and >60 for very sensitive, sensitive, moderately tolerant, tolerant and very tolerant crops.

The Soil Science Society of America (1974) made several modifications to the definition and classification of salt affected soils. Alkali soils are now categorized as sodic soils, because the high pH of these soils is because of excess sodium on the soil exchange complex. Besides, realizing that in most soils studied in USA, SAR and ESP are closely related, the soils are now classified on the basis of SAR rather than ESP, and EC_e omitting the pH for classification. Since SAR of 13 roughly corresponds to ESP of 15, SAR of 13 was used as criteria to classify soils. It eased the process of classification since SAR is easier to determine in the laboratory than ESP. SAR (sodium adsorption ratio) is defined as a ratio for soil extracts used to express the relative activity of sodium ions in exchange reactions with the soil. It is calculated by the formula

$$SAR = \frac{Na^{+}}{\sqrt{\frac{Ca^{++} + Mg^{++}}{2}}} \qquad (1.2)$$

Here the ionic concentrations are expressed as milli-equivalents per liter (me/l). The critical limits of EC_e were also changed from 4 to 2 dS/m. This seems to be more logical as the growth of most sensitive plants, particularly of the horticultural and pulse crops begin to experience the adverse effect of salinity at an EC_e greater than 2 dS/m. The revised classification is given in Table 1.5.

Table 1.5. Revised classification of salt affected soils

Class of the soil	EC_e	SAR	pH_s
Saline	> 2	-	–
Saline-sodic	> 2	> 13	–
Sodic	-	>13	–
Normal	< 2	< 13	–

Indian Scenario

The position of saline and alkali soils in the new comprehensive classification system, proposed by the United States Department of Agriculture in 1960, has been placed in the fourth order, Aridisol because of their close association, chiefly with the arid and semi-arid climatic complex. Realizing that salinization can affect any group of soils, and more so in the marginal arid zones, the alkali soils also figure in the fifth order of *Mollisol* (Agarwal and Gupta, 1968). The different orders, suborders and Great Soil Groups for the saline and alkali soils of India in accordance with this classification are:

Order	4.	Aridisol
Suborder	4.1	Orthids
	4.2.	Argids
Great Soil Groups	4.14	Salorthid
	4.23	Natrargid
	4.24	Nadurargid

Murthy *et al.* (1980) made an attempt to review and classify the salt affected soils in India. The soils are classified at the great-group association level. The legend consists of 12 associations. Salorthid, Natrargid, Halaquept and saline phases of Ustochrept from the major units in the western region; Natrustalf, Natraqualf, Halaquept and saline phases of calciorthid, Haplargid, Camborthid, Ustochrept, Fluvaquept and Haplaquept and saline phases of Haplaquept occur in the eastern region and largely saline and sodic phases of Pellustert, Chromustert, Ustifluevent and Halaquept are found in Peninsular India.

Since India opted to follow the USDA classification (Richards *et al,* 1954), the salt affected soils in India until recently, used to be classified under three heads namely (i) Saline soils, (ii) Non-saline alkali soils, and (iii) saline alkali soils. Bhumbla (1973) prepared a map of the salt affected soils in India and grouped the soils in various categories viz. (i) alkali soils, (ii) saline soils, (iii) potentially saline soils (iv) coastal saline soils (v) deltaic saline soils, and (vi) acid sulphate soils. In this map, saline alkali soils were grouped with alkali soils. The map reported as Fig. 1.1 prepared recently also follows the same criteria and does not include saline-alkali soils as a special class but treats them as alkali soils. Following modifications in the USDA classification are made.

- From management point of view saline-alkali soils are grouped into alkali soils (Table 1.6). In spite of the high EC_e of 12.15 dS/m, the soil is cahracterized as alkali rather than saline alkali.
- SAR is not recommended as a diagnostic criterion because even saline soils have high SAR under the Indian conditions.
- There is no unanimity on the lowering of EC_e limit to categorize soils and still the limit of 4.0 dS/m is widely used as a diagnostic criterion.
- Studies have shown that pH strongly influences the soil physico-chemical behaviour as distinct from the effect of exchangeable sodium on soil properties. For this reason, it has been suggested that pH should be an integral part of the definition of alkali soils.

- Most often an ESP of 15 to 20 is associated with a saturation paste pH of 8.2. For diagnostic purposes therefore, it has been suggested that a saturation paste pH of 8.2 will be more realistic than the value of 8.5 which is nearly always associated with higher values of ESP (Agarwal and Gupta, 1958).

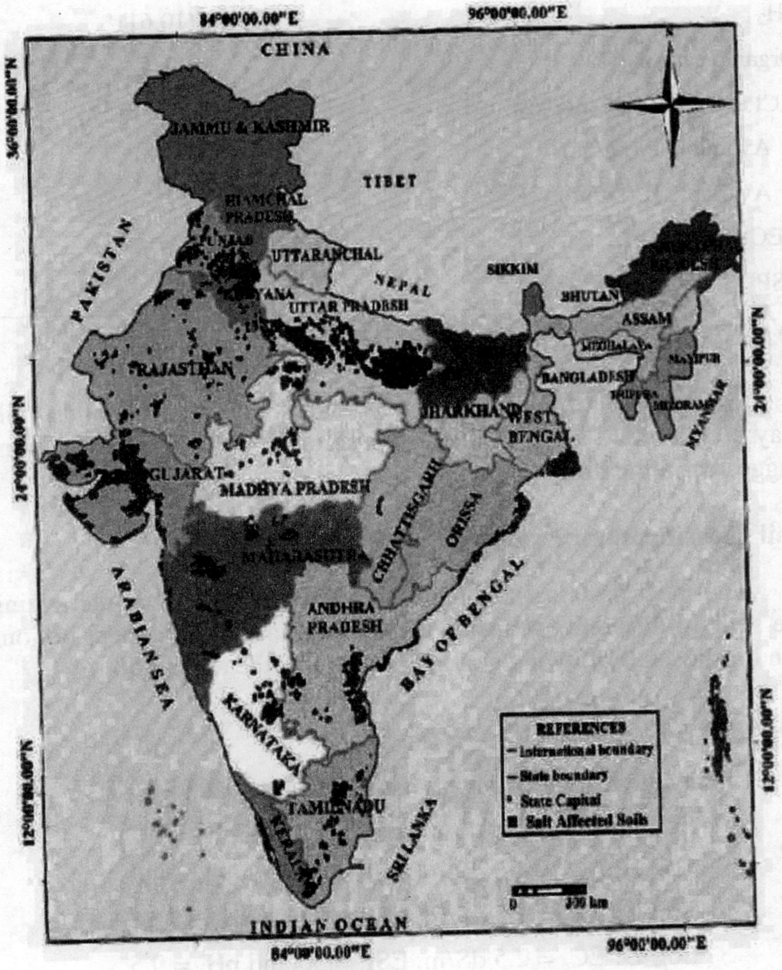

Fig. 1.1. Distribution of salt affected soils in India.

Table 1.6. Physico-chemical properties of saline alkali soil categorized as alkali soil

Soil property	Value
Sand: Silt: Clay (%)	59.3:17.4:23.3
EC_2 (dS/m)	3.09 (12.15)*
pH_2	10.71 (10.61)*
Organic carbon (%)	0.11
KCl extractable N (mg/kg)	317.0
Available P (kg/ha)	37.5
Available K (kg/ha)	174.7
CEC (me/100 g)	27.2
ESP	77.5

*Values in parentheses are EC_e and pH_s

On the basis of these recommendations, it is suggested that soils may be classified as per Table 1.4 preferably with the modifications suggested in the limits of pH_s.

Soil Classification: Example

The most practical application of Table 1.4 is in characterizing the soils in different groups so that appropriate management options can be chosen. Following example illustrates the application.

Example 1.1

Chemical characteristics of 4 soils are given as follows:

Soil A: EC_e = 4.5 dS/m, ESP = 20 and pH_s = 9.2

Soil B: EC_e = 4.5 dS/m, ESP = 10 and pH_s = 8.0

Soil C: EC_e = 1.5 dS/m, ESP = 8 and pH_s = 7.5

Soil D: EC_e = 2.5 dS/m, ESP = 30 and pH_s = 9.5

Classify each soil as per classification criteria.

Soil A

Since the EC_e is 4.5, ESP is 20 and pH_s is 9.2, the values fall in the middle row of Table 1.4. Thus, soil is charac-terized as saline-alkali.

Soil B

Since the values are covered in the ranges indicated in row 1, the soil is saline in nature.

Soil C

It is not covered in any of the rows of Table 1.4, so the soil is characterized as normal soil.

Soil D

Since the values are covered in the ranges indicated in row 3, the soil is non-saline alkali in nature.

NEED TO RECLAIM SALT AFFECTED SOILS

Increasing Food Grains Production

Out of a total geographical area of 329 M ha, net area under cultivation in India has almost stabilized at 140 ± M ha for the last few decades. There seems to be little possibility that the area under cultivation will increase. On the other hand, the population of the country is expected to touch 1.7 billion by 2050. To keep pace with the food requirement of the increasing population, the maximum and minimum demand for Cereals has been assessed at 434 and 381 million metric tons, respectively. To achieve such a target, cereal production must grow by a much higher rate than the current rate of 1.52%. Beyond 2022, even this rate may not be sustainable because of many foreseen problems including climate change (Bale and Jennifer, 2014). Even at the global scale, food production will need to increase by 38% by 2025 and by 57% by 2050 (Wild, 2003) if food supply to the growing world population is to be maintained at the current levels.

The increase in production can either be achieved through multi-cropping on the existing cultivated lands or by bringing additional area under cultivation. The possibility to increase the food grain production to the required extent by these two ways is seriously limited, particularly in India. For example; most of the suitable land is already being cultivated and expansion into new areas to increase food production is rarely possible or desirable. The only feasible alternative is to increase the cultivated area by bringing the waste and barren lands under cultivation. Therefore, the pressing need is to reclaim and utilize the salt affected soils which occupy extensive areas. The reclamation and utilization of nearly seven M ha of the salt affected soils in India can lead to additional production of more than 50 million tons every year. The lands which cannot be put under arable cropping can be put under alternative uses to meet the increasing requirement of fuel, timber and fodder etc.

Social Responsibility

State has the social responsibility of bringing the underprivileged resource poor people in the main stream. In the context of salt affected soils, assuming our land holding to be less than 1 ha, 7 million people have remained out of the development loop because of no or uneconomical productivity of their land holdings. Being a socialistic state, it is the state's responsibility to help these people through land reclamation. Besides, it has been proved beyond doubt that large scale reclamation activities generate employment opportunities for the farmer family and landless people in the village.

Commitments at International Forums

Land degradation neutrality is an idea originated in the United Nations Conference on Sustainable Development (Rio+20) held in Rio in 2012, where member states recognized the need for urgent action to reverse land degradation. It stemmed from the view that complex problem like degradation has cascading effect on natural resources and mankind. India as a signatory to the United Nations Convention to Combat Desertification (UNCCD) is committed to make India land degradation neutral by 2030. It requires concerted efforts by all organizations to become neutral under various land degradation processes including water logging and soil salinization.

As per an assessment made by Central Soil Salinity Research Institute, with the current pace of land reclamation, it may not be possible to achieve this goal for salt affected soils even in 2050 (Table 1.7). Therefore, it is necessary to take steps to increase the current pace of land reclamation efforts. It is all the more necessary, because the anticipated increase in land degradation will be in irrigation commands, where we are making huge investments. It is estimated that current spending to provide irrigation through canals is to the tune of Rs. 1,75000 per ha at the current prices. Once the land is taken out of cultivation, the whole investment gets blocked and the real benefits of irrigation are reduced to that extent.

Table 1.7. CSSRI projections of the extent of salt affected soils in India

Year	Projected salt affected area (M ha)	Reference
Current	6.73	Mondal *et al.* (2011)
2020	11.7*(9+2.7)	CSSRI (1997)
2025	13.0	CSSRI (2007)
2030	15.5	CSSRI (2011)
2050	16.2	CSSRI (2015)

*Salt affected soils and water logging in irrigation commands

■■■■

CHAPTER 2

NATURE AND CHARACTERISTICS OF SALINE AND ALKALI SOILS

INTRODUCTION

The accurate and precise identification of the saline and alkali soils and the diagnosis of the causes of the low productivity are the primary pre-requisites for taking effective measures to economically exploit the productive potential of these lands. Since the term productive potential has been used, it may be worth while to distinguish between the productivity and productive potential of the soils. The productivity of a soil is its initial capability to produce a certain amount of crop per hectare per annum and is expressed as a percentage of the optimum yield per hectare per annum of the same crop grown on the most fertile soils. The potential productivity of a soil, on the other hand, is the productivity which can be attained when all possible improvements, even the most difficult and costly, have been introduced. Potentiality, thus, is the future productivity of a soil. The productivity of salt affected soils is as low as zero (barren) but the productive potential in general is as high as that of a normal soil. It is proposed to discuss the nature and characteristics of these soils so as to understand the causes of low productivity and to devise ways and means to achieve the potential productivity of these soils.

SALT AFFECTED SOILS

All soils contain some amount of soluble salts, which are essential for the healthy growth of plants. If the quantity of soluble salts in a soil exceeds a certain value, the growth and/or the yield of most crops is adversely affected. Similarly, if the sodium on the exchange complex of a soil exceeds the desirable value, the crop productivity is affected. ´Soils, which contain excess soluble salts or high exchangeable sodium percentage or both to adversely affect the plant growth, are called the salt affected soils'. These soils are known by different local terms, such as *Kallar* or *Thur* in Punjab and Haryana, *Usar* or *Reh* in Uttar Pradesh, *Luni* in Rajasthan, *Khar* or *Kshar* in Gujarat and Maharashtra, *Chouddu* or *Uppu* in Andhra Pradesh, *Choppan* in Karnataka etc.

Saline Soils

These soils contain neutral salts (chlorides and sulphates of sodium, calcium and magnesium) in quantities enough to interfere with growth of most crop plants. The pH of saturated soil paste is less than 8.5, soil EC_e is more than 4 dS/m and ESP less than 15. Problem of soil salinity is widespread in the arid and semiarid regions (irrigated lands) and sub-humid to humid regions (coastal salinity). These soils show patchy growth of crops and sometimes, visible signs of salt injury such as tip burn or chlorosis (pale yellow colour) of leaves can be seen. Soluble salts, which are invariably present in large quantities, form a white efflorescence on the soil surface (Fig. 2.1). For this reason, these soils are also called white alkali, a term that has become obsolete. Since the neutral salts present in this type of soil can absorb moisture from the soil surface, these soils look wet although the water is not available to the plants.

Fig. 2.1. Natural view of a saline soil

Following three distinct groups of saline soils can be identified.

- Inland saline soils having shallow water table, water quality being saline in nature. These soils suffer from water logging and salinity.
- Normal soils being irrigated over the years with saline groundwater. These soils normally have deep water table. However, in certain situations the water table could be shallow. Theses soils generally occur in arid and semiarid regions.
- Coastal saline soils where the problem is due to ingress of sea water resulting in high salinity and temporary flooding. The climate is generally humid to semiarid/arid.

Factors Affecting Crop Production in Saline Soils

Plant growth in saline soils is adversely affected due to the following factors:

- Excess soluble salts present in the soil solution depress the water potential of the medium that restricts the water uptake by plant roots. A simple relationship between the osmotic pressure (OP) in atmosphere and electrolyte conductivity of the saturation extract in dS/m at 25ºC can be approximated by the simple relationship:

  $$OP = 0.36 * EC_e \qquad (2.1)$$

  The data in Table 2.1 clearly brings out this effect that plants are unable to take up the water with increasing soil solution salinity (Kelly and Rengasamy, 2006 as reported in Rengasamy, 2010).

- High concentrations of some specific ions (e.g. Na^+, Cl^-, SO_4^{--}, Mg^{2+}, B) can be toxic and can induce physiological disorders
- High salinity may affect several metabolic processes such as CO_2 assimilation, protein synthesis and respiration
- In waterlogged salt affected soils, some elements such as aluminium, iron and manganese may be produced that may accumulate to toxic levels

- Imbalance in the uptake of ions. For example; uptake of K, Ca and phosphorus may be affected because of excess of sodium and chloride ions.

Table 2.1. Percentage of available soil water not taken up by plants in different soil types due to osmotic pressure of soil water salinity (With permission from Rengasamy)

Laboratory measured soil salinity $EC_{1:5}$ (dS/ m)	Gravimetric field soil moisture (%) below which osmotic pressure due to salinity is		Percentage of available soil water not taken up by plants due to osmotic pressure (>1000 kPa) of soil water salinity			
	(>1000 kPa)	(>1500 kPa)	Sand	Sandy loam	Clay loam	Clay
0.25	4.5	3.0	25	0	0	0
0.39	7.0	4.7	50	0	0	0
0.50	9.0	6.0	70	25	0	0
0.72	13.0	8.7	100	50	15	0
1.00	18.0	12.0	100	81	40	4
1.11	20.0	13.3	100	94	50	12
1.25	22.5	15.0	100	100	63	22
1.50	27.0	18.0	100	100	85	40
1.64	29.5	19.7	100	100	98	50
1.64	29.5	19.7	100	100	100	58
2.00	36.0	24.0	100	100	100	76
2.33	42.0	28.0	100	100	100	100

Usually, not one but a combination of the above factors are in operation to affect crop production and productivity in saline lands.

Alkali Soils

These soils contain excess of salts like bicarbonates, carbonates and silicate of sodium, capable of alkaline hydrolysis and have sufficient exchangeable sodium to interfere with growth of most crops. They have pH_s greater than 8.5, electrolyte conductivity of saturation extract less than 4 dS/m and ESP of 15 or more. These soils are also termed as black alkali soils because the organic matter present in the soil solution of highly alkali soils gets dissolved and is deposited on the soil surface through capillary action causing dark black surface. This term is also obsolete now. In extreme cases, the soils remain barren with practically very little or no vegetation (Fig. 2.2).The cropped fields have spotty growth. Crops show scorching and leaf burn injury typical of sodium toxicity. Yellowing of leaves is noticed mainly resulting from poor soil physical properties. Upon drying, these soils show cracking behavior (Fig. 2.2 inset).

Fig. 2.2. Natural view of an alkali soil

Factors affecting plant growth in alkali soils

Plant growth, yield and/or quality of crops in alkali soils is affected by one or a combination of the following factors, the degree of damage depending upon the amount of exchangeable sodium, pH, and nature and kind of crops grown, macro and micro-environmental conditions and management interventions (Gupta *et al.*, 2002).

- High ESP and pH due to high concentrations of HCO_3 and CO_3.
- Poor water and air permeability as a result of high dispersion of soil aggregates and clay particles.
- Toxicity of Na, HCO_3 and CO_3 ions.
- Excess of Na causing imbalance in the uptake of other ions such as K and Ca due to antagonistic effect.

- Decreased solubility and availability of micronutrients like Zn, Fe and Mn due to high pH and presence of $CaCO_3$ and soluble CO_3 and HCO_3 ions.
- Presence of hard $CaCO_3$ layer at or around 1.0-2.0 m depth, which act as a physical barrier for the movement of water and salts as well as penetraton of roots of crops and trees.
- Hard crust on the surface layer which greatly hinders the seedling emergence and reduces the germination percentage and thus plant population.
- Poor drainage causing aeration problems.
- Low organic matter content.

In most situations, a combination of the above factors affects crop production and productivity.

The saline and alkali soils can be visually and qualitatively identified in the field by some visual observations (Table 2.2). However, for accurate characterization, these soils can be distinguished from each other *vis-a-vis* normal soils on the basis of several physical, chemical and biological properties (Table 2.3). Besides, miner-alogical characteristics, and soil water relationships that are affected by physico-chemical properties also needs to be studied. These characteristics are determined either in the laboratory or directly under field situation.

Table 2.2. The visual observations to identify an alkali and a saline soil

Observation	**Alkali soil**	**Saline soil**
Rain water stagnation	Stagnation for longer period	Relatively lesser period
Stagnated water type	Dark coloured, muddy, soapy	Clear
Surface salt crust	Dark brown or ash coloured clay crust	White salt efflorescence
Groundwater	Usually good, occasionally with high Residual Sodium Carbonate (RSC)	Saline, sometimes with high SAR

Observation	Alkali soil	Saline soil
Presence of nodules	$CaCO_3$ nodules are present	Gypsum may be present at some depth
Geographical distribution	Associated with semi-arid and sub-humid climates	Associated with arid and semi-arid climates
Natural vegetation (grasses)	*Sporobolus marginatus, Desmostachya bipinata, Suaeda maritma, Diplachne fusca, Dichanthium annulatum, Cynodon dactylon, Chloris barbata, Brachiaria mutica, Kochia indica, Panicum antidotale*	*Cressa cretica, Cyperus rotundus, Chloris pallida, Sporobolus pallidus, Haloxylon salicornicum, Acluropus lagopoides, Zygophyllum simplex, Dichanthium annulatum, Sueda furuticosa, Butea monosperma*

Table 2.3. Chemical, physical and biological parameters for monitoring soil salinity/alkalinity or overall health

Chemical	Physical including soil water relations	Biological
EC	Effective soil depth	Biological worm population
pH	Bulk density	Bio-diversity
Anions and cations	Soil porosity/Pore space distribution	Crop yield
ESP	Soil tilth/Soil water holding capacity	
SAR	Soil crusting	
CEC	Organic matter	
Calcium carbonate	Soil cracking	
Nutrient status	Soil dispersible clay	
	Infiltration rate	
	Hydraulic conductivity	
	Depth to groundwater	
	Surface stagnation	

PHYSICAL CHARACTERISTICS

Effective Soil Depth

The soil depth that can be effectively exploited by plant roots is an important consideration for plant growth. Generally, excellent yields are produced on normal soils having an effective root zone depth of 45 to 90 cm although shallow rooted crops do fairly well even in a depth up to 30 cm. Root penetration is inhibited seriously in alkali soils, limiting to an effective soil depth of 10 to 15 cm, due to the presence of an impermeable *kankar* (impure $CaCO_3$) layer and/or high ESP. The plant roots are thickened and blackened. In saline soils, the zones of *kankar* and gypsum accumulation, if any, restrict the effective soil depth.

Bulk Density

It is defined as the dry weight of a unit volume of soil and is usually expressed in g/cm^3. The bulk densities of highly productive normal soils usually range from 1.1 to 1.5 (medium to fine texture) and 1.2 to 1.65 (coarse texture). The alkali and saline soils (medium texture) have a high bulk density (1.5 to 1.65 g/cm^3) resulting in the inhibition of root penetration and proliferation and impeded drainage. The infiltration rates and permeability of the medium or fine textured soils are usually low if the bulk density exceeds 1.65 g/cm^3.

Pore Space Distribution

Pore space and bulk density are closely and inversely related. Soils with a low porosity generally contain little air space at field capacity. In highly productive, normal, medium to fine textured soils, the non-capillary porosity, in the plough zone, is usually 10-30%. The alkali soils may have a non-capillary porosity as low as 2% at which the normal aeration of the plants is adversely affected. The optimum values of non-capillary porosity differ for one crop to another. Gupta and Narain (1971) observed that the total porosity and non-capillary porosity in normal soil profile vary from 45.8 to 37.6% and 8.6 to 6.1% as compared to 39.5 to 37.9% and 4.9 to 2.5% in alkali soils.

Soil Tilth

The soil tilth is a product of organic matter content, texture, consistence, etc. which, under good management, combine to form a surface or near surface, physical condition of the soil favourable to plant growth. The highly productive normal soils have a very good tilth. In alkali soils, the soil tilth is very poor because of high pH and ESP. The soils are very hard and compact when dry, and very sticky when wet. These soils have to be tilled very carefully at optimum moisture content. The saline soils generally have satisfactory tillage properties.

Crust Formation

Alkali soils have low stability of structure, disperse and slake when they are wetted by rain or irrigation water and may develop a hard crust as the soil surface dries. This crust presents a serious barrier to the emergence of seedlings. With some crops often this is the main cause of poor crop stand. The crust formation is not only limited to alkali soils but may also occur in normal or saline soils. However, hard surface crusts are typically associated with high ESP (Fig. 2.3), a characteristic of alkali soils, low organic matter, puddling and wetting the soil to zero tension. The chemical analysis

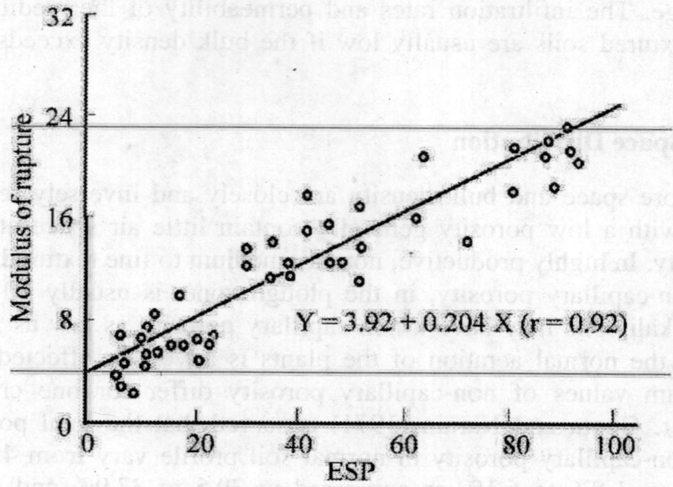

Fig. 2.3. Relationship between ESP and modulus of rupture

of the crust formed on an alkali soil revealed that salt content in the crust is 22.8% out of which sodium carbonate constitutes 48.3%. Other are: sodium bi-carbonate (33.4%), sodium sulphate (13.3%), sodium chloride (3.9%) and sodium nitrate (Traces). Sandhu *el al.* (1980) observed that modulus of rupture linearly increased with increase in ESP of the soil (Fig. 2.3). Earlier, Sharma and Agarwal (1979) observed that modulus of rupture for the same ESP is higher in clay loam than loam and sandy loam soils respectively.

Cracking Behaviour

Due to high amounts of sodium on the exchange complex, the soils are highly dispersed and upon wetting, they get puddled and become hard when drying. As the drying proceeds, thin flakes get peeled off. If drying continues for longer periods, large cracks are formed (Fig. 2.2 inset), which enhance the loss of water from the deeper soil layers. On the contrary, black alkali soils (Vertisols) do not produce large and wide cracks at higher ESP. In fact, the cracking behavior of the alkali soil exhibits development of narrow, shallow, and higher numbers of cracks with increasing soil ESP. Thus, at high ESP, these soils lack in natural deep wide cracks which is the qualifying characteristics of Vertisols.

Water Dispersible Clay

In an aqueous suspension, soil is either dispersed or flocculated depending upon the amount and nature of clay and the nature and the extent of accompanying ions. Generally, calcium and to some extent magnesium, and potassium have a flocculating effect. On the other hand, presence of hydrogen and especially the sodium ions has a dispersive effect, since the forces that bind clay particles together are disrupted when too many large sodium ions come between them. When this separation occurs, the clay particles expand, causing swelling and soil dispersion. Soil dispersion causes clay particles to plug soil pores, resulting in reduced soil permeability. When soil is repeatedly wetted and dried and clay dispersion occurs, it then reforms and solidifies into almost cement-like soil with little or no structure. The three main problems caused by sodium-induced dispersion are reduced infiltration, reduced hydraulic conductivity, and surface crusting.

Dispersion coefficient is defined as the ratio of water dispersible clay to the total clay expressed as percent. The dispersion coefficient of normal soils ranges from 20 to 30%. It is usually very high in alkali soils and may exceed 90% in soils having pH greater than 10.0, increasing with increasing ESP. This is the main factor, which reduces the permeability of the alkali soils by sealing the surface pores under the impact of rainfall or when a water of low salinity is applied. The formation of a layer of high clay content in the profile at some depth is also attributed to high dispersivity.

Khosla *et al.* (1973) and Sandhu *et al.* (1980) observed that the water dispersible clay content increased with an increase in ESP. The dispersion co-efficient linearly increased as the ESP increased from 5 to 93 as per the following equation:

$$Y = 2.74 + 1.36 X \quad (r = 0.896) \tag{2.2}$$

Here Y is the dispersion coefficient and X the ESP.

In a black cotton soil (Vertisol), the dispersion ratio increased steadily up to 15 ESP. Thereafter, the dispersion ratio remained almost similar up to 60 ESP. This suggests that the flock value at 15 ESP is such that the smectitic alkali soil present similar structural degradation as at ESP 60 because of high buffering nature of clay mineral. Most of the fine clay remains in dispersible condition, causing severe erosion losses during the runoff (Singh and Verma, 2016). Similar results were reported by Reeve *et al.* (1954) for eight texturally different soils, varying in cation exchange capacity (CEC) from 10 to 43.5 me/100g and organic matter content from 0.1 to 3.17%.

Diagnosis of Alkali Soils: Physical Properties

Based on some of these physical properties, test have been developed to distinguish the alkali soils from normal or saline soils as well as assess the degree of alkali problem. These tests are based on turbidity, dispersion and swelling of soils (Chinchmalatpure *et al.*, 2015). These tests can be conducted even under field conditions.

Turbidity Test

Take 100 g of surface or subsurface soil in a clear beaker of 600 ml capacity. Gently pour 500 ml of rain water or distilled water.

Gently invert the beaker and return to its original position. Place a white plastic spoon or spatula in the center of soil suspension. Allow it to stand for 4 hrs. Compare the visibility and score the turbidity as follows (Fig. 2.4).

- Clear visible – Non alkali
- Partially visible – Moderately alkali
- Not visible – Alkali

Fig. 2.4. Turbidity test to assess alkali nature of soils

Dispersion Test

Take soil aggregates or fragments of about a size of bean and place them carefully in the beaker. Carefully add enough rain or distilled water from the side to cover the aggregates. Leave the container for 20 hr and carefully assess dispersion. Group it as clear water (no dispersion), partly cloudy (partial dispersion) and very cloudy (complete dispersion). Soil is categorized as non-alkali, moderately alkali and alkali respectively.

Swelling Test

Take two graduated test tubes marked with coloured pencil at each centimeter interval. Fill both the tubes with sieved soil up to 1/3rd level. Add rain or distilled water slowly in one test tube and kerosene oil to the other. Give turn by putting thumb on the top in order to mix thoroughly and finally make soil and water/kerosene oil ratio 1:1. Keep it for 24 hr. Compare the gel thickness in the test tube with water to that of kerosene and categorize the soil as follows:

- No to slight swelling – Non alkali
- 1.5 times swelling of soil in water as compared to that of kerosene oil – Moderately alkali
- More than 1.5 times swelling of soil in water as compared to kerosene oil – Alkali

CHEMICAL CHARACTERISTICS

Salinity (Soluble Salts)

Although some salts are essentially required for the normal growth of the plants, excess of soluble salts in the soil is undesirable as the crop growth is adversely impacted. The total salt content of soil solution or irrigation water is represented in milligrams per liter (mg/l), millimole per liter (mmol/l) or electrolyte conductivity expressed in dS/m at 25°C. Earlier, it also used to be represented by TDS (total dissolved salts), which is the sum of the anions and cations present in units of ppm. Electrolyte conductivity is by far the simplest representation and is a measure of the ease with which current will flow through the soil water or the irrigation water. EC is reciprocal of the electrical resistivity, the resistance in ohms of a conductor, metallic or electrolytic, which is 1 cm long and has a cross sectional area of 1 cm^2. Hence, electrical conductivity was formerly expressed in reciprocal ohms per cm or mhos per cm and now in SI units it is expressed in dS/m. It is equivalent to the earlier unit expressed in millimhos/cm (mmhos/cm) (1 dS/m = 1 mmhos/cm). Milliequivalent per liter is the milliequivalent weight of the anion or compound (1/1000 part of the equivalent weight) in a liter of solution. Various conversions for soil/water salinity are given in Table 2.4.

Table 2.4. Conversion factors for soil/water salinity

Component	Multiply by	To get	Remark
TDS, mg/l	1.0	TDS, ppm	
Salinity, dS/m	1.0	Salinity, mmho/cm	
Salinity, mmho/cm	1000	Salinity, μmho/cm	
Salinity, dS/m (3-30)	0.36	Osmotic pressure, atm	
Salinity, dS/m (< 5 dS/m)	640	TDS, mg/l	
Salinity, dS/m (>5 dS/m)	800	TDS, mg/l	The value of the factor 800 increases with increasing EC and approaches near unity at very high salinity levels.
mg/L	1/equiv-alent weight	me/l	As an example 80 mg/l of calcium is equal to approximately 4 me/l of calcium, the equivalent weight of calcium being 20.04 mg/me.

mg/l = milligrams per liter; me/l = milliequivalents per liter; ppm = parts per million; dS/m = deci-Siemens per meter; μS/cm = micro-Siemens per centimetre; mmho/cm = milli-mhos per centimetre; TDS = total dissolved solids.

In soil testing laboratories, EC is usually determined using an EC meter (Fig. 2.5). It is determined either in the soil saturation extract (EC_e) or in 1:2 soil water extracts (EC_2). Even higher soil water extracts (1:2.5 and 1:5) are used to determine EC especially in heavy textured soils. It has been observed that EC_2 is highly and significantly correlated with EC_e (Gupta *et al.*, 1969 and Singhla *et al.*, 1974). In 1 : 2 soil water suspension, all soluble salts contribute to the measurement of EC equally whether these are neutral or alkaline salts. On the contrary, in the saturation extract of alkali soils, all the soluble salts are not recovered, resulting in low EC_e values. It is owing to the highly dispersed carbonate system wherein the colloidal particles tend to align themselves and form an impervious layer thereby obstructing the passage of the part of the total soluble salt in the soil solution. On the other hand, the chloride salt system facilitates their easy removal in the extract. The salinity appraisal on the basis of EC_2 is very useful particularly when the amount of soil sample is limited or a large number of soil samples are to be analyzed or where repeated samplings are to be made in the same soil to determine the change in salinity with time or treatment. The ratio of EC_e/EC_2 decreases with the increasing saturation percentage of the soil. A calibration curve relating EC_2 and EC_e is used to estimate EC_e (Gupta and Gupta, 2014).

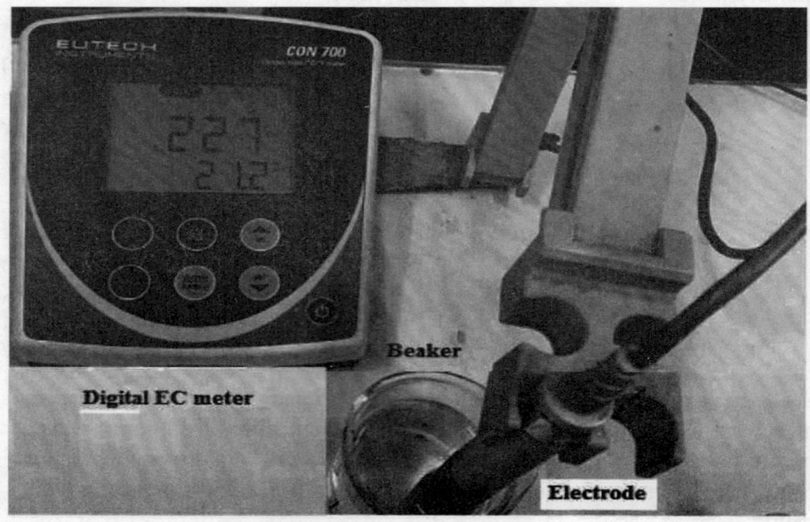

Fig. 2.5. Digital EC meter

EC_e can also be determined using the following relation:

$$EC_e = (\theta_k / \theta_{sp}) * EC_k \quad (2.3)$$

Here θ_k and θ_{sp} are the soil moisture contents in soil-water paste (1: k) and at saturation for a given soil. EC of the soil solution at field capacity (EC_{fc}) and at 15 bar moisture content is usually taken as twice and four times that of EC_e respectively.

The most productive normal soils have EC_e less than 1 dS/m. The most productive normal soils have dominantly chlorides and sulphates of calcium and magnesium salts, followed by sodium salts. Bicarbonates and carbonates may also occur but in small amounts.

The adverse effects of soluble salts on plant growth are governed by the texture of the soil, nature of the salts and nature of plants. At a given salinity, the adverse effects on plant growth are more in a fine than in a coarse textured soil. Chloride salts are more harmful than sulphate salts (Kovda *et al.,* 1973; Table 2.5). As the EC_e of the soil increases from 4 to 20 dS/m in the case of chloridic soils and 6 to 40 dS/m in the case of sulphatic soils, the relative productivity of the soil in most cases decreased from almost 100 to 0.

Table 2.5. Relationship between nature of salts in the soil and plant growth

EC_e (dS/m)		**Relative plant growth (%)**
Chlorides	**Sulphate**	
4	6	100
4-8	6-12	80-90
8-12	12-20	50-70
12-20	20-40	20-50
20	40	0

The soluble salts in saline soils mainly comprise of the chlorides and sulphates of sodium, calcium and magnesium. The primary adverse effect of excessive salinity is due to its adverse impact on the flow of water into the plant by osmosis, which gets reduced and the plant is starved of water even though the soil is moist (physiological drought condition). Besides, these salts are highly deliquescent and therefore absorb moisture from the atmosphere, thus causing the soil to appear moist. Such soils often have a greasy appearance and are almost black in colour. They are likely to be mistaken for alkali soils. If the proportion of magnesium is higher than that of calcium, it may cause harmful effects on the physical properties of the soils and crop growth (Yadav and Girdhar, 1980, 1981). It is because the adverse effects of magnesium on soil physical properties are not as bad as that of sodium but are also not as good as that of calcium.

Bicarbonates occur in very low amounts and carbonates are usually absent or occur only in traces.

The alkali soils, although relatively low in soluble salts, have a composition that differs considerably from that of normal and saline soils. These soils contain mainly carbonates and bicarbonates of sodium, the alkaline salts capable of hydrolysis. Small amounts of chlorides and sulphates also occur. The calcium is highly deficient as it is precipitated in the soil in the presence of bicarbonate ions.

The saline-alkali soils are associated with high salinity and high ESP but their physical properties are not affected because of the high electrolyte concentration due to neutral salts. Upon leaching these soils show all the features of an alkali soil.

The tolerance of plants to soil salinity also differs from one plant to another. The horticultural plants except date palm and pulse crops are sensitive to salinity (EC_e 1.5 - 3.0 dS/m); sunflower, groundnut, maize, sugarcane and rice are the semi tolerant crops (EC_e 3-5 dS/m); and cotton, sorghum, pearl millet, wheat, barley, mustard safflower and sugar beet fall in the tolerant group (EC_e 5-10 dS/m). Amongst vegetable crops, kale, asparagus, spinach are tolerant; tomato, broccoli, cabbage, cauliflower, lettuce, potato, carrot, onion, pea, squash and cucumber are semi-tolerant; and radish, celery, and green bean fall in the sensitive group.

Soil Reaction (pH)

Soil pH in the laboratories is measured using a soil pH meter (Fig. 2.6). The factors that affect the pH values of the salt affected soils are: the total soluble salt concentration, the nature of salts, the composition of exchangeable cations, the nature of cation exchange materials and the presence or absence of gypsum and alkaline earth carbonates. The excessive soluble salt concentration, particularly due to the presence of neutral salts, depresses the soil pH. The highly saline soils have a pH of saturated soil paste around neutral because that is the pH of the neutral salts constituting most of the solutes in the soil solution. As an example, a soil having an EC_e of about 100 dS/m, usually have a pH value close to 7.

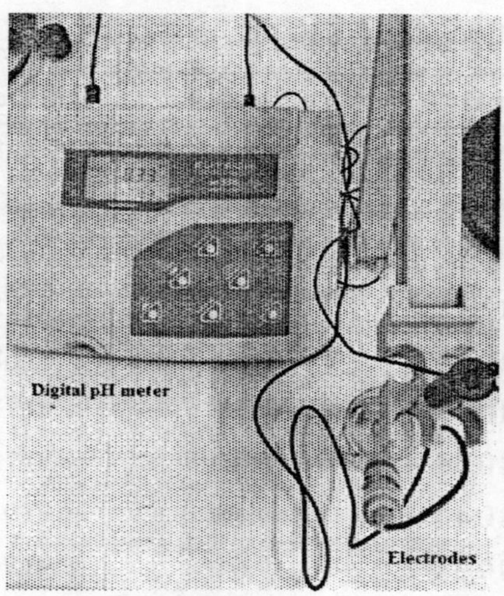

Fig. 2.6. A digital pH meter

The pH of the saturated soil paste (pH_s) has been widely recognized as an index to distinguish alkali soils from normal and/or saline soils. In specific cases, pH of the 1:2 soil water suspensions (pH_2) has also been reliably used for the same purpose (Gupta, 1972). The appraisal of the soil reaction in the saline or alkali soils on the basis of pH_2 is rapid and convenient. It is especially useful when a large number of soil samples are to be analysed and/or where repeated samplings are to be made for the same soil to determine the change in pH with time or treatment. The pH_2 values are usually higher than pH_s. In normal and saline soils, pH_2 values are higher than pH_s by 0.2 –0.5 units. The most striking difference being in alkali soils where pH_2 values may be higher than pH_s by about 1.0 unit (Table 2.6). This corroborates the observation that pH_s of 8.2 will correspond to pH_2 of about 9.0 (Abrol *et al.*, 1980, Table 2.7). The increase in the soil pH values by about 1.0 unit as the moisture is increased from near saturation to 1:2 dilution, is useful in identifying alkali soils. Higher values of pH_2 in alkali soils in comparison to saline soils are due to the nature of electrolytes, former containing predominantly Na_2CO_3, a weak electrolyte, whereas the saline soils contain NaCl, a strong electrolyte. The pH of a weak electrolyte increases sharply with dilution as a result of hydrolysis, whereas changes in the pH with dilution are less pronounced in strong electrolytes (Gupta, 1972).

Table 2.6. Variation of pH_2 and pH_s in alkali soils

pH_2	pH_s	Difference
9.12	8.54	0.58
9.91	9.04	0.87
9.99	9.25	0.74
10.12	9.37	0.75
10.28	9.60	0.68
10.35	9.58	0.77
10.38	9.65	0.73
10.40	9.46	0.94
10.38	9.64	0.74
10.10	9.34	0.76

Source: Datta *et al*. (2017)

Table 2.7. pH of saturated soil paste and approximate ESP

pH_s	Approximate ESP
8.0 - 8.2	5 - 15
8.2 - 8.4	15 - 30
8.4 - 8.6	30 - 50
8.6 - 8.8	50 - 70
8.8	70

To quantify the relative degree of alkali in the soils, the soils having pH_2 values <9.3, 9.3 - 10.0, 10.0 - 10.5 and >10.5 may be called low, medium, high and very high alkali soils respectively. The semi or partially reclaimed alkali soil is one whose original pH_2 value >10 has come down to < 9.3 in the root zone (0-20 cm), but higher values persist in the lower depths (Gupta, 1972).

Cations and Anions

Major cations present are sodium, calcium and magnesium and major anions are carbonates and bicarbonates, chlorides and sulphate. Besides the cations in the soil solution, exchangeable cations also need to be determined. While the cations in soil solutions are used to determine SAR, exchangeable cations help to determine ESP and SAR of the soil. The chemical analysis should be checked and used as follows:

- Sum of cations and sum of anions in the soil solution, when expressed on equivalent basis should be nearly equal. If difference is observed, it may be worthwhile to analyze for nitrate.
- Dominance of chlorides and sulphates indicates the problem of salinity while dominance of carbonates and bicarbonates indicates the problem of alkali.
- Presence of carbonates is an indication of high pH.
- From practical point of view, CEC of the soil can be approximated by the total cation concentration on the exchange complex using the following relation:

$$CEC = Na_x + Ca_x + Mg_x + K_x \quad (2.4)$$

Here x stands for exchangeable cations. Since potassium is usually less, it may be neglected if it has not been analyzed.

Exchangeable Sodium Percentage (ESP)

The sodium adsorption ratio (SAR) and the exchangeable sodium ratio (ESR), ratio of sodium on the exchange complex to the cation exchange capacity minus the exchangeable sodium at the saturation percentage of the soils are significantly correlated. Based upon this relationship, ESP has been observed to be related with SAR by the following equations:

$$ESP = ESR = Na_x+/(Ca_x + Mg_x) \quad (2.5)$$

or

$$ESP = K_g (Na/ (Ca + Mg)/2)^{1/2}) \quad (2.6)$$

Here Na_x, Ca_x and Mg_x are the respective cation concentration on the exchange complex, Kg is the Gapon´s constant. Further, ESP can be found after the chemical analysis of the samples as follows:

$$ESP = \text{Exchangeable sodium/ CEC of the soil} \quad (2.7)$$

Since ESP is related to SAR it can also be estimated by the following relation:

$$ESP = \frac{100\,(-0.0126 + 0.01475\;SAR)}{1 + (-0.0126 + 0.01475\;SAR)} \qquad (2.8)$$

If we incorporate SAR =13 in eq. (2.8), predicted ESP is about 15. Up to an SAR of the saturation extract of about 30 the ESP values are roughly similar to SAR, but above this limit, they diverge and the full eq. (2.8) must be used.

The most productive normal soils have ESP or SAR less than 5 and exchangeable sodium (ES) less than 1.0 me/100 g. The saline soils have ESP or SAR usually less than 15 but SAR values as high as 100 are not uncommon. The alkali soils have ESP or SAR >15 but in highly deteriorated soils ESP may be as high as 100 and SAR may be as high as 500.

A high ESP is often associated with poor soil physical properties, the physical properties becoming increasingly unfavourable with increasing ESP. Besides the ESP, degree in deterioration of physical properties is also governed by the type of clay minerals, the kind of salts present in the soil solution and the soil texture. At a given level of ESP, physical properties are usually poorer in soils with expanding type 2:1 clay minerals (illite, vermiculite, montmorillonite), containing a low electrolyte concentration of salts capable of alkaline hydrolysis, *viz.* sodium carbonate, sodium bicarbonate, silica, *etc.,* than in soils with clay minerals of the non-swelling 1:1 type (kaolinite) or mixed types, containing a high electrolyte concentration of neutral salts, *viz.* sodium chloride or sodium sulphate. The coarse textured soils exhibit less deterioration in physical properties than the fine textured soils at a given ESP.

As the ESP of soil increases from 5 to 60, the relative productivity of the soil decreases from almost 100 to 0. On the basis of ESP, the alkali soils having ESP <30, 30 -60, 60 - 75, and >75 may be called as low, medium, high and very high alkali soils, respectively. From management point of view, the tolerance of plants to ESP is also a factor of practical importance. Horticultural plants are extremely sensitive (ESP 2-10) followed by pulses (ESP 10-15). Wheat, barley, oats, *raya* (mustard), senji, *berseem*, *shaftal*, *dhaincha*, sugarcane, millets and cotton are semi-tolerant (ESP 15-30) and transplanted rice, sugar beet, para grass and Karnal grass are tolerant (ESP 30-60).

Example 2.1

The exchangeable cations concentration in a soil is as follows:

Ca^{2+} = 3.00 me/100 g, Mg^{2+} = 1.75 me/100 g,

Na^{+} = 19.25 me/100 g, and K^{+} = 0.1 me/100 g.

Calculate the exchangeable sodium percentage. Make any assumption that is required in this calculation.

Assumption: Other exchangeable cations are present in negligible quantities.

Cation exchange capacity (CEC) = 3.00+1.75+19.25+0.1 = 24.1 me/100 g

The exchangeable sodium percentage (ESP) is then given by eq. (2.7) as follows:

ESP = (19.25/24.1) * 100 = 79.9

Macronutrients

The testing of the soil for its major nutrients, viz. nitrogen, phosphorus and potassium, is a proven, practical method for evaluating the fertility status of the soils. It helps to formulate recommendations for the applications of fertilizers for higher productivity. The most productive normal soil is high in organic carbon (above 0.75%) which is used as a measure of available nitrogen, high in available P_2O_5 (above 25 kg/ha) and high in available K_2O (above 280 kg/ha). The soil is rated as low in organic carbon, in the available P_2O_5 and in the available K_2O if the respective values are below 0.5%, 10 kg/ha and 110 kg/ha respectively. The corresponding values for soil to be rated medium are 0.5 - 0.75%, 10-25 kg/ha and 110-280 kg/ha respectively. The alkali and saline soils are usually low in nitrogen, medium to high in phosphorus and high in potassium. A common recommendation therefore, has emerged that 25% more nitrogen should be applied in saline/alkali than the normal soil environment.

The indiscriminate use of fertilizers should be avoided in saline and alkali soils. The kind of fertilizers, their method and their time of application as related to the growth stage of the crop, are some of the important considerations which play a vital role in crop response.

The degree of salinity and alkali are also the factors which influence the response of fertilizers. Whereas ammonium sulphate or liquid ammonia is more beneficial to alkali soils, urea is efficacious in saline soils. Cost consideration however, determines the widespread use of urea even in alkali soils. If the degree of salinity is high, the response of fertilizers is limited. In barren alkali soils, P and K are usually high (available P being 5-10 ppm, available K, 500-750 ppm) and therefore, there may not be any response to the application of P and K fertilizers, for few years till these are depleted by few crops in succession. Phosphorus retention increases as salinity increases or as SAR (ESP) decreases (Elmahi and Mustafa, 1980). The fertilizer requirement also depends upon the crops and their varieties.

Micronutrients

The extensive research carried out in plant nutrition has established that sixteen elements are essentially required for optimum plant growth and development. Nine of these elements (C, H, O_2, Ca, Mg, S, N, P and K) are required in relatively larger amounts and are called macro-nutrients. The remaining seven, viz, B, Fe, Mn, Zn, Mo, Cl and Cu, are needed by plants in smaller amounts and are therefore called micro-nutrients.

The deficiency of micronutrients is identified either on the basis of soil and plant analysis or from the visual symptoms appearing on plants, etc. The content of Zn, Fe, Mn and Cu is usually low in alkali soils (Bhumbla and Dhingra, 1964). The availability of micro-nutrients is affected by several factors, the most important being the pH of the soil. As such, deficiency symptoms for zinc, iron and manganese are sometimes strikingly seen on plants growing on alkali soils. Zn, followed by Fe and Mn are limiting factors in achieving higher yields in alkali soils (Table 2.8). The availability of iron and manganese in alkali soils can be increased substantially by submergence and FYM application (Swarup, 1980), but zinc deficiency may still persist. In saline soils, the content of Cu, and Zn is usually less than those in alkali soils and the content of Fe and Mn is generally higher.

Table 2.8. Micronutrients in normal and alkali soils (in ppm) (by the DTPA method)

Soil	Zn	Fe	Mn	Cu
Normal[1]	0.96	15.3	31.7	1.93
Alkali[2]	0.85	4.5	5.5	1.45

Source : [1]Singh, M. (Personal communication), [2]Swarup (1981).

The normal soils, saline soils and alkali soils contains boron in the saturation extract in quantities of less than 1.0, 1 to 2 and 2 to 10 ppm, respectively (Bhargava *et al.,* 1974), 1.0 ppm of boron in soil being generally adequate for the normal growth of most of the plants. The available molybdenum in normal and alkali soils has been reported to be 0.40 and 0.80 ppm respectively (Singh and Singh, 1966).

Toxic Elements

Almost all the essential elements, which are otherwise recognized as micronutrients, may become toxic elements if present in the soil beyond the optimum limits. These elements then adversely affect the growth of certain plants. In addition, some more elements, viz. Al, As, Be, Cd, Cr, F, Pb, Ni, Se and V are also known to cause toxicity to plants, if present beyond permissible limits. The elements which have been recognized to be of practical importance from the toxicity point of view are B, Li, F, Se and Mo. The best productive normal soils are generally free of toxic elements or contain less than permissible limits. Out of these five elements, B, Li and F are phytotoxic whereas Se and Mo may not adversely affect plant growth but the crops, particularly the forages grown on them, may contain an excess of these elements which cause adverse effects to the animals which graze on them. The saline and alkali soils tend to contain one or more of these elements in toxic levels.

Boron

The permissible limits for boron in the saturation extract of soil can be given, tentatively, with respect to the tolerance of plants for boron. The sensitive, semi tolerant and tolerant plants can absorb boron in concentrations below 1.0, 1-5 and 5-10 ppm respectively. The saline soils may contain boron in the range of 5 to 10 ppm but excessive salinity depresses the availability of soluble boron and therefore, highly saline soils having EC_e as high as 100 dS/m may not contain appreciable quantities of boron. In alkali soils, soluble boron increases with pH and ESP; the soils with pH_2 of about 10.8 or ESP near 100 may contain boron to an extent of 10 to 15 ppm. The presence of $CaCO_3$ and gypsum, however, depresses the solubility of boron (Gupta, 1979b).

High levels of boron in saline soils can be reduced by leaching. Contrary to earlier reports, boron may be removed almost in the same proportion as the other soluble salts during the normal leaching process. Additional leaching water may not be necessary to reduce boron to a safe level. In alkali soils, leaching after treatment with gypsum helps to control the adverse effects of boron. In a laboratory study Gupta and Chandra (1972) observed a marked reduction in the water soluble boron content of a highly alkali soil upon addition of gypsum. At high pH and alkalinity, boron is present as highly soluble sodium metaborate which upon addition of gypsum is converted to relatively insoluble calcium metaborate. Leaching also helps to reduce the boron content. It may be noted that gypsum application at the rate of 100% GR may be required to reduce boron so that even boron sensitive crops can be grown during the reclamation period (Table 2.9). The boron can also be leached efficaciously alone by the rainfall of 724 mm (Table 2.10).

Table 2.9. Effect of gypsum treatment and leaching on the boron (ppm) in relation to ESP

Treatment	**B**		**ESP**	
	0-15	**15-30**	**0-15**	**15-30**
Control	3.0	4.4	84.5	95.0
25% GR	2.1	3.9	52.5	87.0
50% GR	1.7	2.5	37.5	82.5
75% GR	1.4	1.8	17.5	83.5
100% GR	0.8	1.3	17.5	43.5

Source : Gupta (1979)

Table 2.10. Effect of leaching by rainfall on the soluble sodium (me/l) and boron (ppm)

Depth of soil (cm)	Sodium (me/l)		Boron (ppm)	
	A	B	A	B
0-30	443	150	13.0	3.0
30-60	135	191	10.0	2.7
60-90	65	61	6.0	4.0
90-120	57	29	4.0	2.0
120-150	21	9	2.0	2.0
150-180	14	8	1.2	1.8
180-210	14	7	1.0	1.0
Groundwater	30	14	1.0	0.8

*A = Before rainfall, B= After rainfall; Source : Bandyopadhya (1971).

Alfalfa, wheat, barley, lucerne, oats, cotton, sugar beet, sorghum and maize are quite tolerant to boron. The oilseed crops such as groundnut and sunflower, leguminous plants, citrus and horticultural plants are known to be sensitive to boron. The boron tolerance increases in the presence of soluble calcium, nitrogenous and phosphatic fertilizers. Unlike tolerance to salinity, the tolerance to boron is more in heavy textured soils than in light textured soils. It is because boron is adsorbed in heavy textured soils.

Selenium

It is not an essential trace element for plant growth but traces of selenium (0.05-0.1 ppm) are required by the animal for normal metabolism. Selenium is the only mineral element known to be absorbed by food and forage plants in sufficient amounts to make the lethal when consumed by animals. Tentative limits of the toxicity of selenium in soils to adversely affect the growth and quality of plants are considered to be equal to 1 ppm. Plants grown on such soils might contain 20-500 ppm of selenium which could prove lethal to the animals. The normal soils or saline soils do not normally contain excessive selenium. However, alkali soils have been reported to contain excessive selenium in certain situations. Narwal *et al.* (1979) did not observe much difference in the selenium content of alkali soils as compared to adjacent normal soils, available Se being in the range of 0.20 to 0.35 ppm in normal as compared to 0.23 to 1.0 ppm in alkali soils. The alkali disease in animals, caused by selenium poisoning, locally called as Degnala or Dekpala in Haryana and Punjab, has been attributed tentatively to the grazing of forages or straw of rice containing selenium. The application of gypsum, sulphur, phosphate and iron can alleviate the selenium toxicity (Singh and Singh, 1980).

Lithium

Lithium is an element which is neither essential for human nor plants. On the contrary, even relatively very small amounts of lithium are toxic to plants. The higher contents of lithium are found in saline soils than normal or alkali soils (Gupta *et al.* 1974, Lavado *et al.,* 1978). The permissible limits for lithium in the soil solution tentatively for sensitive, semi-tolerant and tolerant plants are below 1.0, 1-5 and 5-10 ppm, respectively. Avocados, sour oranges, sugar beet and cabbage are recognized to be semi-tolerant and cotton, maize, wheat and barley tolerant to lithium. Some leguminous plants like gram are very sensitive and may be affected adversely even at 1.0 ppm.

Fluoride

Fluoride like selenium is essential for animals but non-essential for plants. The importance of fluorine is due to its harmful action in animal nutrition even when ingested in minor doses. The water soluble fluorine content in saline or alkali soils have been observed up to 15 ppm. At this level of fluorine, plant growth is not adversely affected. Singh *et al.* (1979) observed that increasing levels of fluoride, above 50 ppm F (as sodium fluoride), affected adversely the growth of rice plant in the alkali soil. This adverse effect of fluorine was, however, nullified by the application of phosphorus.

Molybdenum

The soils with high amounts of available molybdenum result in the accumulation of molybdenum in fodder crops, in concentrations which are toxic to the livestock. The plants may accumulate molybdenum to a level which is toxic to cattle without exhibiting any toxicity to the plant itself. The calcareous alkali soils, particularly waterlogged generally contain excessive molybdenum. Gupta and Dabas (1980) reported total Mo in the alkali soils of Karnal district in Haryana as high (2.32 to 8.0 ppm) and available Mo very high (0.58 to 2.71 ppm). Nayyar (1972) observed the accumulation of molybdenum in the fodder crops ranging from 228 to 538 ppm grown on soils treated with 2 ppm of molybdenum. According to Cunningham, forage containing less than 3 ppm molybdenum is considered safe for cattle and 3 to 10 ppm is safe only if the copper content is normal (5 ppm). Application of gypsum and copper sulphate to the soil has been known to reduce the molybdenum uptake by plants thereby reducing the toxic effects.

BIOLOGICAL CHARACTERISTICS

Organic Matter

Soil organic matter, generally referred to as humus, found in small amounts in productive agricultural soils imparts unique and remarkable properties to the inorganic matrix of sand, silt and clay. It generally represents less than 5% of the dry weight of the soil. Soil organic matter is classified into two main groups namely non-humic and humic substances. The non-humic substances are the unaltered remains of plant and animal tissues and also organic compounds. The humic substances are chemically and biologically modified substances and represent the most active fraction of the organic matter in the soils. These substances resist microbial degradation. Based on their solubility, these can be separated into four fractions, (i) fulvic acid, which is soluble in both alkali and acid; (ii) humic acid, soluble in alkali but insoluble in acid and alcohol; (iii) hymatomelanic acid, soluble in alkali and alcohol and (iv) humin, insoluble in both alkali and acid.

In alkali soils, the alkaline solutions of sodium salts being solvents for soil humus, the latter is highly dispersed in the soil solution. On evaporation, a dark coloured residue is left commonly in the form of a thin crust or rings around small depressions, thereby imparting a dark black colour to the surface of the soil. The organic matter content in alkali soils is therefore, very low, generally in the range of 0.1-0.5%. In saline soils, organic matter is not dispersed as in alkali soils and therefore, the content is higher, generally varying from 0.2 to 1.0%.

The addition of organic matter increases acidity, thus, helping in lowering the pH of alkali soils albeit very slowly. Organic matter is especially helpful where sulphur is added to correct the alkalinity as it supplies food for the bacteria that stimulates the oxidation of sulphur to the sulphate form. The synergetic effect of chemical amendments and organic matter has been documented.

Micro-organisms

The soil is a dynamic medium, constituting the habitat of abundant biodiversity, with unique genetic patterns. Here, one can find the greatest amount and variety of living organisms, which act as a nutrient reservoir. One gram of a normal soil may contain 600 million bacteria belonging to 15,000 or 20,000 different species. These values decrease to 1 million bacteria encompassing only 5000 to 8000 species in desert soils (Informativo Capebe, 2010). Soil degradation by means of physical, chemical, biological and hydrological processes, causes destruction of both the biological potential of the land and of their use to sustain the population connected to it. As the soil becomes saline/alkali, the new regime of the soil affects the qualitative and quantitative composition of the microscopic population. Fungi are reported to be more sensitive to osmotic stress than bacteria. A significant reduction in the total fungal count in soils salinized with different concentrations of sodium chloride has been reported. Similarly, increase in the salinity level to above 5%, drastically reduced the total count of bacteria and actinobacteria (Omar *et al.*, 1994). The total number of bacteria, actinomycetes and fungi are usually the highest in productive normal soil followed by saline and alkali soil (Table 2.11). The highest soil respiration rates were observed in soils with low salinity and the lowest in soils with medium salinity, while the microbial biomass was greater in the treatments with high salinity and lower in those with low salinity.

Table 2.11. Microbial counts in different soils

Nature of the soil	Organism × 10^5/g soil		
	Bacteria	**Actinomycetes**	**Fungi**
Normal	96.00	11.6	0.21
Saline (EC_2 6.1 dS/m)	11.30	0.4	0.045
Alkali (pH_2 10)	2.33	0.04	0.0003

Neelkantan and Pahwa (1977).

Singh (1950) listed the species of blue green algae found on alkali soils of northern India. Through Nostoc commune the annual addition of organic matter to the soil is 37-60% and that of nitrogen 30-38%. Saxena (1960) collected the algal flora of alkali soils and observed that where pH is high (9.0 - 10.9) and the total nitrogen content of the soil is low (below 0.2%), the population of algal flora is poor. In stagnant water ditches, where pH varied from 9.0 to 10.2 and total nitrogen content was about 0.03%, green and blue algae were in equal proportions. At low pH (7.9 - 8.6), the proportion of green over blue algae increased. The development of the algal flora is the most prominent during monsoon. The amount of bacteria, actinobacteria and fungi increased with increasing number of years of reclamation. The activities of urease and alkaline phosphatase also increased, but the activity of invertase altered little (Celia *et al.*, 2012). Batra *et al.* (1997) reported that total organic carbon (%), dehydrogenase activity (µg TPF/g soil), microbial biomass carbon (mg/kg soil), % microbial biomass carbon and DHA/Misc., all were low in a barren alkali soils (pH = 10.6, ESP = 95). The values were 0.5 4.5, 56.7, 2.3 and 0.08 respectively. The values improved to 0.44, 112.1, 203.3, 4.6, 30.55 upon reclamation with Karnal grass and gypsum applied at the rate of 50% GR (pH = 9.3, ESP = 40).

MINERALOGICAL CHARACTERISTICS

Clay Minerals

The kinds of clay minerals present in the soil determine many of the physical and chemical characteristics of a soil. It is well known that the 1:1 layer clay minerals (kaolinite) have only a slight stickiness, a small amount of shrinkage on drying, a small surface area and a low cation exchange capacity. Most 2:1 type clay minerals (illite, vermiculite, montmorillonite) have opposite properties. In general, the most productive normal soils which are dominated with 1:1 layer clay minerals together with iron and aluminium oxides, have excellent soil water relationships and easy workability associated with their high degree of aggregation and non-swelling nature. At the same level of ESP, these soils are more permeable than those containing 2:1 layer clay minerals. However, 1:1 layer clays have low cation exchange capacity and a low capacity for moisture retention. The saline soils in India have been identified to contain high amounts of muscovite, biotite, hornblende and chlorite. The alkali soils are dominated with illite, mixed layer minerals, vermiculite and chlorite etc. The normal soils in association contain the same minerals as adjacent saline or alkali soils, but differ in the contents of their constituent minerals and degree of crystalinity (Chatterji and Sachdeve, 1979).

Kankar (Impure $CaCO_3$)

The *Kankar* or impure $CaCO_3$ commonly accumulates in saline or alkali soils developed under arid and semi-arid climates. The accumulation may be diffused throughout the soil profile, or may take the form of soft concretions or nodules, or may be concentrated in a continuous horizon (caliche, cemented hard clay pan) of different hardness and at different depths below the surface (Gupta and Singh, 1971, Gupta, 1971). Investigations made at the *Usar* Reclamation Farm, Chakri, Kanpur have revealed the occurrence of a compact layer 30-40 cm thick and 30-40 cm below the surface. A zone of *kankar*, often as a thick pan of 1.2 - 1.5 m depth with typical characteristics, exists below this depth (Table 2.12).

The presence of $CaCO_3$ affects both the physical and the chemical properties of a soil. The water movement is restricted and root penetration and proliferation are prevented. The calcareous soils ($CaCO_3$ >2%) usually have pH_s in the range of 7.6 - 8.4. The presence of $CaCO_3$ in the soil profile, particularly the root zone depth, in moderate amounts is helpful in preventing excessive deterioration in saline or alkali soils by checking the rise of ESP. The rain water owing to its acidic nature because of the CO_2 contained in it, is helpful in converting $CaCO_3$ to $Ca(HCO_3)_2$ and replace sodium by the calcium released from $Ca(HCO_3)_2$. The most productive normal soils are slightly calcareous. Highly calcareous soils also suffer seriously with nutrient and micro-nutrient deficiencies. The plants grown on such soils sometimes suffer with lime induced chlorosis.

Table 2.12. Physical and chemical characteristics of clay pans

Depth (cm)	Clay (%)	Bulk density (g/cc)	Aeration porosity (%)	HC (cm/hr)
0-12	18	1.5	5.4	0.039
12-35	22	1.6	3.6	0.014
35-60	25	1.7	2.5	0.004
60-110	24	1.6	2.9	0.009
115-180	19	1.6	-	0.021

HC = Hydraulic conductivity

The soils containing small amounts (2%) of gypsum ($CaSO_4$ 2 H_2O) are considered favourable for crop growth because the calcium replaces sodium on the exchange complex and thereby preserves the soil structure. Excessive gypsum (25%) is harmful to plant growth. Unfortunately, alkali soils do not contain sufficient reserves of gypsum in India. The saline soils, in arid regions and on the sea coast, have been identified to possess gypsum at some depth in the soil profile. Gypsum has also been found to occur in the saline soil profiles of the Rann of Kutch in Gujarat, reported to be in the range of 0.25 - 1.0% of $CaSO_4$ (Satyanarayana, 1961).

SOIL WATER RELATIONS

Infiltration Rate

Infiltration rate is defined as the maximum rate at which a soil, in a given condition at a given time, can absorb rain/irrigation water. Also, the rate at which a soil will absorb water ponded on the surface (shallow depth) when the ponded area is infinitely large or when adequate precautions are taken to minimize the effect of divergent flow at the borders. It is the volume of water passing into the soil per unit area per unit time. The initial intake rate rapidly decreases as the soil is wetted and the intake reaches a steady state after some hours, which is known as the basic or equilibrium rate. The most productive normal soils have basic infiltration rates between 1.0 and 3.5 cm/hr. The alkali soils generally have basic infiltration rate below 0.1 cm/hr characterizing the soils as non-arable except for rice crop. The infiltration rate curve of a normal and alkali soil reveal that the entire curve for the alkali soil falls below that of a normal soil curve (Acharya and Abrol, 1991; Fig. 2.7). The initial and the basic infiltration rates decrease with the increase in ESP (Table 2.13). In a Vertisol, the infiltration rate becomes almost negligible at an ESP of 38 (Singh and Verma, 2016).

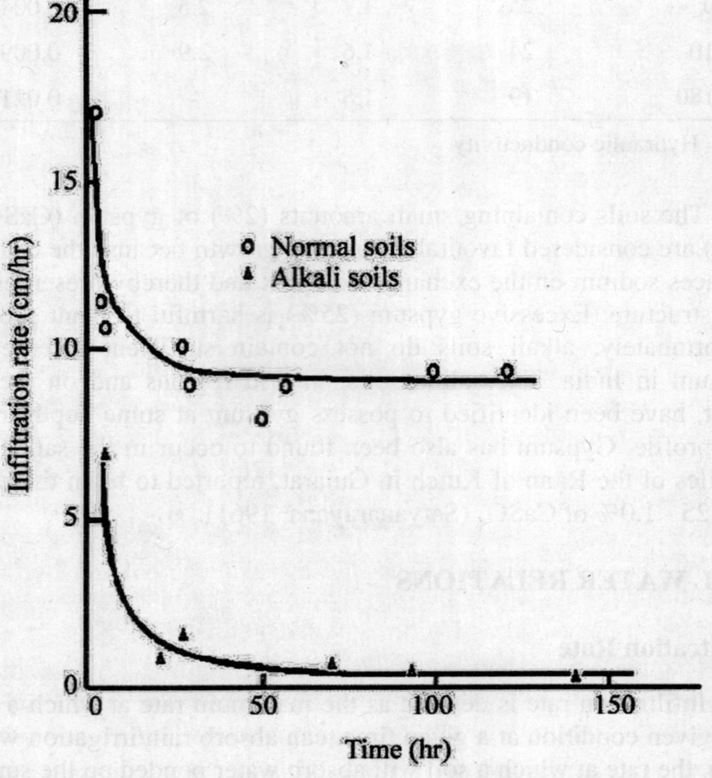

Fig. 2.7. Infiltration rate curves of a normal and an alkali soil

Table 2.13. Infiltration rate as affected by ESP

ESP of the soil	Initial infiltration rate (cm/hr)	Basic infiltration rate (cm/hr)
5	22.1	3.7
14	20.3	2.7
22	11.9	1.4
30	9.9	1.0
44	8.6	0.7
49	5.9	0.3
62	5.4	0.2
65	2.2	0.1
70	2.0	0.06
72	1.8	0.05
74	1.7	0.01

Source : Singh (1980)

Permeability (Hydraulic Conductivity)

Permeability has been defined qualitatively as the quality or state of a porous medium relating to the readiness with which such a medium conducts or transmits fluids. When defined quantitatively the term hydraulic conductivity is used. It is defined as the constant of proportionality between the flux and the total driving force in Darcy's law, or as the flux caused by a unit driving force. The most productive normal soils have such hydraulic conductivity as to ensure that saturated condition, whether from rainfall, irrigation, or both, is unlikely to occur for more than 48 hr in the upper root zone. Several factors, viz. texture, structure and structure stability, consistency, $CaCO_3$ cleavage planes, visible pores and depth to impermeable strata such as bedrock or a hardpan, influence the hydraulic conductivity of the soils. The hydraulic conductivity of a clay soil is as low as 0.13 cm/hr while for sand it is 14 cm/hr. The values for other soil textures are somewhere in between these two values (Hermsmeler, 1978).

Besides soil texture, the hydraulic conductivity for the same texture is also influenced by ESP of the soils (Fig. 2.8). In consequence, the values are lower in alkali soils as compared to normal and saline soils. Hydraulic conductivity in typical normal, saline and alkali soil has been observed as 0.21, 0.29 and 0.11 cm/hr, in the surface 0-15 cm depth (Poonia and Bhumbla, 1973). The unsaturated hydraulic conductivity at 100 cm suction has been observed to be 0.05 cm/d in alkali soil as compared to 0.6 cm/d in a normal soil (Acharya and Abrol, 1975). With further increase in suction, the difference in hydraulic conductivity of two soils generally decrease, values being nearly similar at a suction of 250 cm (Acharya and Abrol, 1991; Fig. 2.9).

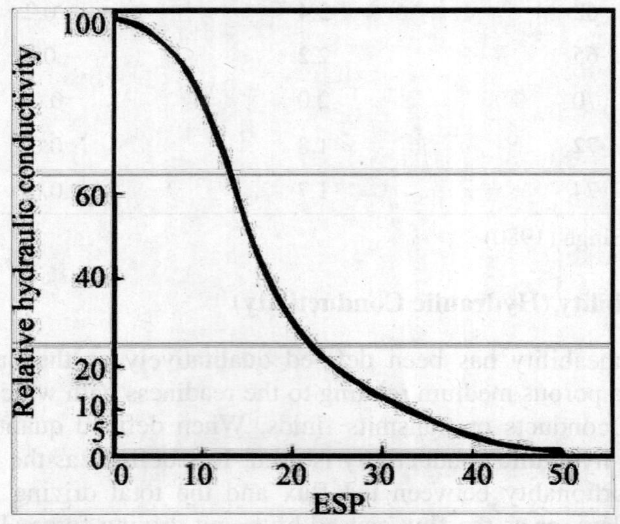

Fig. 2.8. Relative saturated hydraulic conductivity as affected by ESP

Soil Water Availability

The readily available water is that portion of the water in the soil that is readily absorbed by plant roots (about 50-75% of the total available moisture given by the depth of water at field capacity minus the depth of water at permanent wilting point in the root zone). The capacity of a soil to retain water available to plants has a direct bearing on the required depth and the frequency of irrigation. In alkali soils the moisture holding capacity in the moisture available

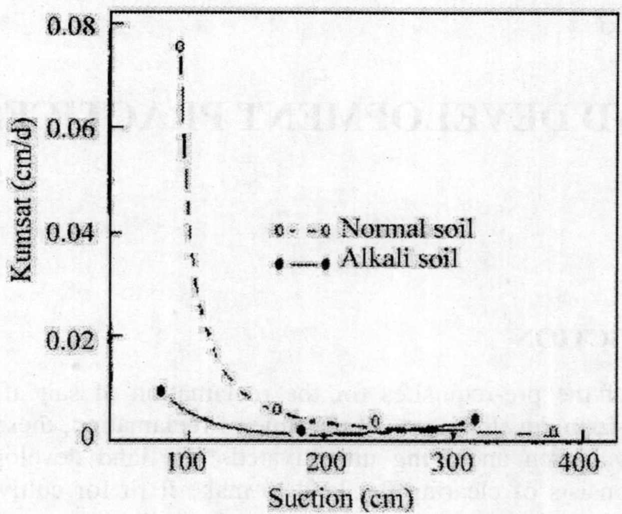

Fig. 2.9. Unsaturated hydraulic conductivity as affected by soil suction in normal and alkali soils

range of the plants may be about 20% less as compared to normal soils. In case of saline soil also the moisture availability to the plants is less due to increase in the osmotic pressure of soil water solution. The osmotic pressure increases at the rate of 0.36 bar per unit increase in the electrical conductivity of soil solution measured in dS/m. The total available water was highest in normal than the alkali (ESP > 50) black clay soil, being 0.31 and 0.26 cm^3/cm^3 water in normal and alkali black clay soil, respectively. The number of days required to reach equilibrium in pressure plate apparatus was also comparatively very high in alkali than normal soil. While the normal soil comes to field capacity within 48 hr, the alkali clay soil took almost a week to attain it (Singh and Verma, 2016). This results in temporary flooding of crop lands in black alkali soils restricting aeration to plant roots.

■■■■

CHAPTER 3

LAND DEVELOPMENT PRACTICES

INTRODUCTION

One of the pre-requisites for the reclamation of salt affected soils is on-farm development. At the time of reclamation, these soils are usually barren and lying uncultivated. The land development package consists of clearing the land to make it fit for cultivation, ensure improved water management and facilitate unrestricted drainage. The degree, number and kinds of on-farm activities will depend upon the kind of salt affected soil, agro-climatic conditions and nature of the land at the time of initiation of land reclamation.

SURVEYS AND LAND CLEARING

Reconnaissance Survey

First and foremost step to reclaim the land is to carry out a reconnaissance survey of the area which should be followed up with a topographic and soil survey. The reconnaissance survey helps in identifying the location of the area boundaries, necessity and extent of the land clearing operation, extent and type of problem and position of the water table. Usually, much larger area is included in this kind of survey compared to the area that is proposed for reclamation. It helps to know the area for locating possible drainage outlets or to assess the possibility of the run-off from the adjoining lands entering the area.

Topographic and Soil Survey

Topographic survey is carried out to prepare a topographic map of the area. Such maps indicate all the permanent structures in the area. The map is used to prepare levelling design, to design the irrigation and drainage systems, and to prepare field's layout. Soil survey reveals the true nature of the problem and helps delineate normal lands from the salt affected lands. Severity and kind of the problem, soil texture, depth of the soil, depth to water table and depth to impermeable layer all could be investigated at this stage.

Land Clearing

Prior to making a detailed survey of the area, it is essential to remove undesirable vegetative growth from the land. In this process, some or all the trees, bushes, vegetation, trash and boulders are cleared from the area. In salt affected soils no major land clearing operation is required. Most operations can be completed manually. In case there are large levelling differences, only then a bulldozer may be required to carry out rough land levelling operation. The operation can also be carried out with a 30-50 HP tractor. For initial breaking of the soil a mould board plough, a disc plough or a sub-soiler can be used.

LAND LEVELLING

In its natural state, ground surface can seldom be suited for efficient and uniform application of water. Land levelling is therefore, an essential prerequisite to establish good agronomic, water, soil and crop management practices. Unevenness of the fields and poor farm designs together result in loss of nearly 10-25% of the irrigation water during its application at the farm. Excess water stagnating at low spots not only delays the tillage operations but has a major impact on the germination, crop stand and yield of crops through nutrient-water interaction, salt and soil moisture distribution pattern and/or accumulation of salts. Fields have uneven crop stands and uneven maturing of crops. All these factors besides reducing yield, impact the product quality which together reduce the potential farm income. In salt affected soils, because too little water would reach the high spots, leaching may be distinctly less than the low spots causing non-uniformity in leaching and hence, patchy crop stand. Therefore, land levelling must precede any reclamation program. Commonly used methods for land levelling operation are the following:

- Manual levelling
- Conventional levelling using animal or mechanical power
- Laser land levelling

Land levelling is commonly misunderstood as the table-top levelling of the fields. The levelling in the real sense of this purpose means to provide a gentle slope to the land surface along the length with nil or too little slope along the width. This would allow the water to move along the length as a uniform sheet. This would also allow to control the depth of water per irrigation. Optimum longitudinal and cross slopes for the borders and furrow methods of irrigation should be selected (Table 3.1). For small areas, levelling operation is usually done by the tractors and the driver's judgment is sufficient to achieve the desired level. In large area planning, however, a complete design of the land levellling operation in blocks of economically feasible sizes is necessary. Before doing so, it is beneficial to complete rough grading with the help of bulldozer. For hauls exceeding 100 m, the efficiency of a bulldozer is reduced significantly. In such cases, use of scraper with a pick up and carry operation can be employed.

Table 3.1. Desirable field slopes for irrigation methods

Methods of irrigation	Slope	Soil type	Slope (%)
Border	Longitudinal	Heavy (clay)	0.05-0.25
		Medium (loam)	0.20-0.40
		Light (sandy)	0.25-0.65
	Cross	For all soils	Zero or as minimum as possible
Furrow	Longitudinal		Maximum permissible
		Sand	0.25
		Sandy loam	0.40
		Fine sandy loam	0.50
		Clay	2.50
		Loam	6.25
	Cross	For all soils	Not exceeding 2.0

Selection of Levelling Implements

Farmers have recognized the value of land levelling since time immemorial and therefore, considerable attention and resources are devoted in proper levelling of fields albeit using traditional methods. The methods are not only cumbersome and time consuming but also lack in precise grade/finish as in most cases farmer/tractor operator judgment overrides in deciding the finishing of the task. Levelling uniformity coefficient (LUC), an index of the effectiveness with which the desired results can be achieved is in most cases low. LUC used to evaluate the performance of the levelling implements can be mathematically described as (Tyagi and Singh, 1979):

$$\mathrm{LUC} = \left[1 - \frac{\Sigma\,|(\mathrm{LD} - \mathrm{LE})|}{\Sigma\,|\,\mathrm{LD}}\right] \tag{3.1}$$

Here LUC is the coefficient of land uniformity, LD is the desired elevation at the grid point, LE is the existing elevation at the grid point and N is the number of grid points. The values of the LUC for the three selected secondary land levelling implements at the desired range of soil moisture indicated that a super leveller with an LUC of 0.96 was the most precise in achieving the desired elevations. It was followed by a combination of blade and float (LUC 0.90). The use of float alone was not suitable because of its low LUC value (0.60). Considering the LUC to be an important parameter for the selection of levelling implements, a combination of terracing blade and float should work well. However, if the choice is to be made for a single implement, the super leveller has an edge over the other implements.

Laser Land Levelling

Although traditional methods appear less expensive yet for the same degree of precision, these may turn out to be quite expensive. With the popularization of laser land levelling more precise results can be obtained. With laser land levelling, skill of the operator becomes immaterial to the levelling operation. The laser levelling equipment essentially consists of four elements (Fig. 3.1):

- The laser emitter
- The laser sensor or receiver

- The control box (electronic and hydraulic control system)
- The tractor and grading implement.

Fig. 3.1. Components of laser land leveller

It has now emerged that laser land levelling through its precision improves irrigation uniformity (Fig. 3.2), irrigation efficiency and as a result the productivity of land and water (Fig. 3.3). Although wide variations have been noticed in the results from various tests, on an average water saving to the tune of 15-20% (17.5%) and a yield increase from 10-15% (12.5%) can be achieved in wheat crop (Ambast *et al.*, 2015). Since the technology is crop neutral, it can also be assumed safely that water saving and yield increase will be achieved in all other crops as well.

Fig. 3.2. Water distribution in conventionally leveled (left) and laser levelled fields

Fig. 3.3. Performance of wheat in a conventionally levelled and laser levelled field

Cost of Land Levelling

The cost of land levelling depends upon the soil type, initial undulations, field size, type of machinery used and the efficiency of the person carrying out the operation. Land levelling cost is not proportional to the differences in the initial and final levels. With increase in precision, cost of land levelling increases almost exponentially. The cost may vary from a few hundred rupees to several thousands of rupees. Laser land levelling is now being undertaken at farmer´s fields both by the departments of agriculture and private entrepreneurs on custom hiring basis. It is usually done on per hour basis. It is assessed that cost of laser land levelling is recovered within one season of wheat crop.

Other Mechanical Treatments

Some of the mechanical treatments that help to improve the physical conditions of alkali lands are the deep ploughing, sub-soiling and the profile inversion. These operations singly or in combination result in a loosened soil mass with improved infiltration rate and hydraulic conductivity. The profile inversion aims to cover an undesirable soil layer with better material from a lower layer or to utilize gypsiferous sub-soil for amelioration of the top layer of the alkali soils. Ramamoorthy (1968) reported that in soils having high concentration of salts on the soil surface and relatively lesser amounts of salts below, soil inversion is quite effective.

FIELD ARRANGEMENT

Laying out fields of workable sizes and shapes is essential for successful irrigation farming. Length and width of the plots are usually decided on the basis of irrigation method to be used. For example, in the case of border irrigation, width of plots may be a multiple of the width of a single border. Although, for the operation of leaching, it is desirable to have small sized plots, yet leaching being a one-time operation, field layout on this basis may be costly and undesirable. The common practice then could be to make temporary dykes (*bunds*) to accomplish leaching.

A well designed irrigation system, which can deliver the required amount of water to all the parts of the area at the required time and in desired amounts, should be established. It is desirable that each field is provided with a drainage channel at the tail end. If the lands are waterlogged and subsurface drainage is to be implemented then more rigorous investigations and design practices need to be adopted.

Assured Supply of Good Quality Water

Timely application of good quality canal or tube well water in adequate quantities is an essential input to any reclamation program. Since reliability of water supply both from the rainfall and canals is poor, it is necessary to have a source of water, which is under the control of the farmers. A tube well for that purpose seems to be the only alternative. Since alkali areas in general are underlain by good quality water and the aquifer characteristics are favourable for the installation of a tube well, this activity could be taken up as a part of the land development. Besides meeting the irrigation needs of the land reclamation program, it should also help in lowering the water table.

Activities Required in Reclamation Program

Some of the on-farm activities that need to be undertaken have been estimated on the basis of experimental and field experiences (Table 3.2). It may also help to understand the components of cost of a reclamation program.

Table 3.2. Envisaged work components in reclamation of alkali/saline soil

Sr. No	Kind of work	Alkali (Quantity/ length/number)	Saline (Quantity/length/ number)
1.	*Bunding* (m/ha)	470	470
2.	Irrigation channels (m/ha)	100	100
3.	Drainage channels (m/ha)	100	100
4.	Outlet (small)	1	1
5.	Land levelling (Tractor/h)	10	10
6.	Tube well	1 per 4 ha	Canal water, 1 per 4 ha
7.	Pump set (no. per tube well or no. per ha)	1 per tube well	1 per tubewell and/or 1 for 200 ha
8.	Soil amendment and application (t/ha)	8-12	-
9.	Subsurface drainage (m)	-	200*

Note: Items 1-7 as per UPLDC norms for alkali lands; * Item 8 is required for alkali lands and item 9 for waterlogged saline soils # If item 9 is required

LAND DEVELOPMENT FOR COASTAL SALINE SOILS

The coastal tract of peninsular India extends from Rann of Kutch in Gujarat to Malabar Coast in Kerala on the western coast, and to Coromandal coast in Tamil Nadu up to Sunderbans delta in West Bengal on the eastern coast. The coastal length given by National Hydrographic Office (NHO) and Survey of India (SoI) in 1970 is 7516.6 km. It comprises of mainland coastline of 5422.6 km and islands coastline of 2094 km for the two islands of Andaman &

Nicobar and Lakshadweep together. Coastal states include Gujarat, Maharashtra, Goa, Karnataka, Kerala, Tamil Nadu, Andhra Pradesh, Odisha and West Bengal. Union territories include Daman & Diu, Puducherry, Andaman & Nicobar Islands (Bay of Bengal) and Lakshadweep Islands (Arabian Sea) (Fig. 3.4). Coastal land salinization and salt water ingress are major hazards encountered along the Indian coast which can hamper the rapid socio-economic growth of the coastal states and the economy of the country as a whole. These soils are formed in the river deltas in a narrow strip of land ranging from a few kilometers close to the sea coast to about 50 km, along the low lying lands, estuaries and in the inland depressions all along the sea coast of India. Salient features of coastal saline soils are reported in Table 3.3.

Table 3.3. Physico- chemical properties of coastal soils of India

State	**AESR No.***	**$pH_{2.5}$**	**EC_e (dS/m)**	**OC (%)**	**CEC {cmol p^+/ kg}**	**BS (%)**
West Bengal	15.1	6.3-7.4	1.5-9.8	0.15-0.52	19.4-22.7	70-80
	18.5	6.5-7.6	4.1-35.0	0.3-0.78	6.6-10.6	74.82
Odisha	18.4	6.3-7.4	2.0-50.0	0.10-0.22	12.8-13.4	75-85
	18.5	4.5-6.3		Tr.-0.20	33.8-55.5	49-69
Andhra Pradesh	18.3	8.0-8.4	4.6-27.0	0.9-1.1	2-6	-
Tamil Nadu	18.1	5.1-6.1	2.0-10.0	0.06-0.09	4.0-6.6	89-100
Kerala	19.3	2.4-3.2	1.0-20.0	9.2-19.9	26.4-58.0	8-57
	19.3	5.1-5.6		0.3-2.6	3.9-7.6	48-58
Karnataka	18.5	3.9-5.7	3.0-10.0	0.07-1.23	0.7-4.6	29-89
Maharashtra	19.3	6.6-7.0	4.0-14.0	0.4-0.7	20.8-28.0	40-45
Goa	19.1	6.3-7.4	4.0-15.0	0.28-0.60	0.5-7.8	48-69
Gujarat	2.2	8.2-8.7	0.20-6.50	0.18-1.10	1.3-3.7	--
	5.3	7.7-8.0	0.46-3.20	0.13-0.15	12.26-19.26	--
Lakshadweep	20.2	8.5-9.1	0.1-0.2	0.11-0.77	0.4-1.5	100

Source: Compiled from several state sources and reproduced from Sen, 2011, AESR, OC, CEC and BS refer to Agro Eco Sub-Region organic carbon, cation exchange capacity and base saturation, respectively.

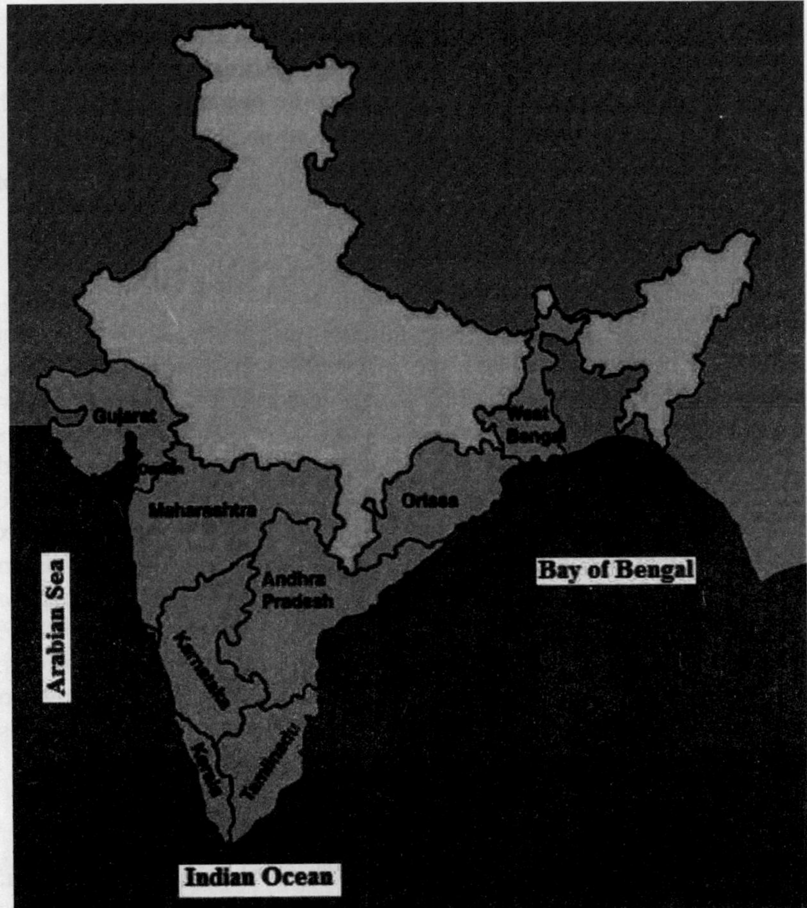

Fig. 3.4. Coastal states and union territories of India

Some of the processes that are most active in formation of coastal saline soils are:

- Sea water inundation during high tides or cyclones resulting in submergence of large cultivable land. The periodical inundations render the land saline and completely unfit for growing any crop
- Use of saline water traveling to interior parts of coastal plains through rivers/creeks for irrigation

- Use of saline groundwater for crop production
- Shallow saline groundwater table during the post monsoon season bringing dissolved salts in the root zone through capillary rise more so during the summer season
- Deposition of wind borne salts, the most active process of soil salinization at Kutchh in Gujarat

STRUCTURES TO MINIMIZE LAND INUNDATION

In order to obviate the re-salinization of land from the first mentioned cause, the reclamation of coastal saline soil is attempted by developing the coastal area through construction of the sea dyke or other structures along the banks of Delta Rivers and creeks (MoWR, 2017). Most of the state governments have adopted different measures to tackle the problem. These vary from simple dykes (*Bunds*), *kharland* Schemes and *nalla* plugs to more complex structures such as *Bandhara, v*ented cross and salt water exclusion dams. From agricultural point of view, dykes or bunds or *kharland* schemes also based on various kinds of dykes are in common use and forms the subject matter of discussion in the following sections. The shallow groundwater table is lowered by subsurface drainage to such a level that the contribution of salts from the groundwater is reduced. This issue is included in Chapter 6.

Dykes

The system of dykes is the oldest, the most common and often, the most economical means of checking inundation of coastal plains in the long run. The system of dykes keeps the deltaic rivers or creeks more or less in their existing course without changing the flood levels from their natural conditions. Earthen *bunds* with stone revetment on the slopes and sluices with wooden needles (wooden plank shutters filled with puddle earth in between planks) are constructed. Very small earthen bunds called *nala* plugs are constructed at various intervals along the channel route. Such embankments have been constructed in Sundarbans (West Bengal), Maharashtra, Karnataka and Odisha etc., usually with a free board of about 1 m above the high tide level. The top width of the embankments ranges from 1 m to 4 m. The sea dykes have a provision for adequate drainage in the form of the tidal gates operating in one direction only i.e. towards the sea.

Factors Affecting Design of the Sea Dykes

The design of dykes includes the proper estimation of certain parameters related to the construction of dykes. The adequacy of the design of a dyke should include (i) elevation of the top of the dyke so that water may not overflow the dyke for the designed floods; (ii) protection against failure due to seepage or piping; (iii) protection against failure of slopes on the wet and the dry side; (iv) Need for revetment on the wet side slope of the dyke such that no damage to the dyke is caused even under sustained onslaught of waves. To further reduce damage from constant erosive action of waves, need of a suitable shelter with trees or grasses around these structures need to be examined to minimize wave action and also to prevent the drifting of saline sands into the inland areas.

Height of the Dyke

The design height of the dyke is an important aspect of the dyke to provide full protection up to a certain flood. In the beginning, the elevation of the dyke used to be determined on the basis of the highest flood stage on record plus a certain amount of free board as a safety factor. This free board was generally kept as 1 m for river dykes but was varied for sea dykes depending upon the magnitude of the wave uprush. A different approach is also followed wherein hypothetical extreme flood stages based upon the probability analysis of the flood events are assessed. However, first approach continues to be popular in India. In deciding the final height of the dyke, due allowance is made for (i) extremely high water levels that are caused by river floods or by sea tides or by a combination of both, (ii) the wind set up, (iii) the wave uprush, (iv) the possible deterioration of the top of the dyke due to erosion or other causes, and (v) the expected settlement of the dyke after construction. These factors provide safety to the dyke against damage due to the cyclonic storms that usually lash the sea coast with a very high wind velocity. A typical cross-section of the dyke is shown in Fig. 3.5, which also has a provision of drain on the land side. A pictorial view of the dyke in Ratnagiri district is shown in Fig. 3.6 (MoWR, 2017).

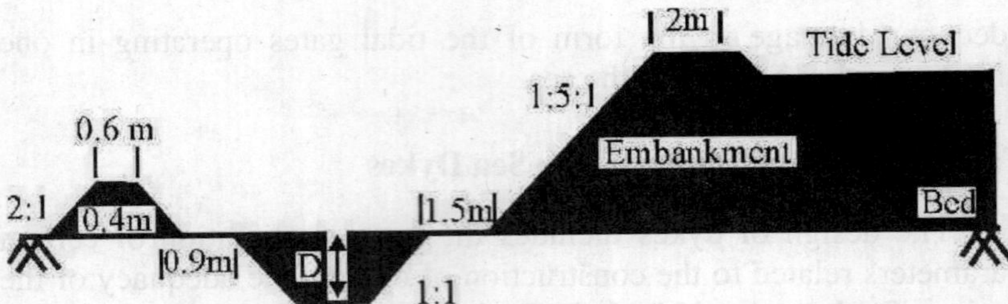

Fig 3.5. A schematic view of an embankment with drain on the land side

Fig. 3.6. A view of the *kharland bund* at Manjare, Taluka Sangmeshwar, district-Ratnagiri, Maharashtra

Side slopes

In general, the magnitude of the wet slope should be decided depending upon the severity of the onrushing wave and the dry slope should be decided based upon the stability of the soil. The height of the dyke, the type of the soil, the foundation conditions and the duration of high water levels against the dyke are other important factors controlling the side slope of the dry side. A relatively low sea dyke, where water levels against the dyke are infrequent and of a short duration, may have a dry slope of 2:1 and for the situation where the water level against the dyke may remain high more frequently, the dry slope should be 3:1. Slope may still needs to be increased if the foundation conditions are weak. The wet slope of the dyke is interrelated with the top of the dyke and the revetment of the slope. With the same type of revetment, a milder slope would reduce the wave uprush and would allow a lowering of the top of the dyke. With the same slope, a more rugged type of revetment would reduce the wave uprush and would also allow a lowering of the top of the dyke. The wet slope of a river dyke is quite often 3:1 while the wet slope of a sea dyke may be as low as 10:1. Sometimes a berm is also incorporated in the slope of a dyke. The purpose of the berm is to act as an energy dissipater by changing the direction of the onrushing water twice and to induce the waves to break at its site. The berm may improve the timing of the recession of one wave so that it will cause the maximum interference with the next onrushing wave. Maintaining a solid, erosion resistant, grass cover over the dyke is a sure prescription of its safety.

The selection of the revetment of a dyke slope depends on degree and the extent of exposure to wave action and availability of materials. The part of the dyke slope that is seldom exposed to water can be protected with a layer of clay and a well maintained grass cover. The part of the dyke slope that is frequently exposed to water so that no firm grass cover can be maintained needs artificial protection. This protection may consist of a layer of riprap that resists the wave action and a sub-layer of gravel that prevents the soil of the dyke from being washed away through the riprap.

Seepage Control in Dykes

Seepage through the dykes makes the dry slope of the dam less stable and may result in piping near the toe of the dyke. In the piping process, the soil from the foundation of the dyke is removed in the shape of a pipe due to the seepage water. The piping starts from the toe of the dyke and continues across the main body of the dyke resulting in the complete failure due to the collapse of the earth material. Piping may also begin with the turbulent water flow in gravel layers underneath the dyke, dislodging particles in underlying or overlying strata of silts and fine sand. Once the dislodging of particles begins, channels are formed that increase in size and length and this may consequently cause the collapse of the dyke. Thus, seepage control assumes significance in the design of dykes.

To check the piping (i) the dyke should be composed partly or completely of impervious material; (ii) in order to control the small amount of seepage that would still come through the dyke and to keep it away from the toe, a horizontal filter can be put underneath the toe; (iii) if impervious material is unavailable at the site, it should be possible to make the dyke of sand and to put a central impervious core, or an impervious layer on the wet slope of the dyke. In doing so, select the impervious material that does not crack and to place and tamp this material carefully so that no big voids are left. The provision of the cut-off wall underneath the dyke also helps in reducing the seepage.

DRAINAGE STRUCTURE : TIDE GATE

When flood waters are excluded from the sea because of the diking system, the drainage problem may get accentuated in the protected area. If outside flood levels are higher than the interior land levels, the natural drainage is impeded. It means some provision is needed either to store the water temporarily inside the dyke or to pump it outside the dyke. To drain the water, tidal gates are provided in the dykes. These tidal gates open only in one direction and allow water to drain from the landside toward the sea side at the time of low tides. During high tides these gates are closed automatically preventing the ingress of water from the sea side. The controlled opening of these gates over flat coastal lands also helps in maintaining the surface water levels in rice crop fields.

Tide gates, or automatic drainage gates as they are often called, provide outlets for the gravity drainage of land side. The typical tide gate structure consists of a pipe or a box culvert through and under the dyke with a gate placed on the ocean end of the culvert (Fig. 3.7). The gates may be circular, square or rectangular in shape and are made of cast iron or steel or aluminum. The differences amongst various sluice gates are due to the shutters and their closing mechanisms to allow only uni-directional flow to river/estuary. They are provided with single or double acting hinges so that they are able to swing free against the culvert opening. The gates may be top-hinged, side-hinged (rectangular tide gate made of aluminum or stainless steel attached to square or rectangular concrete culvert), rubber duckbill made of flexible rubber, bottom-hinged with a pet

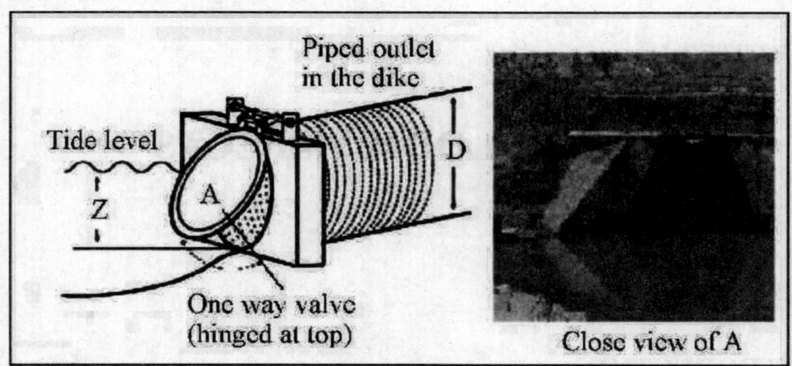

Fig. 3.7. Piped outlet in a dyke with one way valve : Schematic (left) and pictorial (right)

door or self-regulating or buoyant tide gate. In the last type of gate, because of its buoyancy, the lid of the gate remains open, floating above water, most of the time (in other gates, default position of lid is closed position). The only time this type of gate closes is when flood tides reach a level that is high enough to cause upstream flooding (Giannico and Souder, 2005). The gate closes when the tide rises above the water level inside the dyke and opens when it recedes below the water inside the gate. Thus water pressure, caused by the rise and fall of the tide, operates the gate automatically keeping overflow out when tides are in and providing a drainage outlet when tides are out. The size and number of vents are fixed in such a way that the drain discharges fully into the river/estuary with a differential head of at least 0.30 m. Proper location and realistic operational schedules help in effective drainage of the inland areas during the monsoon season.

■■■■

CHAPTER 4

HYDROLOGIC ASPECTS

INTRODUCTION

The water, an important resource for the economic development of a region especially agriculture is becoming scarce. It is because of increasing population, rapid urbanization, increasing industrialization and continuously changing eating habits with improved living standards. As such, any human intervention is now examined in the context of its impact on water resource both in terms of water quantity and quality. Principles of hydrology are often used to assess these impacts.

During the reclamation process many on-farm land development activities are initiated (See chapter 3). These include, construction of earthen embankments to store water for leaching and crop production, construction of drainage channels, installation of tube wells, installation of subsurface drainage for waterlogged saline soils, increasing irrigation requirement all impacting the water resources of the region. In the case of reclamation of alkali soils, application of amendments results in perceptible improvement in infiltration rate of the soils. Clearly both the surface and subsurface components of the water balance of the region will experience a sea change. Since water is key element in any reclamation program, it is proposed to discuss some of the hydrological implications of land reclamation that might affect the water resource in the area upon large scale land reclamation.

HYDROLOGIC NATURE

Hydrology comprising of the terms hydro (water) and logy (study) broadly means the science that deals with the study of water. According to Chow (1964), hydrology, in its broadest sense, is the science that treats the waters of the earth, their occurrence, circulation and distribution, their chemical and physical properties and their reaction with their environment, including their relation to living things. Thus, the domain of hydrology embraces the full life history of water on the earth. To illustrate its practical importance in the reclamation of salt affected soils, one needs to understand the following processes:

- Changes in interception and storage
- Soil moisture and water transmission characteristics
- Overland and channel flow
- Land use changes, water yield and stream flows
- Rainfall-runoff
- Distribution of water in surface and subsurface components
- Increase in water demand
- Salt and water movement impacting land and water resource quality
- Overall water balance

Alkali Soils

The hydrology of alkali land reclamation program has to be seen in the context of the following hydrologic and morphological features of these soils.

(i) The high exchangeable sodium has a marked influence on the infiltration rate, being only a small fraction of the normal land of similar texture group. A major fraction of rainfall therefore, is immediately transformed to runoff. It creates drainage problem in the low lying areas and temporary flooding of the crop lands

(ii) Generally, the alkali area has a flat topography with a slope ranging from 0.1 to 0.3%. In flat lands, during heavy rains, the capacity of the field drains is generally inadequate to remove the excess rain water because of the limited bed slope and low outfalls available at their confluence with the bigger natural streams which also remain in flooded conditions during that period. Even the adjoining normal cultivated fields are affected adversely due to drainage congestion caused by the surface runoff from the alkali areas.

(iii) Partially sodium saturated clay is highly dispersed in the surface layer. It is transported downward through the soil profile, such that it gets accumulated at lower depths where the soil may develop a dense layer of low permeability

(iv) The highly alkali lands are barren without any vegetative growth. As such, negligible resistance is offered to the overland flow following a rainfall event

(v) The groundwater quality is generally good (Table 4.1).

Table 4.1. Average chemical composition of groundwater

EC (dS/m)	Ions (me/l)						RSC (me/l)
	Na	**Mg**	**Ca**	**HCO_3**	**CO_3**	**Cl**	
0.60	2.6	3.9	1.1	3.5	3.0	1.0	1.5

Saline Soils

The hydrology of saline land reclamation program has to be seen in the context of the following hydrologic and morphological features of these soils.

(i) The groundwater in waterlogged saline soils is usually shallow although in arid regions where the rainfall is very low, it may be deep. In the former case, a major fraction of the rainfall is transformed into runoff as the soil has limited capacity to absorb water

(ii) The quality of groundwater is generally poor rendering it unfit for irrigation in normal use. As such, any addition to the groundwater during reclamation without adequate provision of drainage may accentuate the problem of water logging.

(iii) The coastal saline areas, offer several peculiar hydrologic problems besides the normal problems encountered in inland saline soils. Some of these aspects of coastal saline soils are included as a separate section in this chapter.

RAINFALL CHARACTERISTICS

The rain water is the best quality water and its total quantum and distribution together with availability of irrigation water determine the constraints and opportunities for the reclamation of salt-affected soils. Rainfall characteristics such as total annual and seasonal rainfall, rain intensities, and the frequency of rain storms govern the reclamation planning of salt affected soils. Although the salt affected soils are mainly distributed in arid, semi-arid and sub-humid regions, as an example we describe the rainfall, and other climatological parameters of a gauging station in the semi-arid region of the Indo-Gangetic plain. In general, this region has a sub-tropical continental, semi-arid monsoonal type of climate which is characterized by four seasons, viz. (i) the dry summer season from March to June; (ii) the hot rainy season from July to September; (iii) the post monsoon warm season in October and November, and (v) the cold season from December to February. The data in Table 4.2 reveal the variation in rainfall over a period of 16 years (2001-2016). It shows that the rainfall varied from a low of 340.7 mm in 2006 to 1093.3 in 2010 with an average of 674.1 m. Almost 80% of the total rainfall is received during the monsoon season from July to mid-September. During this period some of the storms occur with high intensity creating excess runoff and flooding. The monthly rainfall plotted as a histogram to understand and visualize the monthly variation in rainfall over a year is given in Fig. 4.1. The data reveal that the annual rainfall at Karnal was only 477.1 mm during the year, 2009 and varied widely being 0.00 mm in 3 months (October - December) to a maximum of 215.1 mm in the month of September.

The seasonality of rainfall in the regions of salt affected soils exerts an overall control on the water availability and reclamation practices whilst variability imposes a degree of uncertainty. Important as these aspects are, it is often the characteristics of individual rain storms which are of the greatest importance. The characteristics such as intensity, duration and the frequency of occurrence are also considered in designing of the irrigation and drainage systems.

Table 4.2. Annual rainfall variation at CSSRI, Karnal during 2001-2016

Year	Rainfall (mm)	Year	Rainfall (mm)
2001	835.0	2009	477.1
2002	611.0	2010	1093.3
2003	502.5	2011	707.7
2004	717.1	2012	494.5
2005	671.3	2013	875.9
2006	340.7	2014	713.1
2007	761.8	2015	673.7
2008	682.9	2016	627.5

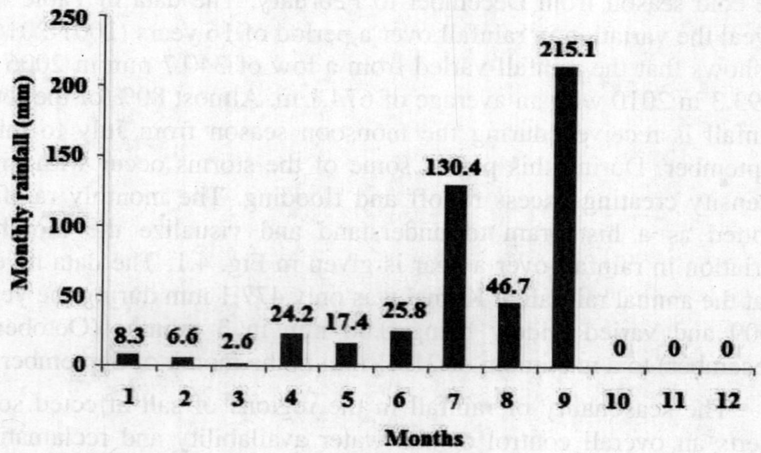

Fig. 4.1. Monthly rainfall distribution at CSSRI gauging station, Karnal during 2009 (values above the bars are rainfall in mm during the month)

The rainfall pattern of several stations located in the areas largely infested with alkali/saline soils and waters has been analyzed by Gupta *et al.* (1973). The data (Table 4.3) for the Karnal gauging station reveal that there is very little difference in 2, 3, 4 and 5 days maximum rainfall values. Therefore, for all practical purposes, a design storm of two days can be selected to design the water management systems for salt affected soils located in this region. It is also seen that two days storms of the five years return period is around 20 cm with a corresponding length of the dry spell of 34 days within the monsoon season. It indicates that conventional drainage comprising of letting the runoff to the river immediately following the rainfall is not a desirable practice. Proper drainage system with adequate provision for the retention of rain water within the catchment area (currently used slogan *´Khet ka pani Khet mein Gaon ka pani Gaon mein´*) will prove to be the perfect solution. It calls for scientific rain water management technologies to solve the irrigation and drainage problems of the region (Narayana *et al.*, 1978).

Table 4.3. Maximum storm rainfall (cm) and dry spells of different return periods

Rainfall duration (days)	Return period (years)					
	1.01	**2.33**	**5**	**10**	**25**	**100**
1	4.1	12.0	15.2	18.3	22.1	28.9
2	5.1	15.5	20.1	23.8	28.5	33.5
3	6.1	17.1	21.9	25.8	30.7	38.1
4	6.7	17.9	22.8	26.8	31.8	39.4
5	6.9	18.5	24.1	28.5	34.0	41.9
Maximum dry spell in monsoon season (days)	15	28	34	39	45	54

Since the water needs of land reclamation and crop production program change within the growing season, probability estimate for the whole season alone may prove to be insufficient. It may be pertinent to have a probability estimate of the occurrence of a particular sequence with rainfall amount throughout the season by different dates. The rainfall probability and reliability data revealed that at 75% probability, the occurrence of dependable rain (> 10 mm) in any five days period during June to September is limited only to certain periods during August. At no time there is the assurance that the rainfall is likely to meet the evaporative demands of the crops in the region. Such a situation also calls for greater emphasis on rainwater conservation not only from drainage system capacity point of view but also to meet the water requirement of crops during the dry spells. Since rice is the recommended first crop for alkali land reclamation, it can be seen that possibility of getting 10 cm of rainfall required for transplanting of the crop is only around middle of July. Transplanting prior to this date will have to be planned with supplementary irrigation.

Besides what has been stated, drainage systems need to be designed for storms of pre-decided durations with certain intensity. Rambabu *et al.* (1979) analyzed the rainfall characteristics for different parts of India. The envelope of intensity-duration-frequency (IDF) curves can be fitted using Horner's equation given as:

$$I(t) = (kT^a)/(t+b)^n \quad (4.1)$$

Such that I is the rainfall intensity, cm/h of a given duration and return period, t is the storm duration (hr), T is the return period, years and k, b, n and m are constants. The range of values for these constants for five zones and five places in India are given in Table 4.4 (Ram Babu *et al.*, 1979). Knowing these values and duration, eq. (4.1) can be used to calculate the design storm. For example, if the rainfall intensity is to be determined for use in Rational Formula to assess runoff from a given rain storm, the time of concentration is used for t in eq. (4.1).

HYDROLOGIC EFFECTS OF ALKALI SOIL RECLAMATION

To assess the complete hydrological influence of any change in land use is to assess how each element of the hydrologic cycle is affected by the change. The important elements of the hydrologic cycle most affected by the land use changes are: (i) interception and storage, (ii) infiltration characteristics, (iii) catchment retention storage, (iv) runoff characteristics, and (v) groundwater recharge. Singh (1980) evaluated the effects of alkali land reclamation on these elements of the hydrological cycle.

Interception and Storage

Reclamation begins with on-farm land development and is followed by cropping. These changes result in increased interception losses and greater storage within the catchment. Increase in these parameters is reflected in reduced runoff.

Table 4.4. Intensity-duration-return period relationships, India

Zone	Station	K	a	b	n
Northern zone		1.50-14.41 (5.92)	0.103- 0.273 (0.162)	0.25-1.40 (0.50)	0.629- 1.296 (1.013)
	Amritsar	14.41	0.1340	1.40	1.2963
Central zone		4.68-11.45 (7.47)	0.108- 0.221 (0.171)	0.15-1.25 (0.75)	0.859-1.121 (0.960)
	Jabalpur	11.379	0.1746	1.25	1.1206
Western zone		3.48- 7.79 (3.97)	0.127-0.209 (0.165)	0.0 – 0.75 (0.15)	0.485- 1.092 (0.733)
	Veravel	7.787	0.2087	0.50	0.8908
Southern zone		5.25-7.14 (6.31)	0.126-0.171 (0.152)	0.5 (0.5)	0.802- 1.128 (0.947)
	Bengaluru	6.275	0.1262	0.50	1.1280
Eastern zone		4.94-16.52 (6.31)	0.115- 0.0.156 (0.135)	0.5-1.50 (0.50)	0.819- 1.073 (0.880)
	Sagar island	16.524	0.1402	1.50	0.9635

Note: The range of values is given followed by average value for the zone in parenthesis

Infiltration

The effect of land reclamation on infiltration rate has been highlighted in Chapter 2 (Fig. 2.7). It revealed that infiltration rate of a reclaimed alkali soil is much higher than a non-reclaimed alkali soil. It is natural that it happens over a time period. The time variation of infiltration rate is discussed in this section based on the Horton (1940) equation. This equation is written as follows:

$$f = fc + \left(f_0 - f_c\right) e^{-kt} \tag{4.2}$$

In eq. (4.2) parameters f_c and f_o represent the basic and initial infiltration rate respectively. The parameter k defines the rate of decrease of the infiltration capacity f from initial infiltration rate f_o

to the final infiltration rate, f_c. During the process of alkali land reclamation, addition of chemical amendment alters the physical properties of soils to a considerable extent including the infiltration rate characteristics. Therefore, the parameters f_c, f_o and k of eq. (4.2) will also change with change in soil ESP. For the alkali soils of the Karnal region, the parameters of Horton's equation were related to ESP by the following equations (Singh, 1980):

$$f_c = Exp(1.90 - 0.07\ ESP) \tag{4.3}$$

$$f_o = Exp(3.46 - 0.04\ ESP) \tag{4.4}$$

$$k = 0.15 \ln(ESP) + 0.24 \tag{4.5}$$

If the values of f_c, f_o and k, given as a function of ESP by eqs. (4.3), (4.4) and (4.5), are incorporated in Horton's basic equation (eq. 4.2), a universal infiltration equation for alkali soils with varying ESPs can be obtained.

Catchment Retention Storage

Singh and Tyagi (1980) proposed a mathematical model that can be used to estimate the rainwater availability upon reclamation of alkali lands. According to this hypothesis two factors contribute to higher availability of rainfall.

- Increase in the basic infiltration rate over a span of several years (Fig. 4.2)
- Increase in watershed retention storage capacity owing to the field *bunding* for reclamation.

Mathematically the two processes are expressed as (Singh *et al.*, 1978):

$$R_w = f(I) + f(s) \tag{4.6}$$

Here R_w is the amount of water generated from rain water, $f(I)$ is the basic infiltration rate function over the span of years until complete reclamation and $f(s)$ is the watershed storage function.

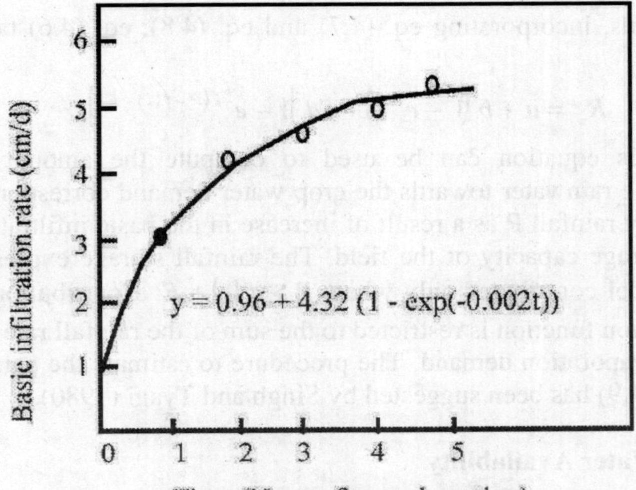

Fig. 4.2. Increasing trend of basic infiltration during reclamation

Improvement in the basic infiltration rate is expressed as an exponential function:

$$f(I) = a + b\left(1 - e^{-ct}\right) \tag{4.7}$$

Here, a is the basic infiltration rate of un-reclaimed alkali soils, cm/d; b is the increase in the basic infiltration rate during the span of years until complete reclamation (cm/day); c is a parameter governing the rate of increase in the basic infiltration rate during the time span of reclamation and t is time in years from the start of the reclamation work.

For the case study presented in this chapter, values of a, b and c have been found to be 0.96, 4.32 and 0.002 (Fig. 4.2).

For the watershed storage, equation given by Linsley *et al.* (1949) has been used:

$$f\ (s) = Sd\left(1 - e^{-KP_e}\right) \tag{4.8}$$

Such that Sd is the maximum storage capacity, cm, K is a constant, and P_e is the rainfall excess (gross rainfall, P minus the sum of evaporation, E and infiltration *f(I)* and interception) and is given as: $P_e = P - E - f(I)$ upon neglecting the interception.

Thus, incorporating eq. (4.7) and eq. (4.8), eq. (4.6) takes the form:

$$R_w = a + b\left(1 - e^{-ct}\right) + Sd\left[1 - e^{-k(P - f(I) - E)}\right] \tag{4.9}$$

This equation can be used to compute the amount of the available rainwater towards the crop water demand corresponding to the daily rainfall P as a result of increase in the basic infiltration rate and storage capacity of the field. The rainfall storage expression in the model contributes only when $P > f(1) + E$. Contribution of the infiltration function is restricted to the sum of the rainfall rate and the daily evaporation demand. The procedure to estimate the parameters of eq. (4.9) has been suggested by Singh and Tyagi (1980).

Rain Water Availability

To estimate the rain water availability due to reclamation, the daily rainfall data for 31 years were analyzed for ascertaining the relative frequency of rain storms of various magnitudes. The average number of rainy days and the average daily pan evaporation values for different months were also calculated (Table 4.5). The basic infiltration rates were calculated following reclamation by fitting the infiltration function f (I).

The quantity of water supply generated by rainfall storms of various magnitudes was computed using eq. (4.8) for an area measuring 350 ha in an operational research project area of the Central Soil Salinity Research Institute, Karnal. The irrigation water supplied from canal and/or tube wells was measured for individual farms. The average water applied for rice and wheat was 70 cm and 36 cm respectively. The water supply generated from rain water for the rice and wheat crops was found to be 55.9 and 10.6 cm, respectively. The total water used for rice-wheat system was 172.6 cm, which is very close to commonly assumed average value of 170 cm of water use in the rice and wheat system for this region.

The reclamation of alkali land has been found to result in increased rainwater utilization within the watershed. Nearly 39% of irrigation water needs of the rice-wheat rotation can be met from rain water alone which otherwise goes as waste besides creating the drainage problem in the low lying area.

Table 4.5. Relative frequency of the rainfall storm depths and average water resource made available due to reclamation

Months	Av. daily evap. (cm)	Av. rainy days	Depth range (cm) of rainfall storms						Av. water supply made available (cm)
			0-3.9	4-7.9	8-11.9	12-15.9	16-19.9	20-23.9	
January	0.22	3	0.960	0.940	-	-	-	-	2.37
February	0.33	3	1.000	-	-	-	-	-	2.38
March	0.53	3	0.963	0.037	-	-	-	-	2.52
April	0.90	3	0.972	0.020	-	-	-	-	0.95
May	1.20	3	1.000	-	-	-	-	-	0.34
June	0.91	6	0.889	0.111	-	-	-	-	4.90
July	0.54	11	0.0876	0.098	0.016	0.004	0.004	0.004	16.50
August	0.41	16	0.842	0.090	0.053	0.015	-	-	26.75
September	0.43	6	0.885	0.083	0.016	0.083	-	0.008	8.65
October	0.37	2	0.909	0.045	0.045	-	-	-	2.74
November	0.29	1	1.000	-	-	-	-	-	0.85
December	0.22	1	0.967	-	-	-	0.032	-	1.22

Av. = average, evap. = evaporation

Retention and Runoff Characteristics

The reclamation of alkali watershed besides altering the retention storage also affects the quantity and time distribution characteristics of runoff. The retention storage was found to increase with increasing area brought under reclamation (Fig. 4.3). From a study on paired identical runoff plots (un-reclaimed and reclaimed alkali) each of 1.26 ha, it was observed that for the average size of rainstorms, the runoff volume as well as the peak discharge values were reduced (Fig. 4.4). The peak discharge reduced by more than 50% whereas the runoff volume reduced by 75% (Singh *et al.*, 1979). These observations indicate that the drainage requirement of alkali soils is reduced to a greater extent after the soils are reclaimed. Thus, a limited capacity drainage network can take care of the drainage needs of reclaimed alkali lands.

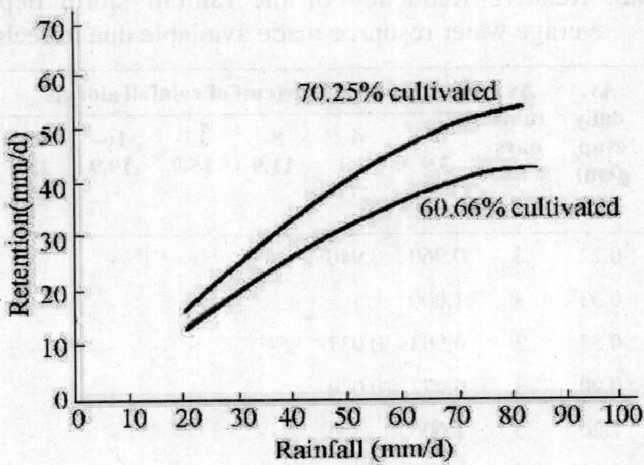

Fig. 4.3. Changes in catchment retention with increase in area reclaimed

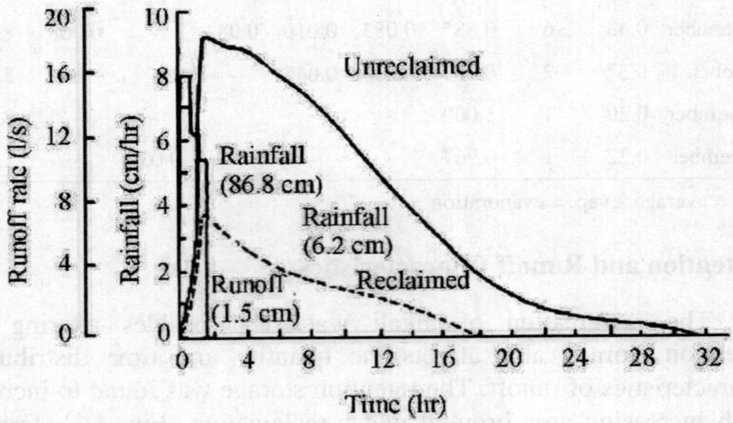

Fig. 4.4. Runoff hydrograph from reclaimed and un-reclaimed areas for a storm on 11.09.1976.

WATER BALANCE STUDIES

Groundwater recharge and discharge components are significantly altered during land reclamation. The principal causes are: (i) rain water retained within the *bunded* fields; (ii) the improved infiltration rate resulting in higher deep percolation losses and (iii)

the increasing use of groundwater for irrigation. The contribution to groundwater recharge as a result of alkali land reclamation program was assessed using the water balance approach. The input sources of all the water within the watershed such as rainfall, canal/groundwater supply and probable groundwater inflow to the study area were observed/assessed. Similarly the output from the watershed comprising of evapotranspiration, runoff, water storage as soil moisture, groundwater rise, and the groundwater export were monitored. The hydrological balance equation was descried as follows (All components expressed in similar units say cm or mm):

$$P = I + \Delta SMS + RO + \Delta SS + \Delta GWS + GWO + ET + SI \quad (4.10)$$

Here, P is the precipitation, I is the interception loss, ΔSMS is change in the soil moisture storage, RO is the surface runoff, ΔSS is the change in the surface storage, ΔGWS is the change in the groundwater storage, GWO is the groundwater outflow, ET is the evapotranspiration and SI is the surface inflow to the watershed. All the parameters for the study site are reported in Table 4.6.

Table 4.6. Monsoon season water balance analysis for an experimental watershed

Water balance components	**Years**			
	1976	**1977**	**1978**	**1979**
Percentage cultivated area	66.66	70.25	75.33	80.03
Rainfall (cm)	86.70	58.90	74.10	50.60
Canal supply (cm)	3.99	12.77	6.19	19.00
Water exported (cm)	4.8	5.4	2.8	7.31
Runoff (cm)	25.8	7.0	5.0	3.3
Evaporation (cm)	55.25	48.92	51.45	61.03
Soil moisture (cm)	7.2	10.0	10.0	6.8
Net local recharge (cm)	-2.36	0.35	11.04	-8.84
Water table depth (June) (m)	6.0	6.72	5.06	5.30
Water table depth (Sept) (m)	4.51	5.34	3.12	6.68
Change in the groundwater storage (cm)	14.9	13.8	19.4	-13.8
Probable recharge from outside (cm)	17.26	13.45	8.36	-4.96

The measured values of various hydrological and meteorological parameters for the seasonal water balance revealed that the net local recharge is increased with increasing area put under cultivation as a result of land reclamation. It can be concluded that hydrological balance is properly safeguarded during the reclamation of alkali lands.

Water Balance: Research Farm

A more rigorous analysis of water balance components was made at CSSRI, research farm of 80 ha (Singh *et al.*, 1979). The June-September water balance components revealed that runoff to rainfall ratio decreased progressively (Table 4.7). Moreover, ground-

Table 4.7. Water balance (June-September) analysis of CSSRI research farm

Particulars	Years					
	1972	**1973**	**1974**	**1975**	**1976**	**1977**
Precipitation (cm)	69.5	58.9	38.4	60.4	86.7	58.9
Canal seepage (cm)*	4.0	4.0	4.0	4.0	4.0	4.0
Total input, I (cm)	73.5	62.9	42.4	64.4	90.7	62.9
Runoff (cm)	12.1	9.4	0.5	1.2	10.0	5.5
Evaporation (cm)	34.2	37.2	37.9	35.8	42.6	26.2
Addition to soil moisture (cm)	9.0	8.0	6.8	8.8	10.9	7.5
Total output, O (cm)	55.3	54.6	45.2	45.8	63.5	39.2
Net local recharge or water surplus, (I-O) (cm)	18.2	8.3	-2.8	18.6	27.2	23.7
Net local recharge or water surplus, (I-O) (% rainfall)	26	14	0	31	31	40
Observed additions to groundwater (cm)	29.1	18.2	17.1	23.9	25.5	22.8
Observed groundwater rise (% rainfall)	42	32	45	40	29	39
Probable recharge from outside, S-(I-O), (cm)	10.9	10.4	19.9	5.3	-1.7	-0-9
Probable recharge from outside, S-(I-O) (% of rainfall)	16	12	52#	8	0	0
Runoff, R (% of rainfall)	17	16	1	2	12	9

*Value adopted from Narayana and Gupta (1974); # The value is unusually high because augmentation tube wells were not operated during the year

water rise almost stabilized at 240 cm, representing about 24 cm of water addition during the season assuming that the drainable pore space is 10%. Besides, the observations emerging from the larger watershed discussed in the previous section could also be established.

Runoff Water Quality

The quality of runoff is impaired in non-reclaimed alkali soils mainly because of excess salts (especially if soils are saline-alkali in nature) and sediments due to the dispersed nature of clay particles. Upon reclamation, substantial improvement in runoff water quality, both on account of reduced salts and reduced sediments, has been observed. Runoff from reclaimed alkali soils may not have any adverse impact on fishes, if stored runoff is used for aquaculture.

HYDROLY OF WATERLOGGED SALINE SOILS

Inland Saline Soils

The basic hydrological principles remain same as are discussed in the context of alkali land reclamation. Without going into these details, few observations are made on the changes in hydrological components. It may be noted that there is no need to apply any amendment to reclaim saline soils. However, to control the water table and for effective leaching, subsurface drainage is an essential pre-requisite to reclaim these soils.

- There will be significant changes in interception losses because of increasing vegetation.
- Runoff in un-reclaimed saline soils is usually high because of lesser water absorbing capacity of these soils. Significant changes in the runoff may be expected for the following reasons:
- On-farm diking of the field to store rainwater for leaching will increase storage significantly impacting the runoff. Besides, a fraction of water will travel through the subsurface drainage system. However, no significant change in infiltration rate will occur due to reclamation.

- Soil and water quality improve over time following reclamation. The disposal of subsurface drainage water in surface water resource may however, impact the quality of water of these resources.

Coastal Saline Soils

The peninsular region of India is bounded by the Arabian Sea on the west, the Bay of Bengal on the east and Indian Ocean to its south (Fig. 3.4). India has one of the most important coastal agro-ecosystem that supports the livelihood security of more than 100 million people. The coastal zone represents the transition from terrestrial to marine influences and vice versa. This zone not only covers shoreline ecosystems, but also the upland watersheds draining into the coastal waters and the near shore sub-littoral ecosystems influenced by land-based activities. The coastal region occupies the deltas, lacustrine fringes, lagoons, coastal marshes and narrow coastal plain or terraces along the creeks. It also includes two distinct major island ecosystems, the Andaman and Nicobar group of Islands in the Bay of Bengal and the Lakshadweep Islands in the Arabian Sea. The coastal agriculture faces several problems, some of them are:

- Climatic hazards (cyclones/floods)
- High humidity
- Saline environment (soil and water)
- Air borne salts
- Shallow saline water table
- Tidal water intrusion
- Heavy rainfall with deep water logging in *kharif* season
- Poor surface and subsurface drainage
- Inadequate good quality irrigation water
- Toxicity of water
- Encroachment of waterways

In spite of these hazards, coastal regions support diverse vegetation ranging from rich tropical rain forests to coastal mangroves. The vast deltaic region on the east coast forms the rice-bowl of the country.

Storms

Two general types of storms affecting coastal regions are - tropical and extra-tropical cyclones while the former originate in tropical latitudes the latter originate in the middle or high latitudes, respectively.

Tropical cyclones are formed only over ocean areas where warm moist air flows into the cyclonic circulation, ascends and cools to the saturation point. The heat released in the subsequent condensation process adds to the energy of storm. A tropical cyclone is an intense low pressure area or a whirl in the atmosphere over tropical or sub-tropical waters, with organized convection (i.e. thunderstorm activity) and winds at low levels, circulating either anti-clockwise (in the northern hemisphere) or clockwise (in the southern hemisphere). From the center of a cyclonic storm, pressure increases outwards. The amount of the pressure drop in the center and the rate at which it increases outwards gives the intensity of the cyclones and the strength of winds.

As per the criteria adopted by the World Meteorological Organization (WMO), India Meteorological Department classifies the low pressure systems in the Bay of Bengal and in the Arabian Sea into 7 classes. Low pressure areas with maximum sustained surface winds of speed between 31 and 61 km/hr (17 to 33 knots, 1 knot = 1.85 km/hr) are called tropical depressions. Once the winds around the low pressure area reach at least 62 km/h, it is called a tropical cyclone and is assigned a name. When wind speed is between 89 and 118 km/hr (48 and 63 kt) it is a severe cyclonic storm (SCS), between 119 and 221 km/hr (64 and 119 knots) it is very SCS and when exceeds 221 km/hr (119 knots), it is a super cyclonic storm.

The extra tropical cyclones develop along frontal zones separating major air masses of different origins. A frontal system in such storms consists of a cold front in the rear of the storm, separating the polar air from the warmer tropical air and a warm front which generally extends from the storm center through the forward part of storm. Most important are the storm that develop as waves along the polar front. Some of these waves develop into intense low pressure systems and cause very severe coastal weather conditions.

The time of occurrence of the storm surge with respect to the stage of the normal astronomical tide is important and can make a lot of difference between serious and minor flooding. Storm surges that coincide with high tide are the most damaging causing extensive destruction and disruption of the agricultural and allied activities in the coastal regions.

In India, cyclonic storms of severe intensities occur in the Bay of Bengal and the Arabian Sea mainly during the pre-monsoon (April-May) and post-monsoon (October-December) periods. Severe cyclones are more frequent in the Bay of Bengal than in the Arabian Sea. Of about 80 tropical cyclones (with wind speeds equal to or greater than 35 knots) forming in the world's waters every year, 6.5% develop in the Bay of Bengal and Arabian Sea, the Bay of Bengal's share coming out to be about 5.5%. The tropical cyclones forming in the Bay of Bengal hit the coast of India (particularly the states of Andhra Pradesh, Odisha and West Bengal). It is observed that 26 of the 35 deadliest tropical cyclones in world history have their origin in the Bay of Bengal. During the past two centuries, 42% of Earth's tropical cyclone-associated deaths have occurred in Bangladesh, and 27% have occurred in India. Bay of Bengal cyclones usually develop near the Andaman Island and initially move in the north-western direction. Later they move north or north east. Cyclones generally move in smooth curved paths but occasionally move zigzag or in loops. On an average, one severe cyclone may be expected to form in the Indian Ocean in the pre monsoon period and one or two in the post monsoon period.

A cyclonic storm with a wind speed of 100 to 150 km/hr and mighty tidal waves of 5-6 m high lashed the coast line on the east coast at about 1700 hours on November 19, 1977, completely washing away several villages in Divi Taluka of Krishna district in Andhra Pradesh. The tidal waves which lasted for several hours swept away thousands of hamlets in the coastal districts of Guntur, Krishna, Prakasam, East Godavari, West Godavari and Visakhapatnam, taking a heavy toll of human lives and cattle heads and colossal damage to property and rich soil. It was followed by widespread and intense rainfalls in the districts of Prakasam, Guntur, Krishna and West and East Godavari on November 20, 1977. Super cyclone crossed Andhra coast about 40 km S-W of Machlipatnam on 4-11 May, 1990. A very severe cyclonic storm crossed Andhra coast near Kakinada on 5-6 November, 1996. A super cyclone remained active during 25-31 October, 1999 and crossed Odisha coast near Paradeep at noon of 29 October. Cyclonic storm, Nilam was the deadliest tropical cyclone to directly affect South India since cyclone Jal in 2010. Originating from an area of low pressure over the Bay of Bengal on October 28, 2012, the system began as a weak depression 550 km (340 mi) northeast of Trincomalee, Sri Lanka. Over the following few days, the depression gradually intensified into a deep depression, and subsequently a cyclonic storm by October 30. It made landfall near Mahabalipuram on October 31 as a strong cyclonic storm with peak winds of 85 km/h (50 mph). In Chennai's Marina Beach, strong winds pushed piles of sand ashore and seawater reached nearly a 100 m (330 ft) inland. Cyclone Ockhi originated from a low pressure area over southwest Bay of Bengal and adjoining areas of south Sri Lanka and equatorial Indian Ocean on 28 November, 2017. It crossed south coast of Gujarat between Surat and Dahanu as a well-marked low-pressure area on December 6. The storm traversed a distance of 2,538 km. In its entire course, Cyclone Ockhi left a trail of massive destruction in Sri Lanka, Lakshadweep, South India, and Maldives as it strengthened from a depression to a mature cyclone.

On the basis of analysis of 525 cyclonic storms recorded from 1877 to 2016. it has been observed that ocean warming from climate change has increased the frequency and average power of regional cyclone storms. Since 1960, a composite index of cyclone frequency and power has increased by about 50% ((Bandyopadhyay *et al.*, 2018).

TIDAL HYDROLOGY AND SALINE SOILS

A tide cycle is the portion of the tide curve between any two consecutive low or two high waters. In large bays and near the mouths of large rivers the fluctuation is very closely represented as cosine curve. Tidal fluctuations are caused by the forces of the sun and the moon acting on the earth's surface. Higher tides occur when the positions of the sun and the moon cause these forces to act together. Neap or low tides will occur when the positions of the sun and the moon are such that their forces act in opposition to each other. Tidal fluctuations near Rajapettai (79°46′22″ and 11°40′55″)

located very near to the coast at 460 m, at full and new moon is shown in Fig. 4.5 (Singaraja *et al.*, 2018). With high tides, saline sea water inundates large areas around the creeks, periodic inundation resulting in the formation of saline soils.

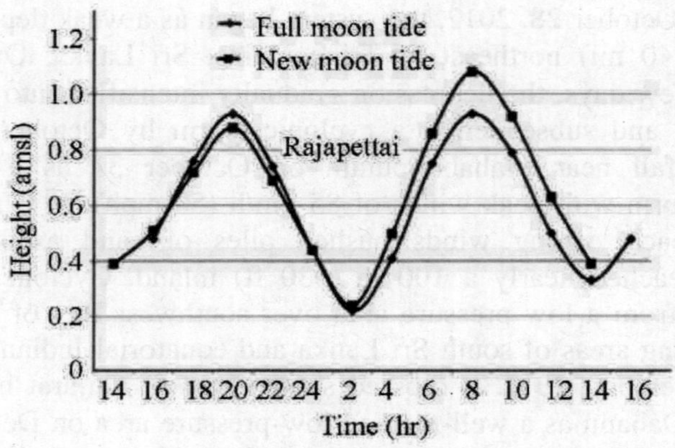

Fig. 4.5. Tidal wave propagation through coastal aquifer

Tidal Effects on Groundwater Fluctuations

Singh and Pandey (1977) studied the effect of tide on the groundwater fluctuations at Parganv *Kharland* farm in Raigarh district of Maharashtra. Observations were made in piezometers 1.5 km away from the sea dyke along the bank of the Shan Bagdi creek where the top 45 cm layer of soil was clay underlain by impure $CaCO_3$ (*Kankar*). The fluctuation of the groundwater table corresponding to the maximum tide of 1.5 m height was only 6.4 cm for an average groundwater table depth of 1.81 m. Similar studies conducted by Rao (1977) at Canning, revealed that the tidal wave in the river propagated similar movement in the aquifer up to a distance less than 532 m only from the river bank (Fig. 4.6). The observed amplitude of the wave at 382 m distance from the river was recorded to be 0.47 m against the amplitude of 2.4 m in the river water level. A time tag of 0.5 hr was also observed during the rising tide and 2.5 hr during the falling tide. However, Singaraja *et al.* (2018) observed that the sinusoidal nature of tide is not well reflected in the water level at Rajapettai being far from the sea. Nonetheless, a minor increase in water level is noticed at the time of high-tide.

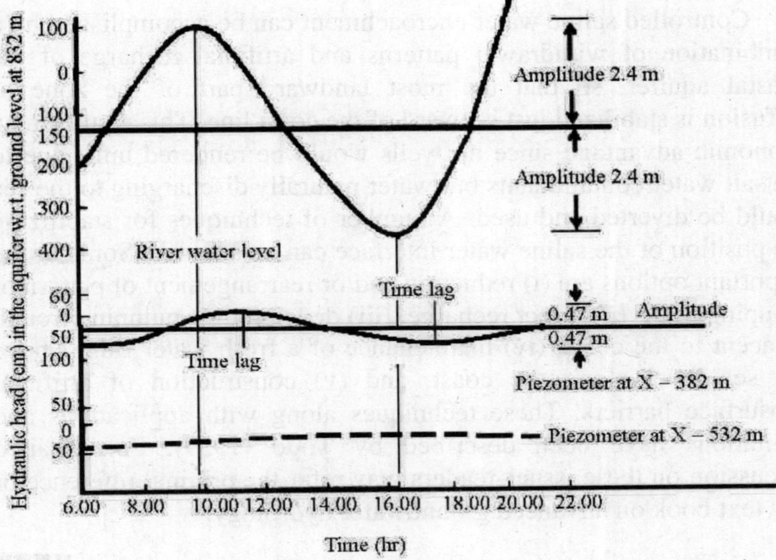

Fig. 4.6. Tidal wave propagation in coastal aquifers

Coastal Aquifer Management

More than half of the world's population lives within 60 km of a coastline. The demand for groundwater in these areas is increasing and in many cases is causing stress on the groundwater resource. A careful and detailed analysis of spatio-temporal variation in water levels reveals useful information on the coastal aquifer system. Seawater intrusion is a natural phenomenon driven by the density difference between freshwater and seawater, but the extent of intrusion has been largely aggravated by the extraction of groundwater. The withdrawal of fresh groundwater from the coastal aquifers reduces the volume of seaward flow of fresh water. Because of this reduction in flow, the zone of diffusion slowly migrates landward to a new position of quasi-equilibrium which reflects the new conditions. In order to have a net withdrawal of water from a coastal aquifer, it is necessary to suffer some landward movement of the saline water interface. This can be considered to be controlled saline water encroachment. Beyond this, intrusion can significantly impact the quality of groundwater due to saline water encroachment into the fresh water aquifers.

Controlled saline water encroachment can be accomplished by a combination of withdrawal patterns and artificial recharge of the coastal aquifer, so that the most landward part of the zone of diffusion is stabilized just seaward of the coast line. This results in an economic advantage since no wells would be rendered unfit due to the salt water contaminants but water naturally discharging to the sea would be diverted and used. A number of techniques for stabilizing the position of the saline water interface can be adopted. Some of the important options are (i) reduction and/or rearrangement of pattern of pumping draft, (ii) direct recharge, (iii) developing a pumping trough adjacent to the coast, (iv) maintenance of a fresh water ridge above the sea level along the coast, and (v) construction of artificial subsurface barriers. These techniques along with applications and limitations have been described by Todd (1959). For detailed discussion on these issues readers may refer the original reference or any text book on advanced groundwater hydrology.

■■■■

CHAPTER 5

IRRIGATION PRACTICES AND METHODS

INTRODUCTION

The main objective of irrigation is to prevent any moisture deficit to the crops during the growing period, so that the crop growth and yield are not adversely affected due to high matric stress. In salt affected soils, besides the matric stress, osmotic stress also impacts the crop growth. Since these stresses are additive in nature, several other aspects come into play while planning and managing irrigation in salt affected soils or irrigation with saline/alkali water. For example, additional water is required in the land reclamation program mainly to meet the leaching needs. Irrigation water management in salt affected soils primarily aims to control the volume and frequency of irrigation water applied to crops, so as to meet the crop needs, minimize salt stress and conserve water resources. It is now known that irrigation water management can increase agricultural production on reclaimed salt affected soils more than any other single agricultural practice. This chapter highlights the principles and practices of irrigation water application in salt affected soils.

WHY IRRIGATION PRACTICES DIFFER IN SALT AFFECTED SOILS?

The salt affected soils differ in several physico-chemical characters than their normal counterparts. As such, saline and alkali soils have altogether different behaviour during irrigation. For example; the electrolytes concentration in the soil solution of saline soil keep the soils coagulated and consequently the water transmission properties such as the infiltration rate and hydraulic conductivity are more favourable than in alkali soils. On the other hand, the high osmotic pressure of the soil solution as a result of the high electrolytic concentration reduces the availability of water to the plants. Besides, the soluble salts move with the water and therefore, may get leached or accumulate at different spots during irrigation. There are certain situations where soluble salts and water do not travel together with the same velocity as is the case of evaporation, where water is lost but salts are left behind. In fact, salts tend to concentrate at places where water evaporates. For example, in furrow irrigation salts will move to the top of the furrow in saline soils or saline-alkali soils. Besides, in surface irrigation methods osmotic pressure increases with each passing day after the irrigation event. On the other hand, high sodium content on the exchange complex in alkali soils exercises a dispersive effect on soil colloids. It results in poor soil physical and water transmission properties. These characters impede the growth of the plants besides affecting the availability of water to the plants. Saline-alkali soils may pose additional problems as both the salt concentration and poor physical properties will influence the irrigation regime. As such, one may need to adopt different irrigation approaches for the management of irrigation water on the two different types of salt affected soils. The objective is to control and/or modify some of the above factors that affect the water availability to the plants.

Besides what is stated, problem is compounded because all irrigation water contains some amount of salts. The degree of problems increases with increasing salinity of irrigation water (See Example 5.1). Since most of the irrigation water is lost through the processes of evaporation and transpiration, only pure water is lost while the dissolved salts remain behind in the soil. Since even the plant uptake of these salts is negligible, the salts accumulate in the soil to adversely impact the crop productivity in the long-term. Such accumulation has to be compensated by application of an excess of irrigation water to leach the salts out of the root zone to maintain a favourable salt balance. The amount of salt removed through drainage water beyond the root zone must at least equal to that applied with irrigation water. It places several other constraints while choosing a particular irrigation method.

In waterlogged salt affected soils, salts are also added to the root zone by the upward movement of groundwater. Higher the salinity of the groundwater, more the salts added to the root zone. The dilemma in this case is whether to allow the crops to draw a part of the crop consumptive use from the shallow water table. In the event plants are allowed to draw water, a major question to be addressed is as to how to manage the likely build-up of salinity with irrigation?

Example 5.1

Calculate the amount of salts added per ha if the EC of the irrigation water is 0.4 (canal water), 1.0 and 4.0 dS/m respectively to grow rice-wheat cropping with an average application of 1200 mm of water.

Since the salinity of three irrigation waters is less than 5 dS/m, we use the relation:

Salts in mg/l = 640*EC of irrigation water

Canal water = 640 *0.4 = 256 mg/l = $256*10^{-6}$ kg

Volume of water applied = 100*100*1200/1000 = 12000 m^3/ha
= $12*10^6$ l/ha

Salt load applied by canal water = $256*10^{-6}*12*10^6$
= 3072 kg/ha

Similarly, salt load applied by 1.0 and 4.0 dS/m can be calculated by simply multiplying the calculated value for 0.4 dS/m by 2.5 and 10 respectively.

Salt load applied by 1 dS/m water = 3072*2.5 = 7680 kg/ha

Salt load applied by 4 dS/m water = 3072*10 = 30720 kg/ha

Clearly the amount of salts added to the soil root zone increases with increasing salinity of irrigation water.

LOW DEPTH HIGH FREQUENCY IRRIGATION

The traditional irrigation cycle for the normal irrigated soils consists of a brief period of infiltration followed by an extended period of water extraction by the crop. This mode of irrigation has been developed primarily because of the ease of replenishing the moisture storage with the minimum cost. To minimize irrigation frequency, the goal is to maximize the quantity of water stored in the root zone as well as the proportion of the water use by the crop before the next irrigation. Since the infiltration and other moisture transmission characteristics in the alkali soils are quite poor, the infiltration period is always a more important part of the irrigation cycle in these soils. Therefore, the irrigation cycle has to be an infiltration dominated process rather than an extraction dominated process. It brings into play a set of laws governing the frequency of irrigation.

Soil Moisture Retention

The amount of water stored in a soil is a function of sizes and volume of water-filled pores. Presence of exchangeable sodium renders the soil aggregates structurally unstable. During and after irrigation or rainfall, there is a breakdown of soil aggregates and the soil particles so dispersed tend to form a surface seal which in turn greatly impedes water intake of the soil. Acharya and Abrol (1975) observed that following prolonged infiltration and subsequent drainage for 48 h, the alkali soil in the subsurface layers held less water; 5 to 6% by weight or 8 to 9% by volume, in comparison to the normal soil (Fig. 5.1). However, higher initial water content in the alkali soil in the surface layers might have resulted either from impeded drainage at the surface or the tendency of Na-saturated dispersed soil to hold water over and above the pore-space saturation. Similarly soil moisture retention at the 7.5 cm depth during the drying cycle indicated that the gypsum treated alkali soils retained more moisture at a given tension as compared to the non-treated alkali soils. Water retention generally increased with decreasing ESP except at or beyond 300 cm tension, where the difference in moisture retention was relatively less (Table 5.1). Clearly, the water retention increases resulting in increased moisture availability range after reclamation, although it will also depend upon the degree of land reclamation.

Table 5.1. Soil moisture retention characteristics

Soil moisture tension (cm)	Soil water content (cm^3/cm^3)	
	Gypsum treated soil	Alkali soil
50	0.35	0.270
100	0.26	0.24
150	0.24	0.215
200	0.22	0.206
300	0.205	0.200
400	0.200	0.185

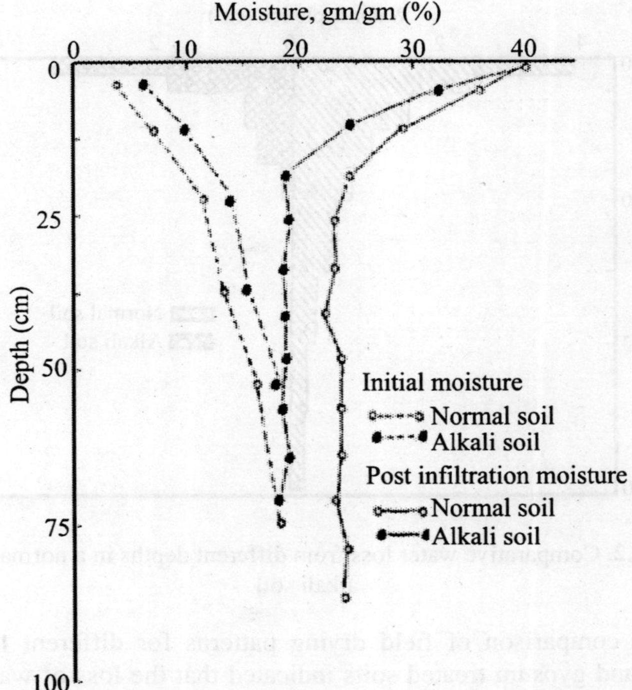

Fig. 5.1. Initial and post infiltration moisture distribution in a normal and an alkali soils

Similarly, as discussed in Chapter 2, the unsaturated hydraulic conductivity decreases rapidly with increasing soil moisture suction in all soils but in the case of alkali soils the hydraulic conductivity decreases rapidly and at the same suction is much lower than that of a normal soil. Therefore, in alkali soils the contribution of water stored in lower soil layers to meet the water needs of crops and that of evaporation from soil surface is restricted. Acharya and Abrol (1975) observed that in the same period of time whereas the losses amounted to about 19.29 cm in the normal soil, only 8.77 cm was lost from the alkali soil (Fig. 5.2). Moreover, there was practically no depletion from layers below 60 cm in the alkali soil, whereas a considerable fraction of the total water was lost from lower soil layers in normal soil. The losses from surface 15 cm accounted for 60% of the total depletion in case of alkali soil, while in the normal soil the losses were only 33%.

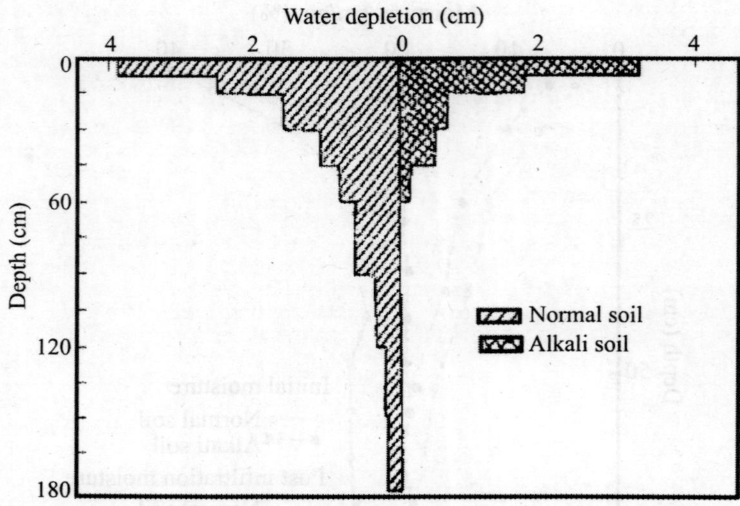

Fig. 5.2. Comparative water loss from different depths in a normal and an alkali soil

A comparison of field drying patterns for different layers of alkali and gypsum treated soils indicated that the loss of water from the surface layer in the case of alkali soil was higher as compared to gypsum treated soils (Table 5.2). However, for lower layers the moisture loss pattern was just the reverse. The ESP significantly affects the magnitude and pattern of the water content changes in the soil profile following evaporation. The rate of drying of two soils having ESP 4 and 38 in the surface 15 cm layer revealed that the surface layers of the high ESP soil with a low upward water flux dried more rapidly than the surface layer of low ESP soil. The results were similar at low and high evaporative demands of 2.5 and 12.5 mm/d (Acharya *et al.,* 1979 and Acharya and Abrol, 1978).

All the factors discussed in the foregoing paragraphs necessitate the frequent irrigations in alkali soils. With increasing irrigation frequency, as long as total water applied remains the same, the low soil water content is eliminated as a factor affecting plant growth. Besides it minimizes deep percolation. Gaul *et al.* (1973) concluded that the grain yield of wheat increased with increasing number of irrigations, subjected to the constant amount of total irrigation water, being nearly the same as under normal conditions. This indicated that the depth of water to be applied at each irrigation must be reduced to obtain good yields without increasing the total amount of water for alkali soils. But depending upon the method of irrigation, water at each irrigation cannot be reduced beyond a certain value particularly in the surface irrigation method.

Table 5.2. Cumulative water loss from different depths of alkali and gypsum treated soils

Time (days)	Cumulative loss (mm)			
	(Soil depth 0-8 cm)		(Soil depth 8-16 cm)	
	Alkali	Gypsum treated	Alkali	Gypsum treated
1	11.0	5.50	0.90	2.70
2	21.25	10.50	1.25	4.10
3	25.20	13.0	1.90	5.0
4	26.60	20.0	1.90	5.75
5	27.50	22.50	1.90	6.50
6	28.0	23.0	1.90	6.50
7	30.50	24.50	3.35	8.50
8	31.0	24.50	3.35	9.20
9	31.4	24.70	3.35	9.15
10	31.6	25.0	3.35	9.85

Similarly, under saline conditions, soil salinity impacts the soil water availability, decreasing it in proportion to its salinity due to the osmotic effect. To withdraw water from a salty solution, the plants must not only overcome the soil water potential but also the osmotic potential, the two potentials being additive in nature. The data reported in Table 2.1 clearly brings out this effect that soils are unable to take up the water with increasing soil solution salinity (Kelly and Rengasamy, 2006 as reported in Rengasamy, 2010). Therefore, to maintain low level of osmotic potential in saline soil, frequent but low depth irrigation needs to be practiced. The irrigations applied at an interval of 10 days gave a grain yield of wheat at 3.1 t/ha as compared to 2.7, 2.4 and 2.1 t/ha, with 15, 20 and 25 days interval respectively, in saline soils in Western

Rajasthan (Sharma *et al.,* 1977). The practical application of the concept in surface irrigation has similar problems as in the case of alkali soils discussed in the previous paragraph.

IRRIGATION METHODS

Commonly used irrigation methods (Fig. 5.3) employed under normal soil and water situations are also employed in salt affected soils and water situations. The proper selection and modifications may however, be required to avoid adverse effects of water stagnation as well as salinization and alkalization of the soils. For example, adoption of flood irrigation and subsurface irrigation may not be desirable under salt affected soils. The later may, however needs to be managed if it is occurring naturally in waterlogged soils having shallow water table.

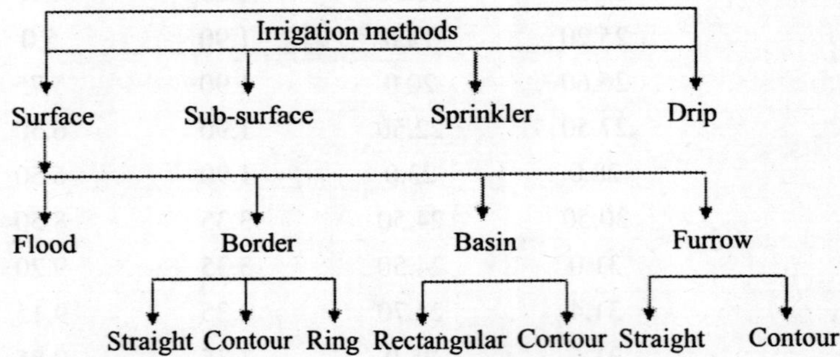

Fig. 5.3. List of irrigation methods

SURFACE METHODS

Surface methods of irrigation include flood, border, basin, and furrow which are further categorized in several groups. All these methods apply water as per irrigation scheduling usually allowing 50% per cent of the readily available water in the root zone to deplete before the next irrigation. The application of these methods is briefly discussed.

Flooding Method

Flooding method has been a common practice until recently as it was passed on from one generation to another generation. Since the irrigations water is applied without any control what so ever, it is also called uncontrolled flooding. In some parts of India, the practice is also known as field to field irrigation. It is an inefficient way of water application as a lot of water is wasted as deep percolation losses with efficiencies being as low as 25% and at the most 40%. The practice results in temporary rise in water table as well as discourages the farmers to go in for land levelling and other improved techniques of crop cultivation. This method is not recommended for salt affected soils.

Border Irrigation Method

In this method, the fields are divided into a number of long parallel strips, generally 5 to 15 m wide and 75 to 300 m in length separated by small border ridges or low dykes of about 15 cm height. Length of the border and slope depend upon the soil type as reported in Table 5.3.

Table 5.3. Border lengths and slopes as affected by soil texture

Soil type	Length of border (m)	Slope (%)
Sandy and sandy loam	60-120	0.25 - 0.6
Medium loam soil	100-180	0.20 - 0.4
Clay loam and clay soil	150-300	0.05 – 0.2

Each strip is irrigated independently. The water spreads and flows down the strip in a sheet confined by the border ridges (Fig. 5.4). After sufficient water is applied to one strip, the irrigation stream is turned into another strip. For uniform flow, it is necessary that longitudinal slope must be uniform, and the transverse slope must be zero or negligible (< 0.03%). The method is most suited to soils having moderately low to moderately high infiltration rates. It is not used in coarse sandy soils that have very high infiltration rates and also in heavy soils having very low infiltration rate. This method is suitable for irrigating wide varieties of close growing crops such as wheat, barley, ground nut, fodder crops, pearl millet and *berseem*. The method has been perfected for rice-wheat cultivation in alkali soils and will be discussed in one of the sections of this chapter. In saline soils, this method is likely to pose problems as the salt moving with irrigation water will accumulate at the tail end of the border to adversely impact the crop productivity at that end.

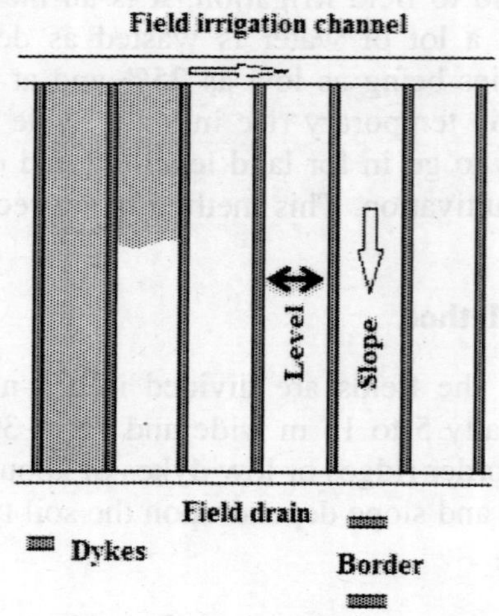

Fig. 5.4. Border irrigation method

Basin Irrigation

In this method, fields are divided into square or rectangular plots each having a nearly level surface (Fig. 5.5). The basins are surrounded by dykes or ridges all around the area forming basins within which the irrigation water can be controlled. The size of the basin varies from 10 m^2 to 25 m^2 depending upon soil type, topography, stream size and crop. For the same steam size, basin area is more in clay than in sand. In some cases (ring basin) the plot may also be circular. In that case it is called ring basin method (see next section). This method is usually practiced in nearly leveled lands and hence the depth of water application is more uniform. The field channels supply water to each basin, during which the basins are filled to desired depth and water is retained until it infiltrates into the soil. It is particularly useful on fine textured soils with low infiltration and percolation rates so that the water is retained on the surface and in the root zone for a longer period of time. Basin irrigation is most commonly used for irrigating crops like groundnut, finger millet, sorghum, and vegetables etc. It is not suited to crops which cannot withstand wet or waterlogged conditions for periods longer than one day such as tuber and root crops like potato, cassava beet and carrot. Basins are most commonly used to leach saline soils because a uniform percolation is pre-requisite for uniform leaching.

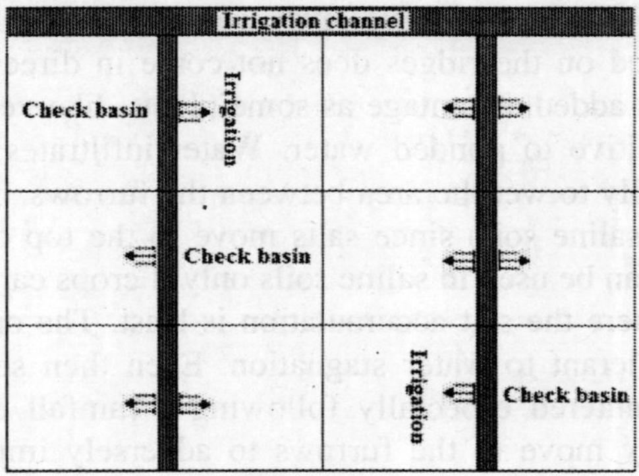

Fig. 5.5. Check basin method of irrigation

Ring Basin Method

This method is a modification of check basin method and is suitable for orchard crops and cucurbits. In this method a circular bund is constructed around each tree/plant or group of plants/trees to create a basin for irrigation. These basins are suitably connected to irrigation conveyance channels such that either each basin is irrigated separately (Fig. 5.6) or a group of basins are irrigated at once by flowing water from one basin to another through inter-connections.

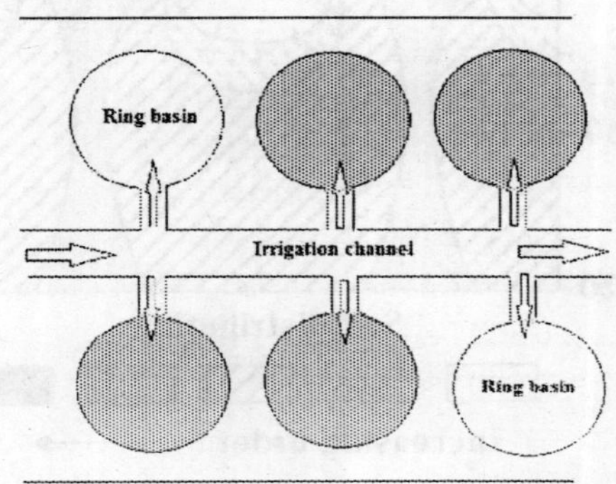

Fig. 5.6. Irrigation of individual ring basins

Furrow Irrigation Method

Furrows are sloping channels made in the soil with the crops being planted on the ridges. The water is applied only in furrows instead of the whole field. This saves water and at the same time the plant cultivated on the ridges does not come in direct contact with water. It is an added advantage as some plants, like vegetable crops, are very sensitive to ponded water. Water infiltrates into soil and spreads laterally to wet the area between the furrows. The method is not suited to saline soils since salts move to the top of the furrows (Fig. 5.7). It can be used in saline soils only if crops can be cultivated in furrows where the salt accumulation is least. The crops however, need to be tolerant to water stagnation. Even then some problems may be encountered especially following a rainfall event. At this time salts may move to the furrows to adversely impact the crop. Therefore, appropriate selection of seed placement is important say somewhere in between the sloping surface. It is suitable to most soils except sand.

Similarly, furrow irrigation has to be managed through seed placement in saline water irrigated lands depending upon whether all the furrow are irrigated or alternate furrow is irrigated.

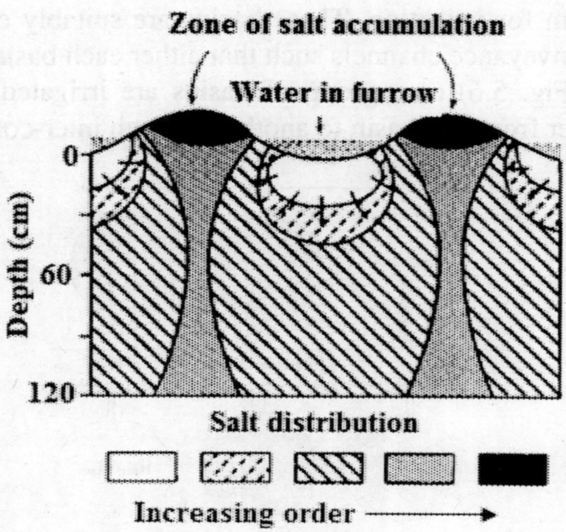

Fig. 5.7. Salt distribution in furrow irrigated saline soils or saline water irrigated land

In a double-row bed system, salt accumulate toward the middle of the bed (Fig 5.8 top). Thus, plants planted away from the middle towards the shoulders of the furrow, encounter relatively low salt stress. Another modification to this method is to have a notch at the center of the ridge and to plant the crop halfway between the wet furrows (Fig. 5.8 bottom). Both these strategies require uniform application of water in the furrows. Otherwise, salts might accumulate closer to one side of the bed near to furrow receiving less water. In that event, the plant growth along that furrow may be adversely affected.

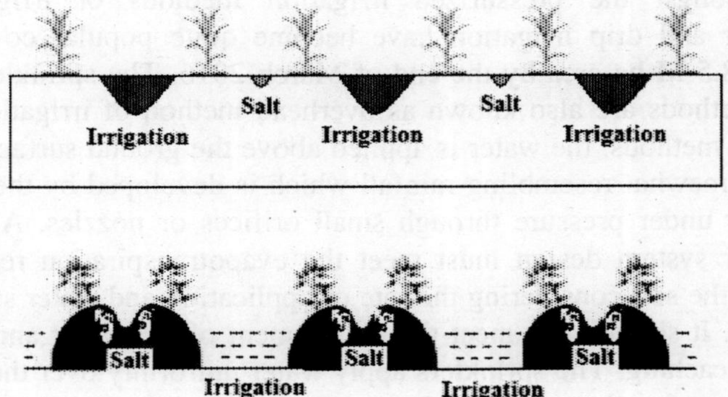

Fig. 5.8. Salt accumulation in double row planting (top) and furrow top with a notch (bottom) where each furrow is irrigated

Alternate furrow irrigation is not an irrigation layout but a technique for water saving. Herein, irrigation is given in alternate furrows. It is further sub-grouped into two classes. In one case, all the irrigations are performed in the same furrow. Dry furrow remain dry throughout the crop growing season. Root distribution in this kind of irrigation is more towards the wet than the dry furrow. This method is sometimes known as skip furrow irrigation. The skipped furrows are sometimes used to raise intercrops requiring less water. Water saving of about 30-35% can be achieved without much yield reduction. With a single-row bed system, salts are pushed across the bed from the irrigated side of the furrow to the dry side. Thus, planting towards the shoulder of the wet furrow is helpful (Fig. 5.9). Care must be taken to ensure that enough water is applied to wet all the way across the bed to prevent build-up in the planted area. This method of salinity management can harm the plant if large amounts of rainfall fill the normally dry furrow and push the salts back across the bed toward the plants. In the second case, dry are wet furrows are alternated i.e. if during the first irrigation, odd number of furrows are used to apply irrigation water, in the next irrigation it will the even number of furrows that will be irrigate. The cycle will continue in this manner ubtil all irrigations are completed. The root distribution in this case will be better than the first case.

Pressurized Irrigation Methods

Amongst the pressurized irrigation methods of irrigation, sprinkler and drip irrigation have become quite popular covering around 8.5 M ha area by the end of March 2016. The sprinkler and perfo-methods are also known as overhead method of irrigation. In both the methods, the water is applied above the ground surface as a spray somewhat resembling rainfall which is developed by the flow of water under pressure through small orifices or nozzles. A good sprinkler system design must meet the evapotranspiration requirement of the soil considering the rate of application and water storage capacity. It should also meet the requirement of the water and crop for salt leaching. The sprinklers apply water uniformly over the land surface keeping the rate of application low enough to prevent runoff. This results in an excellent overall water supply to the crop and adequate and uniform leaching.

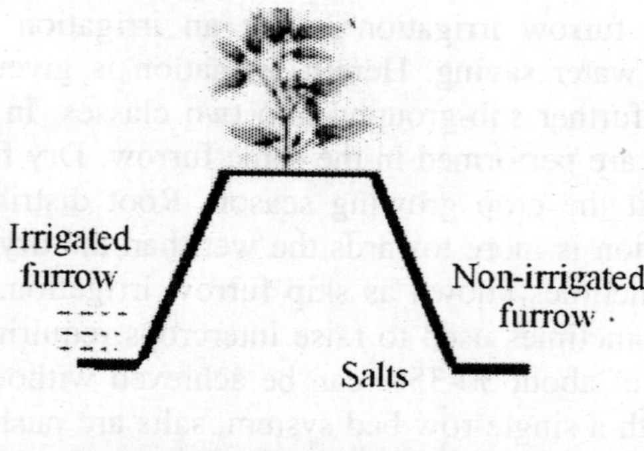

Fig. 5.9. Salt accumulate on the dry furrow side in alternate furrow irrigation

To use saline water with sprinklers, one has to be quite careful. Low saline water say up to 6-7 dS/m may not pose much problem with these methods of irrigation. The crops will grow well depending upon the salt tolerance since crops are able to tolerate more saline water with sprinkler than any of the surface irrigation methods. However, as the water salinity increases crops may be adversely impacted especially under conditions of high temperatures or low humidity levels. Under these conditions, water from the drops will get evaporated both in the air as well as after falling on the leaves resulting in increased salinity of water. Under this situation, plants may be adversely impacted for the following reasons:

- Scorching and leaf burn injury
- Deposition of white salt layer affecting photosynthetic activity
- Absorption of salts by the leaves that may prove toxic to plants. Guidelines on Na and Cl in water used for sprinkler irrigation that may cause foliar injury to crops are given in Table 5.4 (Maas, 1990).

Table 5.4. Susceptibility ranges for crops to foliar injury in sprinkler irrigation

Ion	Na or Cl concentration (mg/L) causing foliar injury			
	Low	**Slight**	**Moderate**	**Severe**
Na concentration	<46	46-230	231-460	>460
Cl concentration	<175	175-350	351-700	>700
Tolerant crops at these levels	Apricot	Pepper	Alfalfa	Sugar beet
	Plum	Potato	Barley	Sunflower
	Tomato	Corn	Sorghum	–

Foliar injury is also influenced by cultural and environmental conditions. As such, the data are presented only as general guidelines for daytime irrigation. Irrigation during the periods of high humidity or at relatively low temperatures such as at night can largely reduce or eliminate these kinds of problem.

Current studies have validated the application of sprinkler irrigation in rice-wheat cropping on semi-reclaimed alkali lands (Fig.

5.10). Here rice is cultivated as direct seeded rice, which requires only wetting of the soil surface as against the standing water in case of transplanted rice. It has been observed that in this way at least 50% water required for rice-wheat cultivation with conventional surface irrigation techniques can be saved (Singh *et al.*, 2014). Sprinkler and LEWA (Low Energy Water Application) methods of irrigation not only gave 10-20% higher grain yield than farmers' practice (surface irrigation) but also saved 30-40% irrigation water indicating their role in water and energy saving in an alkali soil (Sharma *et al.*, 2016).

Fig. 5.10. Sprinkler irrigated direct seeded rice (left) and wheat (right) crops in sequence

Drip irrigation supplies the water to crop on almost a daily basis by means of drippers or emitters. The application is at a very slow rate of 2 to 8 l/hr per dripper. With good quality water, the yields with drip irrigation are mostly higher (up to 50% higher) than other methods under comparable conditions. With poor quality water, yields are generally better with drip due to the continuous high moisture content and the daily replenishment of water lost by evapotranspiration. Salts move to the fringes of the wetting front leaving most of the area under the drip salt free (Fig. 5.11). It is because of this reason; higher yields under relatively more saline soils and irrigation with relatively more saline water are obtained in drip as compared to surface irrigation methods. On the other hand, salts accumulated in the surface soil midway between the drip orifices and at the perimeter of the wetted zone may cause injury if rain washes the salt into the underlying root zone. It is because many fine roots of drip irrigated plants occur in the surface 2.5 cm of soil.

Moderately high salinity in the subsurface to very high salinity in the surface soil can be observed at the midpoint between the emitters. This condition imposes a limit on putting additional plants between the emitters. Further, the zone with a low salt concentration along the row was barely 15 to 20 cm wide. This limits the potential for planting paired rows in most crops. Hence, for the installations of drip irrigation for the use of saline water, there should be a separate lateral for each row of plants and a dripper near each plant. Other benefits such as possible saving in water and nutrients may be large enough in special cases to justify the added investments on the drip system.

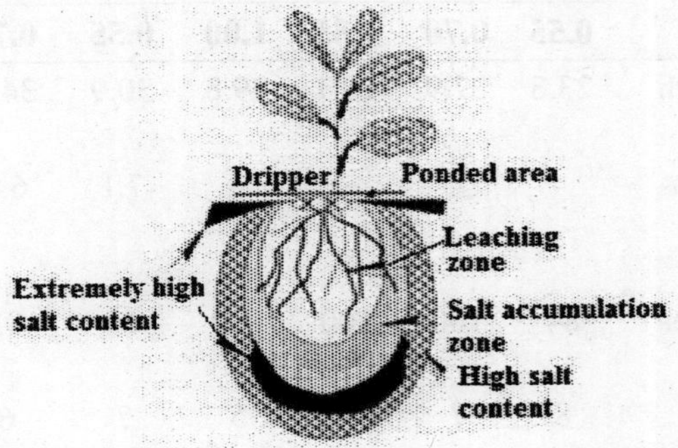

Fig. 5.11. Salt accumulation zones at the wetting front

EVALUATION OF IRRIGATION METHODS

Surface Irrigation

In a comparative study of border, sprinkler, perfo and sub-surface irrigation methods, Pandey and Singh (1977) concluded that for alkali soils the border method of irrigation was at par with the sprinkler irrigation method on the basis of the water use efficiency. However, considering the operation cost of the sprinkler irrigation system, the border irrigation method was best suited to alkali soils.

Tyagi (1976) compared level and sloping borders wherein prevailing land slope was advantageously incorporated in the graded border. As shown in Table 5.5, the treatments included four cut off ratios of 0.55, 0.7, 0.85 and 1.0 for both level and graded (0.91% slope) borders. For a stream size of 10 l/s, the irrigation was scheduled on the basis of pan evaporation with a pan coefficient of 0.7. It was concluded that the graded border with 70% cutoff ratio (for the first irrigation it was common as 85% except for 100% throughout) gives the highest application, distribution and water use efficiencies of 89%, 96% and 122 kg/ha-cm respectively as compared to 55%, 54% and 80 kg/ha-cm respectively, in case of the level border.

Table 5.5. Irrigation depth, number of irrigations, efficiencies and crop yield in level and graded borders

Items	Graded border cutoff ratio				Level border cutoff ratios			
	0.55	0.70	0.85	1.00	0.55	0.70	0.85	1.00
Irrigation depth (cm)	23.8	27.9	29.1	29.8	30.9	34.5	35.5	32.5
Excess application in level borders (cm)	-	-	-	-	7.1	6.6	6.4	2.7
Percentage area covered	84	100	100	100	65	81	93	100
Number of irrigations	8	7	6	5	7	6	5	4
Application efficiency (%)	83.0	78.0	64.0	57.5	69.8	56.0	52.6	42.7
Distribution efficiency (%)	59.2	82.0	64.6	58.0	39.3	51.0	61.7	70.9
Crop yield (t/ha)	3.3	4.2	4.1	3.8	3.1	3.4	3.5	3.4
Water use efficiency (kg/ha-cm)	131	150	141	131	106	98	108.3	105

Tyagi (1978) also attempted to quantify the yield loss in the case of wheat crop due to water stagnation at the tail ends. The effects of water ponding on crop yield as affected by varying cut-off ratios is given in Table 5.6. The overall maximum crop yield of 4.33 t/ha was obtained at cut-off ratio of 0.7 for all irrigations except the first irrigation for which the cut-off ratio was 0.85. The crop yield decreased by as much as 22% in the last 15-25 m length on downstream end of the borders at cut-off ratios of 0.85 and 1.00.

Narayana *et al.* (1980) evaluated the optimum stream size for the design of borders for alkali soils. The irrigation efficiencies computed on the basis of the observations of the soil moisture and the wheat crop yields are presented in Table 5.7. The crop yield data confirmed that the border dimensions with length of 50 and 70 m and width of 6 and 8 m gave higher irrigation efficiencies and crop yields. From the same study, a general comparison of the advance and the recession curves for different sizes of borders indicated that the opportunity time was more uniform and probably in the desirable range for borders of 50 m and 70 m long and 6 and 8 m width.

Table 5.6. Effect of water ponding on crop yield (1976-77)

Cutoff ratio	Length of ponded water from D/S (m)	Grain yield in ponded length of border (kg/ha)	Grain yield in un-ponded length of border (kg/ha)	Reduction in grain yield (%)
0.55	-	-	3300	-
0.70	11	3686	4330	12.6
0.85	17	3335	4306	22.5
1.00	26	3150	4010	22.0

Table 5.7. Efficiencies in borders of different dimensions

Length (m)	Width (m)			
	4	**6**	**8**	**10**
Distribution efficiency (%)				
50	58	87	70	50
70	45	72	64	44
90	41	69	73	70
Application efficiency (%)				
50	69	79	79	65
70	42	74	63	64
90	61	58	63	67
Yield of wheat (t/ha)				
50	4.1	5.2	5.4	4.6
70	4.6	5.4	5.4	5.0
90	4.6	4.7	4.9	4.5

Borders for Rice-Wheat Cropping Sequence

As described previously, borders used in wheat crop require a certain minimum degree of slope along the length of the field for efficient application of irrigation water. On the other hand rice, the second crop in sequence, requires a certain depth of ponding. Therefore, to achieve uniform ponding, a level field is essential. Both the conditions are difficult to meet in the same field. For this situation, the compartmentalization of long sloping borders has been recommended (Pandey *et al.*, 1977). For an average and a minimum depth of ponding at the head end, an equation was developed for the depth at the tail end for a compartmentalized border:

$$H_n = \left(d_1 + \frac{LS}{2}\right) + \frac{LS}{2n} \tag{5.1}$$

Here, H_n is the depth of water at the tail end of each compartment when the original border has been divided in n equal compartments, n is the number of compartments, d_1 is the depth of ponding at the head end of the border, L is the total length of the border and S is the slope of the border in fraction, m/m.

An equation for the depth at the head end, d_n of each compartment was also given as

$$d_n = d - \frac{LS}{2n} \tag{5.2}$$

Here d is the average depth of ponding. The uniformity of ponding is then given as:

$$u_p = 100\left(1 - \frac{LS}{2d_n}\right) \tag{5.3}$$

These theoretical expressions can be used to fix the tail end bund height of each compartment so that the irrigation or rain water in excess of the average depth of ponding can be drained out from the upper compartment to the lower compartment and then to the field drains.

Sprinkler, Drip and Surface Methods

Bernstein and Francois (1973) evaluated the drip, furrow and sprinkler irrigation methods. They concluded that the drip irrigated plots out yielded the furrow and sprinkler irrigated ones by about 50% when the same amount of low salinity water (EC 0.7 dS/m) was applied. The saline irrigation water (EC 4 dS/m) caused only 24% reduction in yield with drip irrigation but 54 and 94% reductions in yield for furrow and sprinkler irrigated plots respectively. When the irrigation frequency was increased for furrow and sprinkler treatment, the yield differences decreased and virtually disappeared with the low salinity water while the saline water caused 18 and 59% yield reduction for the furrow and sprinkler irrigated plots, respectively. Low yields and injury by sprinkling with saline water were attributed to osmotic shock caused by the flushing salts accumulated between the irrigation into the root zone and at the soil surface rather than to the foliar absorption of salts.

Agarwal and Khanna (1983) conducted a trial at Hisar to grow radish crop with saline water (EC = 6.5 dS/m) and good quality canal water (EC = 0.25 dS/m). The yield was higher with drip being maximum in subsurface than surface drip. Moreover, the yield reduction with saline water was much less in drip (10.3% in surface drip) as compared to surface irrigation (39.6% in best treatment). Recent studies have shown that benefits of subsurface drip are not always forthcoming and the usefulness has to be investigated considering the soil, crop, salinity of the soil and water etc. Field experiments conducted at CCS HAU Hisar for tomato, cauliflower, cabbage, brinjal, watermelon, grapes, cotton, and sugarcane revealed that drip irrigation performed better under deep water table (> 5m) conditions but the performance under shallow water table (<1.5m) conditions was mixed (Singh and Kumar, 1989, Singh *et al.*, 1990 and Singh and Kumar, 1994).

Yadav (2006) studied the effect of saline water irrigation on cauliflower with drip and surface irrigation methods. Higher yields were obtained at all salinity levels in drip than with check basin. The yield obtained at EC_{iw} of 10 dS/m in drip was about the same as obtained at EC_{iw} of 2.5 dS/m in check basin irrigation (Table 5.8). A study conducted at Central Soil Salinity Research Institute, Karnal, revealed that the yield of okra (*Abelmoschus esculentus* L. Moench,

Var: Mahyco-10 Hy) reduced by 22.9% at EC_{iw} of 8 dS/m compared to only 5.9% at 6.0 dS/m with drip irrigation (Saxena *et al.*, 2013). Clearly Okra can be grown with saline water having EC of 6 dS/m with drip. On the contrary severe yield reductions are reported with surface irrigation methods at this salinity level.

Table 5.8. Effect of method of irrigation on cauliflower yield with saline water

EC_{iw} (dS/m)	**Drip irrigation**	**Check basin**	**Mean**
0.0	17.5	13.0	15.3
2.5	17.7	13.1	15.4
5.0	15.0	10.0	12.5
10.0	13.0	7.5	10.2
Mean	15.8	10.9	

IRRIGATION SCHEDULING

While planning to design an irrigation system, the two questions arise:

- How much water is enough at a particular time?
- When is the best time to apply it?

Irrigation scheduling is used to make decisions on these aspects of irrigation. The area of interest can be a small field, a farm or a region/basin. Irrigation schedule should ensure that neither plant faces any water stress nor the water is applied in excess of the crop requirement. Prihar *et al.* (1975) critically reviewed the common approaches for scheduling irrigation. A list of various methods commonly used for irrigation scheduling is given in Fig. 5.12. All of these are not included since only few of these have relevance to salt affected soils. These are discussed in the following sections.

In salt affected soils, the importance of irrigation scheduling is magnified because of the need for extra application of irrigation water for leaching and shallow depth high frequency regime of irrigation. It assumes greater significance when the water supply is limited. Over irrigation results in water logging, generally reducing the crop yields due to poor aeration and loss of nutrients. Feel and appearance and visual approaches often pose several difficulties in irrigation scheduling in salt affected soils. It is because of the adverse physico-chemical properties of the soils as well as appearance of similar visible symptoms due to reasons other than the moisture stress. Similarly, time approach especially rotational water supply fails to take care of the need for frequent applications of water in salt affected soils. Soil water regime, meteorological approaches and critical growth stage approach especially under managed deficit irrigation are therefore, more commonly used approaches for irrigation scheduling in salt affected soils.

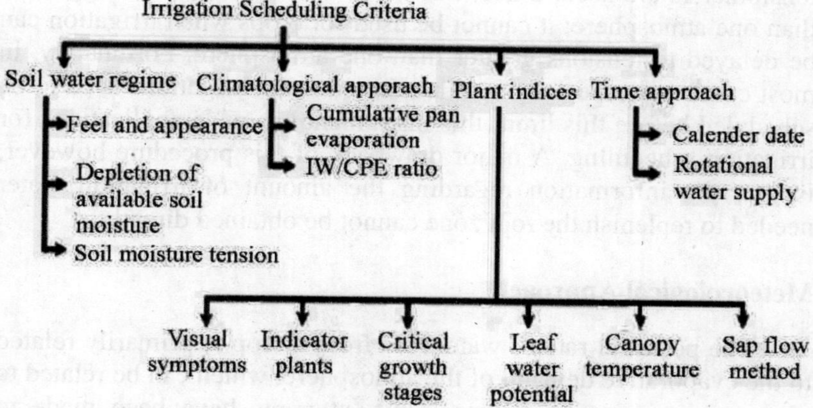

Fig. 5.12 Approaches for irrigation scheduling

Soil Water Regime Approach

This concept for scheduling irrigation is based upon the depletion of available soil water or increase in the soil moisture tension. In this approach, it has been assumed that plant available water is the water held by the soil between field capacity and the permanent wilting point. Since these two parameters are controlled mainly by soil texture, the amount of water available is less in a light than a fine textured soil. Initially, it was considered that water between these two limits is equally available to plants, but later studies proved that it is not so. In a comprehensive review, Stanhill (1957) reported that in 80% experiments plant growth was affected by the extent of soil water depletion from the root zone prior to irrigation. In most instances the optimum yield was obtained when the irrigation is given at or before half the available water in the root zone is depleted. Later studies especially under managed deficit irrigations, brought out that while it is true for the initial growth stages, irrigations can be delayed at later growth stages by allowing the plants to draw 60-80% of the available soil water. Apparently, one can control the frequency of water application using this approach.

It was also recognized by some workers that the plant response to irrigation is better correlated with the soil water suction than with the soil water content. To measure soil water tension, generally, tensiometers are used. Since, this instrument fails at tensions greater than one atmosphere; it cannot be used for crops when irrigation can be delayed to tensions greater than one atmosphere. Fortunately, in most cases and especially in salt affected soils, irrigations need to be scheduled before this limit; this instrument has wide applications for irrigation scheduling. A major drawback of this procedure however, is that the information regarding the amount of irrigation water needed to replenish the root zone cannot be obtained directly.

Meteorological Approach

The potential rate of water loss from a crop is primarily related to the evaporative demand of the atmosphere, which can be related to open pan evaporation values. Thus, attempts have been made to work out empirical coefficients to describe the relation between water loss from the open pan and the crop and to compute potential evapotranspiration from meteorological data. For the practical application of this approach, irrigation can also be scheduled, based upon the allowable water depletion from root zone and potential evapotranspiration rates (PE) of crops over short time intervals of growth. Prihar *et al.* (1974) suggested a practical approach based upon IW/CPE, i.e. the ratio between fixed irrigation water depth (IW) and the net cumulative pan evaporation (CPE) since the previous irrigation. The response of wheat to a few irrigation treatment based on IW/CPE is shown in Table 5.9. The higher yield was obtained with IW/CPE equal to 1.0 followed by 0.75 throughout the growing season.

Table 5.9. Effect of irrigation treatments based on IW/CPE ratio on the yield of wheat

IW/CPE	Grain yield (t/ha)		Remarks
	1969-70	1970-71	
1.0 throughout	4.79	4.12	Wheat was planted in the third week of November and first irrigation was applied to all treatments 28 days after sowing.
0.75 throughout	4.65	3.80	
0.50 throughout	4.05	3.65	
0.50 till mid- Feb. and 0.75 later.	4.12	3.25	
0.50 till- mid-Feb. and 1.00 later	4.16	3.50	
L.S.D. (P = 0.5)	0.45	0.31	

Critical Growth Stages

Critical growth stages are known to provide great opportunities in irrigation scheduling especially when water is limited. Under these situations, one can schedule irrigations at growth stages that are more critical than others in increasing the crop yields. In a alkali soil having pH 9.2, and ESP 38, an increase in the irrigation frequency resulted in greater relative growth rate, rooting density, productive tiller per meter and grain yield of wheat. The three irrigations given at crown root initiation, tillering and milk stages gave significantly higher grain yield as compared to other treatment with three irrigations (Sharma *et al.*, 1990). In a alkali land under reclamation (pH 8.8 and ESP 23), rosette stage (28-30 days after sowing) was found to be critical for mustard. Irrigation at this stage gave significantly greater relative growth rate (RGR), branches and pods/plant, seed and straw yields as compared to one irrigation at pod formation stage (Sharma and Singh, 1993). Critical growth stages for some of the crops are given in Table 5.10.

Table 5.9. Critical growth stages for various crops in order of sensitivity to water stress

Crop	Crop duration (days)	Critical stages for irrigation
Wheat (dwarf)	130-140	Crowing, flowering, jointing, milk and dough
Rice (dwarf)	130-140	Flowering, grain filling, tillering
Rice (Basmati)	140-150	Flowering, grain development
Summer maize	90-110	Tasseling and silking, and grain development
Pearl millet	75-85	Heading
Sorghum	90-110	Flowering, seeding, grain filling
Chickpea	130-140	Pre-flowering and pod development
Groundnut	110-135	Pegging, pod-setting and pod filling
Sunflower	100-120	Flowering, seed filling, late vegetative
Sugarcane	280-330	Tillering and grand growth
Mustard	140-150	Pre-flowering and pod development

■■■■

CHAPTER 6

DRAINAGE METHODS

INTRODUCTION

The removal of excess surface or subsurface water from the field for optimum agricultural production is called drainage. The drainage helps in the transforming swamps, bogs, deltas and other waterlogged soils, which are generally too wet for crop production, into areas of high productivity. The necessity of drainage even in arid and semi-arid regions has been recognized for reclaiming the salt affected soils by leaching the salts from the root zone depth. The improvement of such problematic soils by means of surface or subsurface drainage can lead to the creation of an environment for sustainable productivity, conductive to human settlement, economic development and social progress. Thus, drainage is clearly one of the most important reclamation and water management practices for sustainable irrigated agriculture.

There are two basic types of drainage systems viz. (i) surface drainage system which removes the excess surface water from the fields, and (ii) subsurface drainage system which removes the subsurface excess water from the soil root zone depths by lowering or stabilizing the water table at a particular depth (Fig. 6.1). Sometimes, the subsurface drainage system is also helpful in removing the surface water from isolated low lying spots in the field to a limited extent. Based on the functional aspects, the third kind of drainage system is termed as interceptor drainage system although basic principles used are similar to surface and subsurface drainage systems. It is also known as across the slope system. In this system, open/tile lines run across rather than down the slope. It intercepts the flowing water before it resurfaces on the ground. Based on functionality, materials used and their orientation with respect to ground surface, subsurface drainage systems have been classified under various groups (Fig. 6.1).

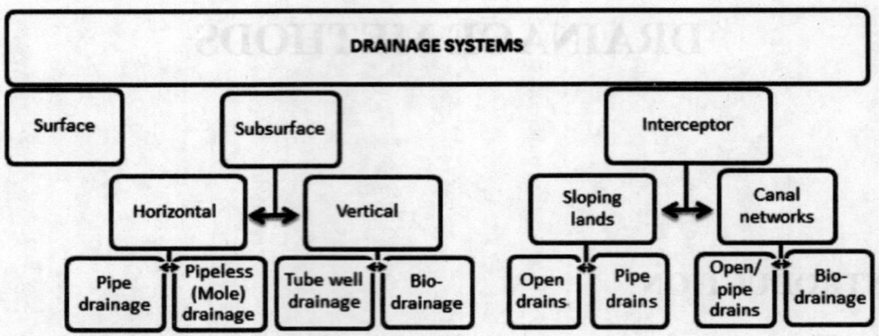

Fig. 6.1. Categorization of drainage systems

The surface drainage systems should:

(i) Remove surplus water from the field or other areas:

(a) in the case of a row crop, every row should be drained. Water after collecting in the rows should move without ponding to a field drain,

(b) If the land is in the form of a pasture or grain or hay field, or where the surface is smooth, the water should flow as sheets flow over a smooth ground surface without pocketing in a field drain,

(c) all collection drains should be so built that water can enter them without impounding and eroding,

(ii) Remove water within a certain period of time (design rate of runoff); this will vary according to soil, climate, land use, topography, economic feasibility, engineering feasibility, and crop types. Water should be removed before the soil profile is saturated and the crop is damaged. It will need to be removed faster for some crops than for others;

(iii) It should not cause any hindrance in farming or farming operations;

(a) each system should be designed so that it fits the crops and crop rotations,

(b) the system must be flexible enough so that the farm mechanization is conveniently possible,

(c) have a minimum grade which is steep enough to prevent pocketing of water on the field or in those drains which are to be crossed,.

(iv) It should be easy to adequately maintain it;

(a) require a minimum of maintenance,

(b) regular cultural operations should not materially impair effectiveness,

(v) Designed considering all phases of soil and water management;

(a) erosion control needs,

(b) irrigation needs,

(c) soil structure and fertility needs besides, system should not expose excessive amounts of harmful subsoil.

Similarly the subsurface drainage system should;

(i) facilitate the control of salinity in the soil by providing the improved opportunity for effective leaching;

(ii) control the water table at desired depth by lowering it during rainy season at a rate such that crops are not affected due to lack of proper aeration;

(iii) promote the beneficial soil bacteria activity and improve the tilth of the soil;

(iv) increase the crop growing period of *rabi* crop by bringing the soil in workable condition just after the rainy season.

(v) provide the fringe benefits for controlling the variety of health hazards associated with excess water on the land or in the subsurface.

The need for surface drainage in semi-reclaimed alkali lands for 8 crops is revealed in Table 6.1. The data show that even one day of water stagnation in wheat crop can reduce the crop yield by 8%.

Table 6.1. Yield losses in various crops due to water stagnation of various durations

Crop	Yield losses# over drained (%) with water stagnation (days)			
	1	**2**	**4**	**6**
Sorghum	3	11	16	20
Pearl millet	6	15	22	27
Pigeon pea	4	14	18	21
Wheat	8	17	27	39
Barley	4	7	13	25
Mustard	8	16	22	29
Berseem (seed)	2	21	35	48
Sunflower	13	19	26	30

Yield losses are relative to drained plots

Suitability of Drainage Systems

Not all kinds of drainage system will suit various conditions encountered in salt affected soils. The suitability of various drainage systems is given in Table 6.2.

Table 6. 2. Application of drainage systems for various situations in land reclamation

Type of soil or situation	Drainage system in order of preference
Saline soils	Surface, subsurface pipe drainage (For water-logged saline soils), pipe less drainage (for heavy textured saline soils)
Alkali or saline-alkali soils	Surface drainage, vertical drainage (Tube well), subsurface pipe drainage (after thorough investigation in waterlogged alkali lands)
Interception (Canal network)	Bio-interceptor, pipe drainage as seepage interceptors

Bedding System

It is one of the oldest of all the known drainage system. The bedding system is used on fields which are practically flat (slopes of

0 to 1.5%). It is used on soils which are slowly permeable and where it is not feasible to install the tile drainage systems. Bedding started as a practice because it can be easily performed with the farm equipment commonly available with the farming community. It is accomplished by the continuous ploughing of rectangular shaped lands, developing dead furrows and back furrows in the same location each year. Gradually a crown is built, and the dead furrows are deepened into ditches. In earlier works the lands had strips of 10 to 12 m spacing and dead channels of about 20 cm depth.

The beds are established with lengths parallel to the prevailing slopes. Sometimes collection ditches are spaced from 100 to 300 m apart to intercept the water from the dead furrows. Rows generally run parallel to the dead furrows.

The beds in use these days are often smoothened to remove surface irregularities and eliminate pockets where water can stagnate. They are also syste-matically cross drained so that the row drainage does not become excessive. A few permanent pastures or hay crops do not require as narrow beds as general rotation crops do. For the bedding system to function properly, all dead furrows must have a continuous grade with no obstruction or low points that would interfere with the complete drainage of the furrows.

The crops planted and the tillage implements used have a direct bearing on the degree of drainage provided by this system. When the crop is grain or hay, the surface of the bed is smooth and the lateral drainage to the dead furrow is good. When the bed is planted to a row crop, ridges are formed by the tillage tools and the lateral drainage is either poor or nonexistent, depending somewhat upon the crown of the bed. Therefore, bedding is limited in its usefulness for row crops unless each, of the rows on the bed has some slope to a drain without pockets.

Random System

The random drain system is adapted to a depression type topography having wet pockets or depressions too deep or large to be filled up by land smoothening, or where the farmer does not wish to spend money to grade the land. These types of drain meander from one low spot or depression to another so that the water flows towards the lateral outlet ditch. The drains must be of sufficient size and depth to drain off the impounded water rapidly and completely.

These drains receive water either from crop rows or by sheet flow over smooth surfaces. Generally, they should be designed and built so that they can be crossed by farm equipment. The drains should have a minimum of 24 cm depth, and should have side slopes not steeper than 8 to 1. Generally, they are built with a 2 or 3 m bottom with crossable, 10 to 1 or flatter slopes. The minimum longitudinal grade should be at least 0.05%, but 0.1% is preferred. It may be noted that the flatter the grade, the more difficult it is to build and maintain the drains without pockets and back fall.

Parallel Field Drain System

The parallel system is used for flat, poorly drained uniform surface soils, which may have few or numerous small depressions. The channels are spaced farther apart than a bedding system and the channels have greater capacity to carry runoff than the dead furrows (Fig. 6.2). The distance separating ditches can vary depending upon the soils; maximum lateral flow length to ditch may be 180 m for non-erodible soils but may be reduced to 90 m for erosive soils. The depth of the drain should be minimum 20 cm having a minimum cross section area of 0.45 m^2. The side slopes are 8:1 or flatter. The row or field drains are perpendicular to the field lateral. The system is most suitable where farm mechanization is practised. Land smoothening of field surface between laterals is necessary so that land remains well graded.

Diversion System

The diversion system of surface drainage uses one or more diversions to divert the surface and subsurface flow. A diversion is an individually designed channel constructed across the slope. It serves the purpose of intercepting the surface runoff and the sub-surface flow to reduce erosion or improve drainage, or both. The intercepted flow is then conducted to an adequate outlet at a safe velocity. The drainage department of Haryana has constructed numerous diversion drains which take off the water from the salt affected areas.

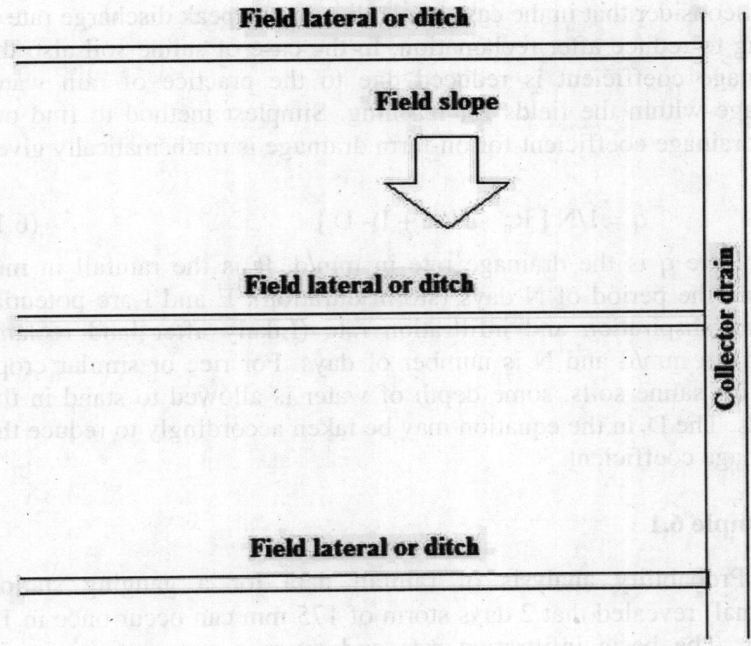

Fig. 6.2. Parallel field drain system

Parallel Lateral Open Ditch System

The parallel lateral open ditch systems besides providing surface drainage also help to control the groundwater table. It is why; many times the system is referred as open subsurface drainage system. The layout of the system is similar to the field drain system but ditches are deeper. These drains cannot be crossed with farm machinery. The minimum size for open ditches is 0.3 m deep and side slopes are steeper than 6:1. When financial or other constraints do not allow construction of tile/pipe drainage, this system of drainage is recommended till finances are generated or made available.

DESIGN OF SURFACE DRAINAGE SYSTEM

The surface drainage system is designed to have sufficient capacity to carry the designed runoff usually given as surface drainage coefficient. To select the design value of the peak discharge, one must consider that in the case of alkali soils the peak discharge rate is going to reduce after reclamation. In the case of saline soil also the drainage coefficient is reduced due to the practice of rain water storage within the fields for leaching. Simplest method to find out the drainage coefficient for on-farm drainage is mathematically given as:

$$q = 1/N\,[\,R_n - N(E + I) - D_r] \qquad (6.1)$$

Here q is the drainage rate in mm/d, R is the rainfall in mm during the period of N days (storm duration), E and I are potential evapotranspiration and infiltration rate (Likely after land reclamation) in mm/d and N is number of days. For rice or similar crops and for saline soils, some depth of water is allowed to stand in the fields. The D_r in the equation may be taken accordingly to reduce the drainage coefficient.

Example 6.1

Probability analysis of rainfall data for a gauging station (Karnal) revealed that 2 days storm of 175 mm can occur once in 10 years. The basic infiltration rate and average evaporation are 54 mm/d and 6 mm/d, respectively. Calculate the drainage coefficient if it is proposed to remove the water from the crop land within 2 days. Assume that 3 cm of water may be allowed to stand in the fields for leaching. What capacity collector needs to be designed to take acre of land drainage for a farm of 5 ha.

Given : N = 2 days, R_n = 175 mm, E = 6 mm/d, I = 54 mm/d and D_r = 3 cm or 30 mm. To assess q, incorporate these parameters in eq. (6.1)

$$q = 1/2\,[175\text{-}2(6\text{+}54) - 30.0]\ \text{mm/d} = 25/2 = 12.5\ \text{mm/d}$$

The capacity of the collector, Q is given as

$$Q = q{*}A = (12.5/100) {*}5 {*}100{*}100 = 6250\ m^3/d = 0.072\ m^3/s$$

Once the peak discharge rate value is selected for design purposes, the cross section of the drains can be decided using Manning's formula. It computes water velocity in the channel:

$$V = \frac{1}{n}\,R^{2/3}\,S^{1/2} \qquad (6.2)$$

$$\text{or} \qquad Q = \frac{A}{n} R^{2/3} S^{1/2} \tag{6.3}$$

Here, V is the average flow velocity (m/s), A is the cross-section flow area (m^2), n is the Manning's roughness coefficient, S is the hydraulic gradient (m/m), and R is the hydraulic radius given as: cross section area divided by the wetted parameter (m)

Generally rectangular, trapezoidal, V shape or parabolic sections are selected for the drains, trapezoidal and parabolic being more commonly adopted. Since the designs are made to achieve the most efficient hydraulic cross section that results in minimum volume of excavation for the same discharge, the bottom width of the trapezoidal drain is decided using the following relationships:

$$b = 2d \tan \theta / 2 \tag{6.4}$$

where, b is the bottom width, d is the depth of flow which depends upon the outlet conditions and θ is the angle given by channel side slopes. Side slopes are decided on the basis of soil texture. For salt affected soils it should not be steeper than 45°. Approximate relation R = d/2 can be used, although it is valid only if trapezoidal channel is half hexagon.

For a parabolic cross section, the area is given by 2/3 t d and the wetted perimeter P = t+8 d^2/3t. Therefore, R = t. 2d/ (1.5 t^2 +4d^2). Here t is the top width. The approximate relation d =1.5 R is commonly used in the design of parabolic channels.

Once the side slope, depth and the bottom width are fixed for the open surface drainage, the velocity V is computed using eq. (6.2). The channel bed slope must follow more or less closely the slope of the ground surface in the direction of the ditch line. The value of Manning's roughness coefficient is generally recommended as 0.035 although it also depends upon soil type and type of channel roughness. The velocity computed by this formula should always be less than 1 m/s for salt affected soils. If, it comes more than the permissible velocity, the section of the drain should be increased by adopting either the larger values of the depth or the bottom width or both. In such situations the criteria of the most efficient section may be overlooked. After deciding the cross sectional area of the channel with the permissible velocity, it should be checked whether it can carry the design discharge, otherwise the cross sectional area needs to be further increased.

NON-CONVENTIONAL DRAINAGE SYSTEMS

Conventional surface drainage systems comprising of mains, sub-mains or collector drains implemented by the states, mostly as flood control measure, have failed to improve agricultural productivity. It is attributed to the following factors (Gupta, 2018):

- Drains fail to function when these are required the most
- Leaving the construction of field drains to the farmers who fail to construct the field drains
- No link to the systems for land locked areas.

As a result some of the non-conventional drainage systems can be used. The two major systems are:

- Construction of on-farm ponds for drainage and reuse for irrigation and/or for aquaculture
- Construction of artificial groundwater recharge structures especially in alkali lands

SUBSURFACE DRAINAGE

System Description

A subsurface drainage system mainly comprises of the following (Fig. 6.3):

- Laterals
- Collectors/Main drain
- Manholes and other structures
- Pump/gravity outlet (Sump)

Laterals

Horizontal subsurface drainage system comprises of a network of buried pipes as lateral drains (Fig. 6.3). These are the heart of a subsurface drainage system and are laid at a pre-decided depth and spacing. The laterals collect the water from the soil and empty it into the main or the sub-main. These are made of cement concrete or clay tiles or corrugated plastic pipes with factory drilled holes. Latter are now commonly used laterals, having outside diameters (OD) of 80 and 100 mm. The spacing varies from as low as 20 m for heavy textured soils to as high as 200 m for sandy soils. A filter is put around the drains to prevent or restrict fine-grained materials from entering the drains along with the groundwater. Earlier gravel was used as envelope but now synthetic filters are used for this purpose.

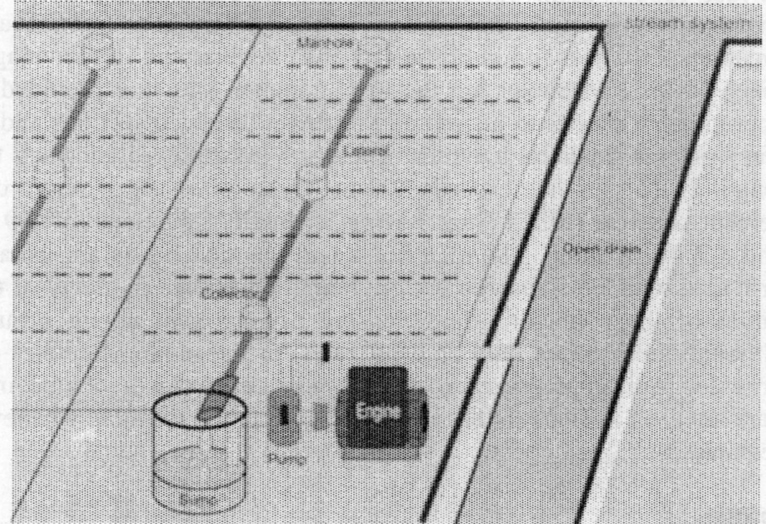

Fig. 6.3. Assembly of various components of subsurface drainage system

Collectors/main drain

Laterals are connected to buried collector drains or sometimes referred as sub-mains leading to open ditches serving as outlet drains (Fig. 6.3). Collector drains carry the drain water to the main drain and to the drainage outlet. The collector drains are laid deeper than the deepest lateral drain in a tile network. The collector drains are of larger diameter in order to collect the flow from many lateral drains. Collector drains are usually non–perforated pipes but in areas having shallow water table, perforated collectors are used so that pipes do not float during construction. A main drain is the one into which several lines empty their water through an appropriately constructed outlet. It is often an open drain that takes the water to a point of disposal using natural/constructed drain networks.

Manholes/other structures

A subsurface drainage project cannot simply be constructed by laying out perforated pipes. It must have a number of engineering structures for disposal of drainage effluent, to maintain the system, maintain the infrastructure of the project area, and transport the surface runoff. Some common structures used in a drainage system are: manhole/junction box, drops, surface water inlets, relief and breathers, crossings, drain bridges, closing devices and flow management devices. Manholes are the most common structure provided at appropriate locations to inspect and to detect damaged sections and to carry out maintenance activities. These are invariably used when the length of the pipe line exceeds 300 m. These are made of reinforced cement concrete (RCC) rings having an internal diameter of 0.9 m and 1.2 m for sump. The man holes and junction boxes are usually provided with silt traps, the bottom being lower than the level of the outlet of the collector. Manholes are usually kept above ground although there are no hard and fast rules for the same. Manholes are provided with appropriate covers. In most cases plastic-coated iron bars are used as steps so that a person can enter the manhole whenever required.

Outlets

Commonly, two kinds of outlets are used. If the natural slope permits, the disposal of drainage water from the collector or main drain into a natural water body is accomplished through a gravity outlet (Fig. 6.4left). On the other hand, in flat lands, possibility of having a gravity outlet is minimal. In such cases, a sump is constructed. It is a prefabricated RCC structure (Fig. 6.4 right) with several rings joined together one above the other or constructed *in-situ*. The collector is connected at the bottom. Water is pumped from it to the natural drain. If the water quality permits, pumped water can also be used to irrigate the crops.

A map of each field that shows the location and size of all tile lines and the depth and grade should be prepared at the time of laying out the tile lines. Such a map is helpful in the long run in formulating perfect operation and maintenance schedules. It also helps in deciding the location of the new lines to be put in future depending upon the need.

Fig. 6.4. A view of an gravity outlet (left) and a RCC ring for construction of manhole, junction box or sump (right)

Layout

There are eight different layouts of the subsurface drainage systems as described herein. Amongst these first four are more commonly employed, while others are used under specific conditions.

Natural or Random System: This system consists of placing the subsurface drains in such a way that all points of wet lands are connected. It is preferred when there are a few depressions which hold water and remain wet while the rest of the land drains well (Fig. 6.5 left).

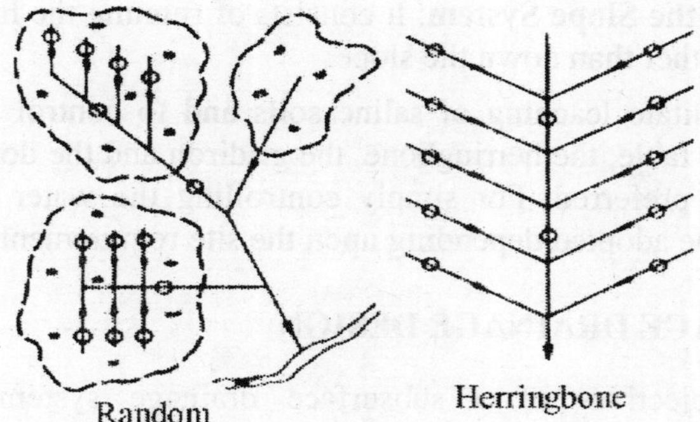

Fig. 6.5. Layout of random system (left) and herringbone system (right)

Herringbone System: In the herringbone system, parallel laterals drains enter the main drain at an angle from one or both sides (Fig. 6.5 right). Varying the angle of confluence with the main drain helps in achieving the desired grade of the lateral drains. The main/collector drain is aligned along the main slope.

Gridiron System: The grid iron pattern or the parallel system is the most preferred system of land drainage. This system is mostly used on flat, regularly shaped fields having soils of uniform permeability. Here, parallel lateral drains are aligned perpendicular to the collector/main drain. Laterals can enter either from one or from both the sides of the collectors.

Double Main System: It is like two gridiron systems placed together with parallel sub-mains or main lines leaving some distance between the two lines.

Cutoff System: This is used on sloping sites where the water comes to the surface and makes the slope side as well as the bottom, wet and un-productive.

Elkington System: It is similar to the cutoff system except that at regular intervals some wells are constructed and are interconnected with the tile lines so that the water may enter into the wells and is removed by tile lines. Sometimes when the earth is not firm, the caving of wells is checked by filling them with gravel.

Grouping System: It consists of collecting the water from a number of small systems and discharging it into the ditch through one outlet.

Across the Slope System: It consists of running the lines of the tile across rather than down the slope.

To facilitate leaching of saline soils and to control the saline groundwater table, the herringbone, the gridiron and the double main systems are preferred. For simply controlling the water table any system can be adopted depending upon the site requirement.

SUBSURFACE DRAINAGE DESIGN

The objectives of a subsurface drainage system for the reclamation of waterlogged saline soils are two folds:

- To facilitate the leaching of salts from the root zone by the application of excess water
- To keep the groundwater level relatively deep so that leaching can take place as well as the salts from the saline

groundwater body are prevented from accumulating at the surface due to the evaporation of saline water by capillary rise.

To realize the above objectives, a specific combination of drain spacing and drain depth is used. The problem in drainage design therefore, is to find out the appropriate drain depth and spacing. Generally the depth of the drains is decided on the basis of the outlet availability, crop root zone and other conveniences during the construction of the drainage system.

To determine and decide on the proper spacing, it is necessary that the desired function of the drainage system be expressed in those terms that can be used as input information in the mathematical formulae for calculating drain spacing. These formulae relate the performance of the drainage system to the geometry of the system and the pertinent soil factors.

The drainage flow theories have been classified under three heads, namely steady state, non-steady state and fluctuating water table theories. A steady state condition exists when a system, its flow rates and boundaries do not change with time, i.e. when the system is in dynamic equilibrium. Otherwise a non-steady state condition exists. In this case, usually a falling water table theory is used. Fluctuating water table is also non-steady state theory but is more realistic than the falling water table theory. It has been observed that for the Indian conditions, steady state formulae give reasonably good design for the reclamation of waterlogged saline lands. Therefore, only this theory is introduced in the next section.

Steady State Theory

The steady state condition is seldom reached under field conditions. It may be established during periods of long duration of low intensity rainfall. The simplicity of the steady state theory makes it attractive for routine use in designs. The drain spacing formulae are relatively simple and convenient to use.

Ellipse equation

One of the earliest attempts to find a mathematical solution to a steady state problem resulted in the development of an elliptic equation for the shape of the water table over the drains (Fig. 6.6). Colding (1872), a Danish engineer was the first to develop this equation. Later on, Hooghoudt (1937) derived the same equation giving a different reasoning. This equation is presently the most well- known and widely used equation. It is known as Hooghoudt's equation. Mathematically, the equation is written as:

$$L^2 = \frac{4K\left(H^2 - h^2 + 2dH - 2dh\right)}{q} \tag{6.5}$$

where, L is the spacing between parallel drains (m), K is the hydraulic conductivity of the soil (m/d), H is the height of the water table in the middle of the drains (m), h is the depth of the water in the subsurface drain or the height of the water level above the tile line (m), d is the depth of impermeable layer below the drain bottom (m), q is the replenishment rate or rainfall intensity (m/d).

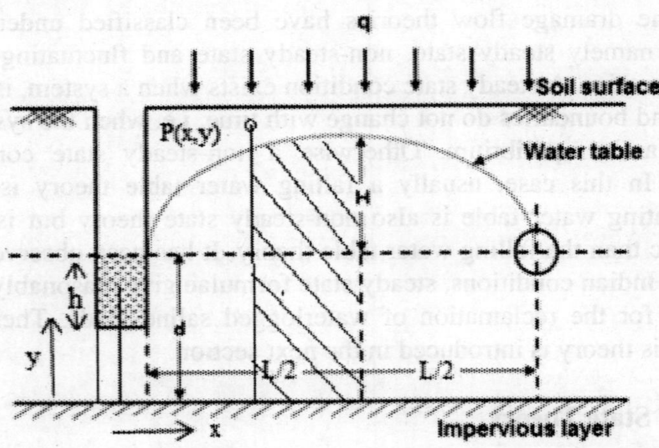

Fig. 6.6. Geometry and symbols for sub surface drainage system.

For practical purposes, if the drain is considered empty the Hooghoudt's equation is reduced to:

$$L^2 = \frac{4KH}{q}\left(2d + H\right) \tag{6.6}$$

For the layered soil with hydraulic conductivity K_a above the drain line and hydraulic conductivity K_b below the drain line the Hooghoudt's drain depth and spacing formula is given as:

$$L^2 = \frac{4}{q}\left(K_a H^2\right) + \left(\frac{8}{q} K_b d\ H\right) \qquad (6.7)$$

The multilayered soil above or below the drain line can be treated by the above formula by considering the weighted mean of the horizontal conductivities as per the formula:

$$K = \frac{K_1 L_1 + K_2 L_2 + \ldots\ldots + K_n L_n}{L_1 + L_2 + \ldots + L_a} \qquad (6.8)$$

Here K is the weighted mean of the hydraulic conductivities, $K_1, K_2, \ldots K_n$ of layers having thicknesses $L_1, L_2, \ldots L_a$ etc.

To overcome various assumptions made in deriving the equation, Hooghoudt suggested replacing the depth d, with equivalent depth d_e, which takes care of the radial flow below the drains. Various formulae, nomographs and Tables have been given to calculate the equivalent depth (Gupta, 2018) but are being left out being beyond the scope of this book.

VERTICAL DRAINAGE

The drainage of irrigated land by wells can prevent the concentration of soluble salts at the soil surface or in the root zone and can prevent the prolonged saturation of the root zone. In many irrigated areas pumped wells system for drainage is the most feasible approach having advantages over closed drains particularly for the regions having drain outlet problems and even restricted natural surface drainage. Vertical drainage is defined as ´the removal of groundwater by means of a borehole (or well) either under pressure (artesian conditions) or through pumping´. Vertical drainage may not be feasible at all places. Therefore, site conditions must be evaluated before planning a vertical drainage system. A comparative statement is given in Table 6.3 that describes the site characters where vertical drainage may prove to be successful The vertical drainage system practically includes one or more deep/shallow tube well(s). The network of open or closed drains to dispose of the water also constitutes a part of the system. The group of tube wells having sufficient overlap of their individual cones of depressions controls the groundwater table or artesian groundwater conditions. The overlapping of cone of depressions ensures uniform lowering of the water table at all the points. Lowering of water table in many parts of India ascribed to overexploitation through tube wells attests to its feasibility in lowering the water table. The pumped water can be used for irrigation either alone or in conjunction with fresh water, provided the proper salt balance is maintained. Water balance studies as described in Chapter 4 must be conducted to assess the number of tube wells required to maintain water balance in the area.

Table 6.3. Favourable and non-favourable site features for vertical drainage

Favourable features	Non-favourable features
Good quality aquifers are available	Non-availability of good aquifers; low transmissivity does not allow construction of tube wells of sufficient capacity
Good quality groundwater	Groundwater of high EC, SAR or RSC resulting in problems for its disposal. Tube well components are more prone to wear and tear/corrosion due to salinity resulting in higher maintenance cost
Cost of tube well construction is less and electrical power is available	Lack of fresh water without which possibility of conjunctive use of drainage water may pose problems
The tube wells are installed by the farmers. The government inputs are restricted to priority connections of electricity	High operational and maintenance costs when systems are installed under government set-up
Upward seepage pressure from a confined aquifer is cause of drainage problem and the cost of horizontal subsurface drainage is prohibitive	
The areas to be reclaimed are only waterlogged with no or minimal salinity	
Potential of using drainage water for irrigation or minimum disposal related problems.	
One window system for agriculture and groundwater development is available	

BIO-DRAINAGE

Biological drainage or bio-drainage is defined as ´the application of a biological material to absorb, utilize, transport and evapo-transpire water stagnating on the land surface or in the root zone to drain the land´. Quick growing plants are employed as bio-drain materials as these are capable of absorbing and transpiring huge quantities of water. Since water is transpired straight into the atmosphere, there are no disposal problems. In essence, bio-drainage involves three processes (Fig. 6.7):

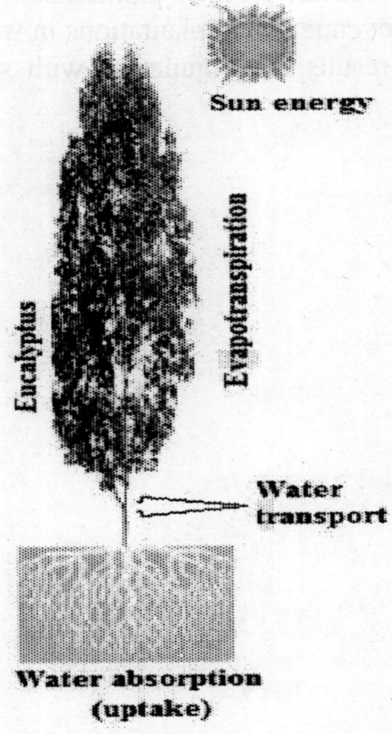

Fig. 6.7. Processes involved in bio-drainage

- The uptake of water from the soil medium by the roots (absorption)
- Transportation of water from root to plant parts
- Transpiration of water by the plants through the tree leaves.

Apparently, a good bio-drainage friendly plant must be able to absorb, transport and transpire water under stresses of water logging and soil salinity (Gupta, 2005). A critical part of bio-drainage is to select suitable plants that are tolerant to water logging and soil salinity. Dagar (2011) listed several species for block plantations, important amongst them being *Eucalyptus*, *Casuarina glauca*, *Acacia ampleceps*, *Terminalia arjuna* (*Arjun*), *Pongamia pinnata* (*Papri*) and *Syzygium cuminii* (*Jamun*). *Eucalyptus* with its quick growing characteristics fits well into any bio-drainage scheme. A major problem with this system arises due to the fact that bio-drainage is not capable of taking care of salt balance since plants do not absorb salts. Therefore, these systems may be used as a preventive measure for salinity control. Practically, block plantations in alkali lands and on *killa* lines (as per revenue maps) plantations in waterlogged saline lands have given good results in conjunction with some vertical drainage inputs.

■■■■

CHAPTER 7

LEACHING AND SOIL SALINIZATION

LEACHING

Leaching of saline soils is essentially a process whereby water of low salt concentration is applied to displace the soil solution of relatively high concentration. The application of the excess water to pass through the root zone with the aim of pushing the salts below the root zone is defined as leaching. The leaching of salts is carried out to meet one of the following objectives:

- To remove or to reduce the salts in the root zone commonly known as one time leaching for reclamation
- To maintain the salt balance in the reclaimed or irrigated lands so that crops do not suffer due to excess salts at any time in future.

This Chapter deals with these issues besides the issue of soil salinization/alkalization of reclaimed lands.

LEACHING FOR RECLAMATION

Since the initial salinity in the soils is too high for crop production, the same has to be leached down by applying additional quantities of irrigation water. The process of extra application of water to leach down the salts is known as leaching. During leaching, salts are dissolved in the water passing through the soil and are removed beyond the root zone depth in natural or artificial drainage systems. Leaching may occur either naturally as a result of rainfall or may be achieved artificially through application of additional water specifically for leaching purpose. In both cases, water must pass beyond the root zone so that excess salt are pushed beyond the root zone. Sometimes and especially in heavy textured saline-alkali soils, flushing method is used to remove the salts. This method is principally different than the leaching process.

Flushing for Salt Removal

In flushing method of salt removal, salts are removed by surface washing of the soil as compared to pushing of soluble salts below the root zone in leaching. In flushing, salts move with runoff or irrigation water as the water moves down the sloping land surface. The water with washed away salts is drained through a surface drain provided at the tail end. Reeve *et al.* (1955) reported that salts removed by flushing in a 100 m border strip constituted only 1% of the total salt content of the 60 cm deep soil profile. Yadav and Agarwal (1959) observed that the volume of leachate and the quantity of sodium salt removed from a saline-alkali soil increased when the pots were given flushing in between two irrigations. Gupta (1985) observed that except for the first storm or storms following long dry spells, no appreciable amount of salt is present in the runoff occurring due to heavy rain over saline soils. The flushing method is useful only when soil permeability is low or in soils where impermeable layer exists near to the soil surface. Elgabaly (1971) recommended that in areas where there exists a salt crust of highly hygroscopic salts which may prevent use of levelling or other heavy equipment because of low bearing capacity; it is useful to flush the salts. It will permit better drying of the surface soil. A combination of flushing and leaching has been successfully used for the reclamation of lands in several projects of *Kharland* Development Board, Gujarat.

Flushing proved effective in reclamation of saline-alkali soils of the central India. For effective flushing, water is kept ponded over the land surface for 2-3 days followed by puddling to dissolve soluble salts followed by drainage of the ponded water. Two flushing with an interval of about 7 days in between the two flushing is recommended. Since the soluble carbonates and bicarbonates of sodium can be effectively flushed, gypsum application after the flushing proved relatively more effective than without flushing. It is because of the fact that in the latter case gypsum reacts with soluble salts to adversely impact the gypsum efficiency.

A well designed surface drainage system is pre-requisite for the efficient functioning of this technique. Thus, in land locked fields, flushing may not be practicable. Due to high water requirement, inefficient salt removal and several technical constraints; use of this technique as a means of reclamation is limited.

Leaching

The only reasonable way to accomplish reclamation seems to be application of adequate quantities of water to the soil surface, such that the excess salts present in the root zone are dissolved and transported below the root zone with percolating water. For efficient leaching, following steps should form the part of a leaching program.

Field Preparation

On-farm land development activities should be completed especially the levelling and dyking the fields since inadequate levelling would lead to spotted desalinization and may cause delay or even failure of the reclamation program. To enhance uniformity of leaching, smaller plots with temporary bunds should be constructed such that the plots can later on be adjusted within the overall layout plan of the farm. Tillage of land before and in between leaching is beneficial because it would reduce the time of reclamation as well as improve the leaching efficiency.

Time of Leaching

Since adequate availability of water at the time of leaching is the essential pre-requisite for a successful leaching program, the time at which such quantities of water can be arranged would be most appropriate time for leaching. If there are no constraints of water, the best time for efficient leaching is when the water table is deepest and evaporation is minimal. Leaching of sodium sulphate, however, might be retarded during winter season when temperatures are low. It is better to choose a period when water table is at least 1 m deep. In case leaching is to be accomplished through rainfall, it is natural that one has to wait for a period when excess rainfall is expected. For example, in the Indian context, at most places in arid and semiarid zones, excess rainfall is only experienced during the monsoon season. Therefore, July-August period appears to be the best time for leaching. If, some water is available towards the beginning of this period, it could be used for leaching as it would enhance the efficiency of leaching through rainfall.

Depth of Reclamation

The depth of reclamation depends largely upon the potential land use. It is 60 cm for shallow rooted crops or pastures to more than 2 m for trees. For diversified cropping, the depth of reclamation usually varies from 1 to 1.5 m. It may be noted that these are the minimum recommended depths and appropriate measures against resalinization should be adopted.

How Much to Leach?

The salt status of the soil following leaching is normally determined by the tolerance of the crop to be grown soon after the reclamation. The level of salts should be such that for the planned cropping sequence and water management strategies, salt constituents of concern (salinity, specific ion concentration or alkali) should not exceed the permissible limit. If the level achieved after leaching does not permit cultivation of the proposed crop(s), either a more tolerant crop is selected or additional water is applied for leaching. Most comprehensive data on salinity tolerance of crops have been published and often this data form the guidelines for crop selection in reclaimed lands (Maas and Hoffman, 1977; Maas, 1980). Crop tolerance table provides to some extent similar information on some crops for some selected stations in India (Table 7.1). The threshold EC i.e. the EC_e up to which there is no decline in yield, slope which represents the decline in yield per unit increase in soil salinity beyond the threshold are reported (Table 7.1).

Crop productivity at the end of the season depends upon the initial crop stand. Therefore, most favourable conditions must be created during the germination and initial establishment stages. In one time leaching for reclamation, amount of salts to be leached is usually expressed as a fraction (Lf) of the amount of salts present in the profile at the start of leaching. Mathematically,

$$Lf = 1 - (\text{Threshold salinity of the crop} / \text{Initial salinity of the soil}) \quad (7.1)$$

Table 7.1. Crop tolerance to soil salinity for working out Lf

Sites	Soil type	Crop	Threshold salinity (dS/m)	Slope (%)
Sampla	Sandy loam	Wheat	4.0	29.0
		Mustard	6.0	75.0
		Barley	7.0	19.0
Karnal	Sandy loam	*Moong* bean	1.8	20.7
		Mustard	3.8	6.9
		Sorghum	2.2	10.6
Agra	Sandy loam	Wheat	8.2	19.8
		Mustard	6.1	20.7
		Berseem	3.5	12.5
		Potato	4.4	16.1
		Tomato	1.3	6.5
Dharwad	Black clay	Wheat	2.3	20.5
		Safflower	2.8	20.7
		Sorghum (W)	2.1	3.9
		Italian millet	6.1	50.0
		Sataria	1.1	2.2
Indore	Black clay	*Berseem*	2.0	11.2
		Safflower	2.8	5.0
		Maize	0.5	7.9
Chakuly	Clay	Rice	0.6	2.7

When some yield reduction is permissible, then threshold salinity in eq. (7.1) can be replaced by the salinity at which desired yield reduction (EC_d) can be expected. It can be calculated by the data reported in Table 7.1 utilizing the following relationship:

$$EC_d = (\text{Desired yield reduction/Slope}) + \text{Threshold EC} \qquad (7.2)$$

If tolerance of the crop at the germination stage is different than the mean values in Table 7.1 (Threshold), it would be proper to use the threshold salinity levels at the germination stage.

LEACHING CURVES

In planning reclamation programs, the leaching curve is a useful tool to decide whether desalinization is technically and economically feasible and how much water needs to be used to reclaim the land? Leaching curves are developed with sufficient accuracy on the basis of results from simple and rapid leaching tests. While conducting these tests, same process of leaching is adopted that is likely to be used to leach the salts in the reclamation project. A simple, straight forward way to prepare leaching curves is by running a test similar to the one used for the evaluation of the infiltration rate. Two concentric cylinders are used for this test (Fig. 7.1). The water level in the inner cylinder is kept at a constant level and water infiltrated is measured till the pre-determined amount of water has entered the soil. The experiment is usually a replicated trial. In these experiments, 3 or 4 depths of water are passed and are taken as the treatments. Soil samples in the inner cylinder are taken before and after the completion of the test for evaluating moisture content and salt status. For salt status appraisal, the most convenient approach is to measure the bulk soil electrolyte conductivity of different layers.

Fig. 7.1. A view of the set-up for preparing leaching curves

The amount of water drained below a given depth, D_w, is found by subtracting from the total depth of water that has infiltrated, the amount of water evaporated and the amount of water absorbed in the soil layer. The EC_{eq} is commonly used to make the leaching curve independent of the salinity of irrigation water and the existing evaporation and drainage conditions. It is obtained by measuring the soil salinity in the 0-5 cm layer in the treatment where largest amount of water has entered the profile. It is expected that the salinity of this layer reaches a practically constant minimum soon after the leaching commences.

The dimensionless plot of $(EC - EC_{eq})/(EC_o - EC_{eq})$ (EC_o is the initial soil EC_e) versus D_w/D_s (D_s is the depth of the soil) provides the desired leaching curve. This curve can be made use of in future for the test soil and conditions under which leaching tests have been conducted. Data illustrating this approach for preparing a leaching curve are shown in Fig. 7.2 left. It shows that water requirement for leaching a light textured soil is relatively less than a heavy textured soil. A plot on a log-normal paper results in a straight line (Fig. 7.2 right)

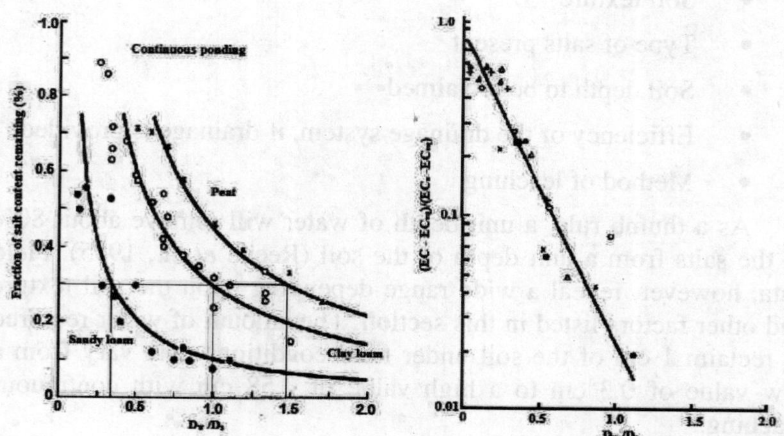

Fig. 7.2. Leaching curves for different soils (left) and on log-normal paper (right)

The leaching curves so prepared may not be applicable to drained lands. The main limitation of a leaching curve so prepared relate to the type of flow encountered in the drained lands. In constructing leaching curves by the method illustrated, flow is vertical and therefore, will be applicable to soil depths at which water movement is vertical or practically vertical. Thus, the leaching curves prepared are not applicable to strips near the drains or 1 m below the soil surface in widely spaced laterals with lateral drain depths of 1.5 m or more, where flow is likely to be radial.

Amount of Water Required for Leaching

In an ideal soil system without bypass, without dissolution or precipitation of salts, without salt diffusion or dispersion, the salt concentration of the soil water passing a given depth in the soil profile should drop to the concentration of the applied water when one pore volume (1 PV) of the applied water equals one pore space or one pore volume of the soil volume to be leached. The actual amount of water required under practical leaching will, however, depend upon many factors. Important among them are:

- Amount of salt initially present in the profile
- Desired level of salt in the root zone
- Soil texture
- Type of salts present
- Soil depth to be reclaimed
- Efficiency of the drainage system, if drainage is provided
- Method of leaching

As a thumb rule, a unit depth of water will remove about 80% of the salts from a unit depth of the soil (Reeve *et al.*, 1955). Field data, however, reveal a wide range depending upon the soil texture and other factors listed in this section. The amount of water required to reclaim 1 cm of the soil under field conditions may vary from a low value of 0.3 cm to a high value of 2.58 cm with continuous leaching.

LEACHING EFFICIENCY

When water is applied for leaching purpose, some of the irrigation water will move rapidly through the larger pores and reach the lower boundary of the root zone with little increase in the salt content. The water moving through the smaller pores may displace the soil water already present, as piston flow, so that the drainage water from the smaller pores will have about the same salt concentration as that of the soil water in the root zone. Thus, the water draining from the lower boundary of the root zone can be considered as a mixture of irrigation water that has passed unchanged through the root zone and soil solution that has been directly displaced by irrigation water. The leaching efficiency (E_t) has accordingly been defined as the hypothetical fraction of the drainage water consisting of the displaced soil solution. Thus, the concentration of the water draining from the root zone is calculated as (Bouwer, 1969):

$$C_d = E_t C_s + (1 - E_t) C_i \tag{7.3}$$

$$C_d = C_i + (C_s - C_i) E_t \tag{7.4}$$

In which C_s is the salt concentration of the soil water in the root zone and C_i is the salt concentration of leaching water applied. The value of E_t depends mainly on the size distribution of the water filled pores, which in turn is governed by the texture, structure and swelling properties of soils. Generally the value of E_t varies from 0.2 for heavy soils to 0.6 for light soils. The soil cracks, root holes, worm holes and other large diameter pores add to the inherent non uniform distribution of water application in farm irrigation systems, causing leaching efficiency appreciably less than one. The original formulation of Boumans and van Der Molen (1964) that states that leaching efficiency depends upon soil texture seems to be based upon this concept. On the contrary, it is apparent that the parameter E_t is more likely to be a bypass factor or mixing efficiency. Since leaching efficiency decreases with increasing amount of water passed through the soil profile, leaching efficiency can be defined as the amount of salt drained as a fraction of the amount of salt present in the profile per unit pore volume after pre- decided pore volumes of water have been drained (Gupta and Pandey, 1980).

Several ways and means to improve the leaching efficiency are discussed in the following sections.

Continuous and Intermittent Leaching

The common method of leaching is to store water within the dykes of the saline fields. The applied water passes rapidly through the large pores and flushes out the salts. The small pores conduct water less rapidly and take much longer time to be flushed. As long as water is ponded on the surface it continues to move through the large pores resulting in inefficient leaching of the soil profile. Leaching efficiency can be improved by creating unsaturated conditions during leaching and by reducing the average rate of water movement. Intermittent ponding or sprinkler irrigation can create such situations and result in sometimes thorough leaching of salts from the soil profile. Intermittent leaching minimizes loss of water through large pores and provides greater opportunities for diffusion of salts from immobile or less mobile pockets to the mobile zone. Intermittent leaching has been found to be useful in field trials. It has emerged that as much as 30 to 70% water required for leaching in continuous ponding can be saved (Dahiya *et al.,* 1981). Nielsen *et al.* (1966) reported that intermittent ponding may not be useful in cases where surface layer is more restrictive than the lower layers, since in such a case even under continuous ponding, unsaturated conditions will prevail in the profile. Intermittent ponding has not been found superior to continuous ponding in alkali soils (Abrol and Bhumbla, 1973), sandy loam soil (Lefellar and Sharma, 1977) and for reclamation of saline soils formed as a result of 2.5 years of application of saline water (Jury *et al.*, 1979). Similar results have been reported for leaching of boron (Beyce, 1972) reported that nearly 50% saving in water can be made with intermittent leaching compared to continuous leaching in a sandy loam soil. Elgabaly (1971) reported that although intermittent ponding saved water, it took nearly 40% more time for leaching compared to continuous leaching. Surface evaporation of water occurring between leaching may sometime overshadow the increased leaching efficiency brought about by intermittent leaching. Synthesis of present day data reveal that intermittent leaching may be useful in soils with large bypass (with cracks or structured soils), poor soil permeability, deep groundwater table and during periods of low evaporative demand. It may prove to be the most efficient method of leaching where water costs are high. On the other hand continuous ponding may give reasonably good results in soils with good permeability, unstructured soils, areas with shallow saline groundwater and during periods of high evaporative demand.

Initial Moisture Content and Application Rates

Studies by Nielsen and Bigger (1961) revealed that the degree of saturation and the rate of water movement in a given soil also affect the transport of salts through the soil. Field studies by Wilson *et al.* (1961) and Nielsen *et al.* (1966) also demonstrated the influence of the soil water content and the flow rate on the leaching efficiency. In these studies, comparison was made between (a) continuous ponding and intermittent ponding, (b) ponding and intermittent sprinkling, and (c) continuous ponding and rainfall. In each study, the slower movement of water achieved by intermittent ponding, sprinkling or rainfall resulted in significantly higher leaching efficiencies. All these techniques suggested that leaching efficiency can be improved by decreasing the application rate. While studying the effect of initial moisture content, flow velocity and redistribution time on the movement and distribution of salts in soil columns, it emerged that the salt front coincides with the water front in the initially dry soil and lags behind it in the initially moist soil. Immediately after infiltration and after redistribution, the salt peak occurs at a depth above which the total water storage equaled infiltration. But the salt spread in the profile increases as the initial water content increases. It also emerged that immediately following infiltration, the salt is displaced to deeper soil layers at a rate slower than that of water. Later on, the movement of the salt is faster than that of water. But when the application along with the redistribution time is matched, the salts move deeper with water added at faster than at slower rates. It is concluded that the slower rates of water application may not increase the leaching efficiency of water under field conditions. In fact, more water was required to leach the same amount of salt when water was applied at 1 cm/d compared to when it was applied at 2 or 3 cm/d (Thokal, 1990).

Conjunctive Use of Saline and Fresh water

If good quality water is either limited or is unavailable at the desired time, conjunctive use of saline and good quality waters for leaching has immense potential especially under the following conditions (Reeve and Doering, 1966).

- The salinity and SAR of the saline water is less than the salinity and SAR of the soil

- Gradually low EC and SAR water is used for leaching followed by fresh water in the final stage of leaching. Conjunctive use requires that last stage be synchronized with availability of good quality water.

A number of studies have validated the application of this concept. Attempts have also been made to quantify the saving in fresh water with conjunctive use in comparison to when fresh water alone is used (Gupta and Pandey, 1983; Ranade, 1984; Chawla, 1985). Gupta and Pandey (1983) observed that although 18-33% more water (saline+fresh) is needed in conjunctive use, the fresh water requirement is reduced by 15–40% depending upon the salinity of the groundwater, initial salinity of the soil and fraction of the salts drained with saline water. Chawla (1985) also observed that in sandy loam soils fresh water saving in the range of 48.3 to 61.2% can be made although the total water requirement is increased by 37.9 to 63.8%. Typical leaching behavior of a saline soil with conjunctive and fresh water alone is shown in Fig. 7.3.

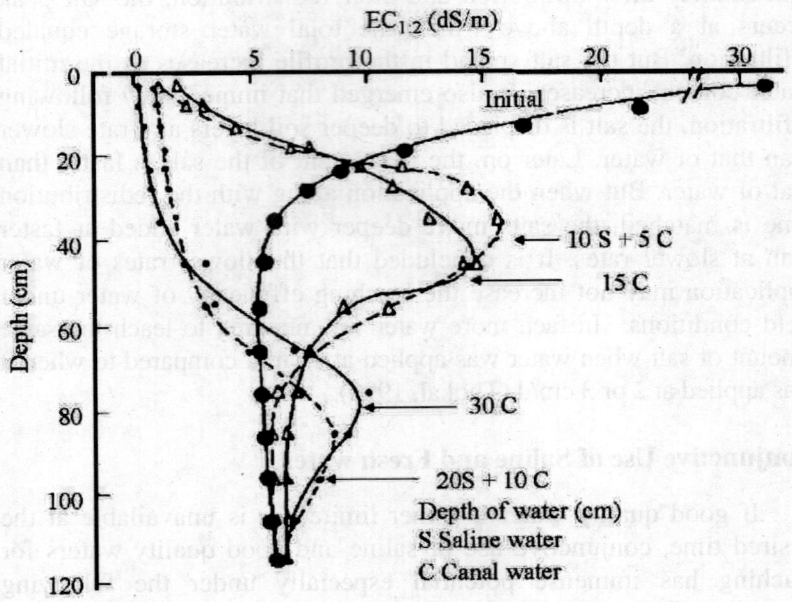

Fig. 7.3. Leaching behavior of saline soils with conjunctive use of saline and fresh water

Additional Practices

Leaching is the only practical means to reduce soil salinity and bring saline soils under the plow. To reclaim the soils with limited amount of water or to achieve early reclamation, a number of practices can be used. Some useful practices are summarized in Table 7.2.

Table 7.2. Practices to modify leaching behavior

Practice	Effect
Sub-soiling	• Improves infiltration rate • Removes/reduces natural channels
Tillage with inversion	• Improves infiltration rate of the soil • Physically moves the surface layer with high concentration of salts to lower depths
Chemical amendment	• Improves infiltration rate • Controls surface crusting • Maintains soil structure
Breaking hard pan (Auger hole piercing)	• Improves water movement through the hard pan • Reduces chances of water table build-up
Irrigation practices	• Improve leaching efficiency by maintaining high moisture content • Situations resembling intermittent leaching can be created

RAINFALL AND SOIL DESALINIZATION

One of the fortunate situations with continental monsoon climate of India is that nearly 75% of the total annual rainfall occurs in a short span of 2–3 months (July-September). Normally water penetrating into soils during this period exceeds evapotranspiration demand of crops. As a result, it induces leaching of salts from saline soils or soils irrigated with saline water. Intensity and duration of the rainfall, duration of dry spells between two rain events and soil conditions such as soil texture, initial salinity and its distribution in the soil profile, levelling and *bunding*, condition of the top soil surface and moisture content of the soil profile at the time of rainfall, all influence the process of desalinization/leaching. In general, duration and degree of salt leaching over the seasons would greatly be dictated by the monsoon rains and the evapotranspiration. Monitoring of soil profiles before and after the monsoon season has revealed that leaching due to rainfall is substantial (Gupta and Abichandani, 1970). They reported that after the rains, say about 35 to 45 cm, the initial high salt concentration in the saline soils is greatly reduced rendering the root zone completely non-saline. The hydrolysis of Na-clay appeared to be of small significance and leaching did not result in high alkalinity. On the other hand, in highly calcareous soils, ESP was also reduced significantly along with the soluble salts. The surface soils of Na^+ = Mg^{++} - Cl^- - SO_4^{--} type before rainfall became Na^+ - Ca^{++} - Cl^- – HCO_3 type after rainfall.

Dhir (1977) compared the effect of soil texture on leaching of salts during the rainy season at several locations in Pali district of Rajasthan. Results showed that in those years which received only 60% of the mean annual rainfall, the complete leaching from one year's rainfall occurred only in the loamy sand soils. However, only 80% of the salts were leached from the top 40 cm layer in the sandy loam soils. On heavier soils, desalinization was of the order of 50-80 per cent for this depth. In a year which received 142% of the average rainfall, the complete leaching of the accumulated salts was achieved in all soils except in the silty clay loams.

The effect of natural leaching by rainfall on alkali soil at Karnal was studied by Bandyopadhya (1971). It was reported that soluble salts were leached beyond the 60 cm depth in the soil profile during the monsoon but in the dry season the extensive evaporation increased the salinity of the soil. The presence of enough calcium carbonate did not help in reducing ESP of the soil due to high pH and lack of Ca^{2+} ions in the soil solution.

Minhas and Gupta (1992) estimated the desalinization of soils irrigated with saline water due to rainfall under normal cultivation practices. The following equations were fitted to the heavy, medium and coarse textured soils respectively.

$$EC_a/EC_b = 0.1916\,(D_s/D_{rw}) + 0.0962 \qquad r = 0.75 \qquad (7.5)$$

$$EC_a/EC_b = 0.1919\,(D_s/D_{rw}) - 0.0017 \qquad r = 0.95 \qquad (7.6)$$

$$EC_a/EC_b = 0.1577\ (D_s/D_{rw}) + 0.0068 \quad r = 0.66 \tag{7.7}$$

Here, EC_a and EC_b are the soil salinity after and before the rainy season, dS/m; D_s is the depth of the soil, m and D_{rw} is the depth of the rainfall, m. The above relations reveal that about 1.85, 0.95 and 0.76 cm of rain water will be required to leach 80% of the salt from each cm of the soil. On the other hand, for medium textured soils, about 0.5-0.6 cm of leaching water per cm of soil is sufficient to leach about 80% of the salts.

On the basis of these and many more studies, it has emerged that soils have low salinity after the monsoon or even at the time of sowing of next *rabi* crop (Fig. 7.4). Such regularities influence the way in which these soils have been utilized by the farming communities in India for centuries. However, this scenario may vastly change if summer/monsoon season crops are cultivated with saline water. Moreover, for the same EC_i, salinization and desalinization cycle will also be influenced with SAR of the water resulting in more salinization and less desalinization with increasing SAR (Fig. 7.4).

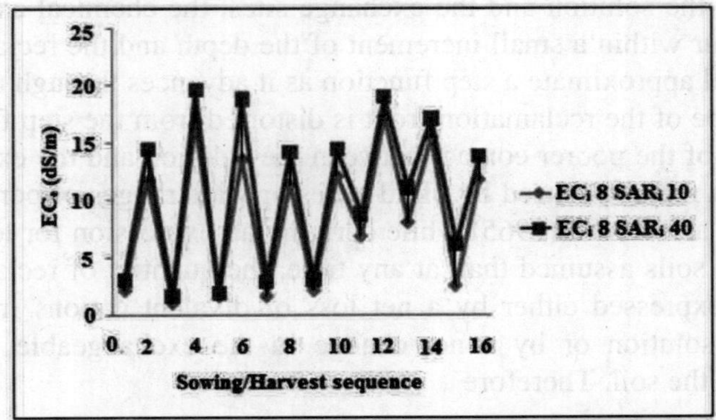

Fig. 7.4. Mean profile salinity at sowing and harvest of *rabi* crops at Agra

LEACHING OF ALKALI SOILS

Electrolyte Concentration of Leaching Water

The hydraulic properties of alkali soils are functions of both the electrolyte concentration of the leaching water and the exchangeable sodium status of the soils. For the reclamation of alkali soils by leaching it is essential to maintain good hydraulic conductivity for the water to percolate through the soil profile so that the divalent cations (Ca^{++} and Mg^{++}) can be transported to the cation-exchange sites for exchange with the sodium. Whereas the waters of low electrolyte concentration cause alkali soils to disperse and reduce its permeability, the waters having high electrolyte concentration have a flocculating effect on the soil particles resulting in the maintenance of higher permeability during the leaching process. The leaching water must also have a high calcium and magnesium concentration for the cation exchange process to take place. A leaching water is considered as reclaiming water relative to a particular soil if its SAR is less than the SAR of the soil water. Reeve and Bower (1960) demonstrated that even the water containing a high sodium ion concentration, such as ocean waters and dilution thereof can be used to reclaim an alkali soil because high permeability can be maintained during the process of leaching. As a reclaiming solution passes through a soil, calcium and magnesium are exchanged for sodium until the SAR of the solution corresponds to the SAR of the equilibrium solution for the soil at that ESP. If there is ample contact between the solution and the exchange sites, the chemical exchange will occur within a small increment of the depth and the reclamation front will approximate a step function as it advances through the soil. The shape of the reclamation front is distorted from the step function because of the poorer contact between the solution and the exchange sites that may be caused by blind pores or wide ranges of pore sizes. Doering and Reeve (1965) while deriving an expression for leaching of alkali soils assumed that, at any time, the quantity of reclamation can be expressed either by a net loss of divalent cations from the applied solution or by a net change in the exchangeable sodium status of the soil. Therefore

$$D_w \Delta(Ca+Mg)\, F = D_s \rho\ CEC \Delta(ESF) \qquad (7.8)$$

In which D_w is the depth of water in cubic centimeter per square centimeter, Δ(Ca+Mg) is the reduction of the divalent cation concentration of the solution between inflow and outflow in me/ml, F is the efficiency factor (always <1) that considers the contact between the solution and the exchange sites, D_s is the depth of soil in cm in which the exchangeable sodium fraction has been reduced by Δ(ESF), ρ is the bulk density (g/cc), and CEC is the cation exchange capacity in me/100 g. Eq. (7.8) can be written in the following form:

$$\frac{D_w}{D_s} = \frac{\rho CEC}{F} \frac{\Delta(ESF)}{\Delta(Ca + Mg)} \tag{7.9}$$

Here, D_w / D_s represents the depth of water per unit depth of soil. As an approximation, F can be taken as unity and all other quantities on the right side of Eq. (7.9) can be determined for each dilution step in the reclamation process. For homogenous soil, $\rho CEC / F$ is constant, and the total depth of water needed can be estimated by adding the depths of water needed for each reclamation step by eq. (7.9). For the nth leaching treatment, the depth of water per unit depth of soil D_w/D_s can be obtained in metric units by the following expression:

$$\left[\frac{\Delta D_w}{\Delta D_s}\right]_n = \frac{1000\,\rho_s\, CEC\,(\Delta ESF)_n}{\rho_w\, F\,[\Delta\,(Ca + Mg)]_n} \tag{7.10}$$

As for the homogenous soil, the product $\rho_s\, CEC$ is constant. Thus, for N successive dilutions

$$\left(\frac{D_w}{D_s}\right)_n = \sum_{n-1}^{n=N} \left(\frac{\Delta D_w}{\Delta D_s}\right)_n$$

$$= \frac{1000\,\rho_s\, CEC}{\rho_w\, F} \sum_{n=1}^{n=N} \frac{(\Delta ESF)_n}{[\Delta\,(Ca + Mg)]_n} \tag{7.11}$$

The change in exchangeable sodium fraction $(\Delta ESF)_n$ can be calculated from the algorithm

$$(\Delta\, ESF)_n = ESF_{n-1} - ESF_n \tag{7.12}$$

The equilibrium exchangeable sodium fraction for any dilution n can also be expressed in term of the SAR of water for the nth dilution and a soil constant k whose value depends mostly on the surface exchange density of soil ranging from 0.01 to 0.025 $(me/l)^{-1/2}$ for most soils.

$$(\Delta\, ESF)_n = 1/[1/(KSAR_{n-1}) + 1] = 1/[1/(KSAR_a) + 1] \tag{7.13}$$

From the above equation, SAR can easily be calculated for any step in a dilution series. The reclamation of alkali soil by the high salt water dilution method depends mainly upon the availability of high salt water in which the divalent cation concentrations are more than 0.3 of the total salt concentration because the amount of water that must pass through the soil profile increases exponentially as the divalent cation fraction decreases.

Leaching Amended Alkali Soils

While saline soils can be reclaimed through leaching alone, alkali soils require application of an amendment to replace Na ions by a more favourable ion, the Ca on the exchange complex. Therefore, leaching in amended alkali soils is not only required to transport the applied chemical constituent of the amendment to the site of exchange, but also to transport the exchanged ion and soluble salts below the root zone. Commonly, gypsum and pyrites are applied as amendments to reclaim alkali soils. Leaching practice for the two amendments are different as discussed in the following sections.

Leaching Gypsum Amended Soils

Since gypsum is sparingly soluble in water, it was believed that soils amended with gypsum would require huge quantities of water for leaching. Based on the solubility of gypsum in pure water, it was estimated that 120 cm of water must pass through the profile to dissolve 15 t/ha of gypsum commonly applied to reclaim highly degraded alkali soils. Recent trials, however, do not support this contention since gypsum solubility in the soils increases several fold because of the presence of a sink for the calcium ions (Calcium is taken out of the solution as it gets adsorbed replacing sodium). It is estimated that to dissolve 99% of gypsum in the size range of 0.5-2.0 mm in top 30 cm of the profile would require only 36 cm of water. More recent studies indicated that water requirement for leaching gypsum amended soils may not be much different than what would be needed to remove soluble salts. A good field practice for leaching gypsum amended soils is to continuously pond the fields after gypsum application for about a week or so before transplanting the rice crop. Remaining water requirement for leaching would be met through deep percolation losses from the rice fields.

Leaching Pyrite Amended soils

Pyrite should be applied to the soils at relatively high moisture content. After the amendment is applied, fields are left as such for about a week without initiating the process of leaching. It allows time for the pyrite to oxidize. Only after the oxidation has taken place, water is ponded on the soil surface for about a week either in the continuous ponded mode or in intermittent ponding mode. Total time required for leaching will be slightly higher if intermittent leaching is adopted. Since leaching is followed by rice transplanting as in the case of gypsum amended soils, it takes care of the remaining leaching process.

LEACHING FOR SALT BALANCE

Once the land is reclaimed, a major pre-requisites for sustainable crop production is to maintain the salt balance (salt leaching is equal to or more than the salt applied). It is all the more necessary, if saline water is used for irrigation. In this case, some leaching is always desired so as to restrict the salt build-up within the desired limits. The process is principally different than the one time leaching for land reclamation discussed in previous sections. The water required for this process is known as leaching requirement. The leaching requirement has been defined as the fraction of irrigation water that must be leached through the root zone to keep the salinity of the soil below a specific value. Under the steady state condition, the leaching requirement (LR) to maintain a proper salt balance in the root zone can be mathematically expressed as:

$$C_i D_i = C_d D_d \tag{7.14}$$

Such that

$$D_d/D_i = C_i/C_d \tag{7.15}$$

The water drained as a fraction of the water applied for irrigation is termed as LR. Moreover, salt concentration can be expressed in terms of respective EC of the water, equation for LR can be written as:

$$LR = D_d/D_i = C_i/C_d = EC_i/EC_d \tag{7.16}$$

In these equations, D_i is the depth of irrigation water applied (mm), D_d is the depth of water draining from the root zone (mm), C_i is the salt concentration of the irrigation water, which can be replaced by EC_i (dS/m) and C_d is the salt concentration of the soil water draining from the lower boundary of the root zone or EC of the drainage water, EC_d (dS/m). This equation takes into account only the salt contained in the irrigation water and the salts removed from root zone by the drainage water.

To determine LR, one needs to select an appropriate value of C_d for a given quality of irrigation water. A concentration of the soil saturation extract that produced a 50% decrease in the yield was used as the value of C_d by Bernstein (1964) and Bower *et al.* (1969). Benstein and Francois (1973) observed that the crops are generally insensitive to the bottom root zone values of the saturation extract and even C_d values about four times more than the previously selected value can be taken for C_d in a field situation. Boumans and Van der Molen (1964), on the basis of field observation, introduced the concept of leaching efficiency to predict the average C_d values.

Irrigation, Leaching Fraction and Soil Salinization

In general soil salinity is related to the salinity of irrigation water being high with high salinity water and low with low salinity water. The soil salinity and salinity of irrigation water are related as per the following equation:

$$EC_e = EC_i * X \qquad (7.17)$$

Here X is the concentration factor, which depends upon the leaching fraction (LF). The steady state root zone soil salinity (EC_e) likely to develop upon long-term use of saline water at various LF and a water uptake pattern of 40:30:20:10 is shown in Fig. 7.5. This water uptake pattern assumes that plants draw water in the ratio of 40:30:20:10 from each 25% of the root zone beginning from the top. It can be seen that at low leaching fraction (0.05), steady state salinity is about 3.2 times the salinity of the irrigation water while at high leaching fraction it is even less than the salinity of the irrigation water as is evident from leaching fraction exceeding 0.30. It is apparently clear that there is ample scope to control the mean root zone salinity at a pre-decided level by controlling leaching fraction. Moreover, for the same amount of water applied, the leaching fractions is more in a light than a heavy textured soil resulting in higher salinity build-up in the later. Clearly, the effect of amount of irrigation water applied and the soil texture can be lumped together in leaching fraction. It may be emphasized that LF should not be confused with LR. Whereas LR is theoretically calculated value to maintain salt balance, LF is the actual fraction of the irrigation water that leaves the root zone. If rainfall contribution is neglected, LF can be easily related to the average water application efficiency. An application efficiency of 70% means that leaching fraction is around 0.30.

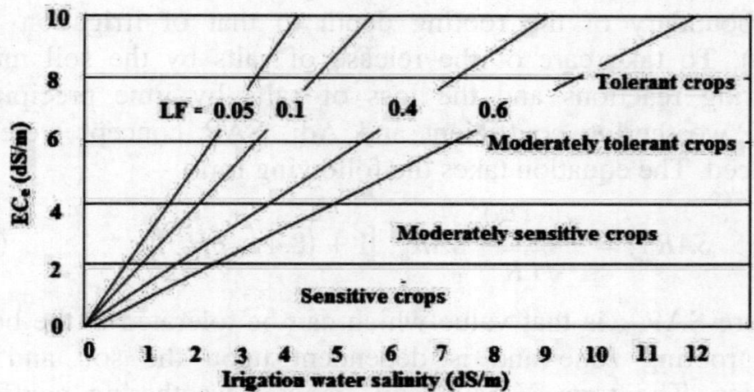

Fig. 7.5. Steady state root zone soil salinity (EC_e) likely to develop with applied irrigation water salinity (EC_i) at various LF

The depth of irrigation water required to maintain a given level of salinity in the soil in terms of the evapotranspiration of the crop can be expressed as

$$D_i = D_e\left[1/(1-LR)\right] \tag{7.18}$$

Here D_e is the equivalent depth of water representing evapotranspiration or consumptive use. If LR is written in terms of EC (eq. 7.16) and incorporated in eq. (7.18) then the depth of irrigation water required is equal to:

$$D_i = \left[EC_d / (EC_d - EC_i)\right] D_e \tag{7.19}$$

LR for SAR Control

Rhoades (1968, 1977) presented a method to estimate the leaching requirement to keep the sodium absorption ratio (SAR) at the bottom of the plant root zone below a limiting value. Since there is a well-established relationship between the exchangeable sodium percentage of the soil exchange complex and the SAR of the soil saturation extract, the leaching requirement needed to prevent the excessive build-up of the exchangeable sodium in the soils was designated as LR_{SAR} but commonly used notation was LR in this case as well. The theory behind the calculation of LR_{SAR} is analogous to the leaching requirement to control EC discussed in the previous paragraph. An equation was developed to calculate LR_{SAR} that relates the sodium adsorption ratio of drainage water $(SAR)_{dw}$ from the lower boundary of the rooting depth to that of irrigation water (SAR_{iw}). To take care of the release of salts by the soil mineral weathering reactions and the loss of salts by lime precipitation mineral weathering coefficient and Adj. SAR concept were also introduced. The equation takes the following form:

$$SAR_{dw} = \frac{Y^{(1+2\,LR)}}{\sqrt{LR}}\,SAR_{iw}\,\left[1 + \left(8.4 - pH_c\right)\right] \qquad (7.20)$$

Here SAR_{dw} is that value which can be tolerated at the bottom of the rooting zone and is dependent upon the soil and crop properties. The term Y is a soil mineral weathering coefficient usually taken as 0.65. A method for its determination was presented by Rhoades (1968). The pH_c is the theoretical pH at which the water of a given (Ca^{++} + Mg^{++}) and (CO_3^{-} +HCO_3^{-}) content would have an equilibrium with solid lime. The term (8.4 – pH_c) as such, is an index of the tendency of the water to either precipitate or dissolve lime, when reacted with a calcareous soil. This equation in this form is limited to calcareous soils and waters containing bicarbonates. It may be noted that the concept of adjusted SAR used in deriving eq. (7.20) has become obsolete and is no more recommended. Considering this fact, eq. (7.20) has also become obsolete but is included since it is a part of literature on leaching requirement.

For waters that do not contain bicarbonate but have sodium hazards owing to their SAR_{iw} values, the following equation can be used to estimate LR_{SAR} values:

$$SAR_{dw} = \frac{Y^{(1+2\,LR)}}{\sqrt{LR}}\,SAR_{iw} \qquad (7.21)$$

It should be noted that the leaching requirement for salinity control may or may not be simultaneously satisfied for the exchangeable sodium control depending on the water properties and the allowable SAR_{dw}. For a relatively low value of SAR_{dw} tolerated at the bottom of the rooting zone, the leaching requirement required for salinity control of relatively low concentrations, and high SAR_{iw} waters will be unsatisfactory for the exchangeable sodium control.

On the other hand, for waters having high salt contents or for relatively high SAR_{dw} values, the leaching requirement estimated for salinity control may also be adequate for exchangeable sodium control.

SALT BALANCE OF THE ROOT ZONE

The water and salt balance of the root zone are so intricately woven that it is not possible to separate them out. The salt balance of the root zone of irrigated lands can be worked out by various components of water balance (Fig. 7.6) multiplied with their respective salt concentrations (Gupta and Khosla, 1996). The resulting salt storage or leaching from the root zone can be mathematically put into the form of an equation as follows:

$$S = R\,C_r + I\,C_i + G\,C_g + S_1 - DC_d - S_2 \tag{7.22}$$

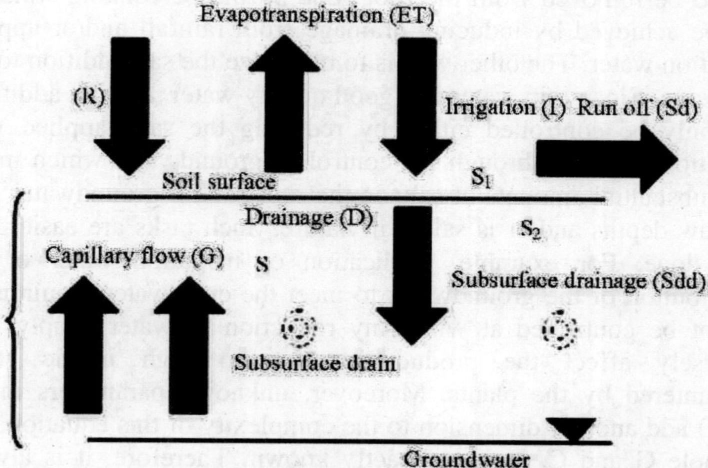

Fig. 7.6. Water balance components of the root zone of irrigated lands

Here salt storage (+) or leaching (-), S, in the root zone is given by the components which add salt to the root zone (+) or those which remove the salt (-). These components are added together with their respective signs. In eq. (7.22) R, I, G and D are the amounts of rainfall, irrigation, groundwater contribution and drainage below the root zone and C is the concentration of water with subscripts describing the respective components. In this equation salt are added to the root zone through rainfall (RC_r), application of irrigation water (IC_i), groundwater (GC_g) and direct addition of salts such as through the application of amendments/fertilizer or generated in the root zone through mineral dissolution (S_1). On the minus side are the salt removed due to drainage (DC_d) and the salt precipitation (S_2). In most cases, the terms S_1 and S_2 are neglected leaving only four terms on the right hand side of eq. (7.22). It may be noted that irrigation induced mineral dissolution is significant and it may not be possible to neglect this component in irrigated lands without causing significant differences in the calculated and actual salt balance. Besides, what is discussed, eq. (7.22) when simplified by incorporating S, C_r, G, S_1 and S_2 all equal to zero, it turns out to be the simplified salt balance eq. (7.14) since $I = D_i$.

Since the salt uptake by the crops is negligible, the only way to avoid soil salinization is that an equivalent amount of salt added should be removed from the root zone by the percolating water. It can be achieved by inducing drainage from rainfall and/or applied irrigation water. The other way is to minimize the salt addition to the root zone. Since rain water is a good quality water, the salt additions can only be controlled either by reducing the salts applied with irrigation water or through the control of groundwater which might add substantial amounts of salts in the root zone if groundwater is at shallow depths and/or is saline in nature. Such tasks are easier said than done. For example, application of irrigation water or the contribution of the groundwater to meet the crop water requirement cannot be controlled at will. Any reduction in water supply may adversely affect the productivity due to high matric stress encountered by the plants. Moreover, unknown parameters in eq. (7.22) add another dimension to the complexity of this equation. For example G and C_d are not exactly known. Therefore, it is always useful to monitor the salt status of irrigated lands over a few years' time to work out the salt regime. It will not only help to diagnose the problem but will also help to identify factors likely to accentuate salt accumulation or leaching of the salts.

Factors Affecting Salt Balance in the Root Zone

Soil salinity being dynamic, has large spatial and temporal variability. Some of the factors that influence the salinity of the root zone are the amount of water applied, the salinity of the irrigation water, depth to water table, soil texture and other soil characteristic which affect its permeability, evapotranspiration, total rainfall and its distribution, irrigation methods employed, setting of the land in respect of the adjoining area (topography) and above all the overall management inputs that goes with any agricultural activity. The following salt balance scenarios are anticipated:

- Balance in favor of leaching (S in eq. (7.22) is negative)
- Stable salt balance (S in eq. (7.22) is zero)
- Balance in favor of salt accumulation (S in eq. (7.22) is positive)

Salt balance with respect to time is determined for short durations (say between irrigations or a month), medium durations (say a season) and long durations (say annual) over a few years' time. Similarly, salt balance is established over small terrestrial units (say a farm) or over an irrigation command or on a regional basis. Usually, larger areas are chosen for salt balance studies, since salt move from one area to another and normally operate on regional scale. While preparing the salt balance sheet, causes, rate and degree of accumulation or leaching needs to be established.

WATER QUALITY AND SOIL SALINIZATION

Soils, which are otherwise salt free or are reclaimed with huge investments, may be rendered unproductive by soil salinization. In monsoon climatic conditions processes of soil salinization and desalinization operate over time and one can observe the following cycle:

- The period of soil salinity build- up during November to April when saline water is used for irrigation and/or through capillary action if the shallow groundwater conditions exists after the monsoon season
- Period of salt redistribution within the root zone from April to mid-June or till the onset of monsoon. During this period, soil salinity is usually maximum in the surface layer (0-15 cm)

- Period of desalinization during monsoon season when rainfall is in excess over the potential evapotranspiration.

Clearly, two situations can be identified for soil salinization. One is the capillary rise from the shallow water table and the other is the use of saline water for irrigation.

Critical Depth to Water Table

Under shallow water table situations, yield is reduced due to poor aeration, limited root extension and salt build-up especially when the groundwater is saline. The capillary rise/flow is related to soil texture, depth to water table and water gradients. While capillary rise increases with increasing silt content of the soils, the rate of evaporation increases with decreasing depth to water table (Fig. 7.7, van Hooren, 1979). The concept of 'Critical Depth to Water Table', first advocated by Talsma (1963), has been in vogue for quite some time. According to this concept, the process of secondary salinization is quite active once the water table is above this critical depth. The critical depth, for the purpose of this concept, is defined as the depth to water table at which relative evaporation is 0.1 or the rate is about 10% of its value when the water table is at or near the soil surface. For practical standpoint, the evaporation rate should not exceed 0.5 mm/d to 1 mm/d, the maximum allowable flux to maintain favorable water and salt balance. On this basis, critical depth may vary from a meter in light textured soils to a few meters in fine textured soils.

Recent studies have revealed the limitations of this concept and arguments contesting the applicability of this concept have been put forth. It is argued that if the flux at the bottom of the root zone is kept downward, there should be no opportunity for soil salinization to occur even under shallow water table situation. In the Indian context, subsurface drains installed at shallow depths performed well and maintained water and salt balance because flux of the water at critical periods is downward (monsoon season, when the water table is at shallow depths). During the summer season, when evaporative demand is high, water table is deep enough to allow excessive upward flux. Thus, applicability of the concept of critical depth to water table needs to be examined before it is applied to field situations. Otherwise, we may end up in costly solutions in land reclamation programs.

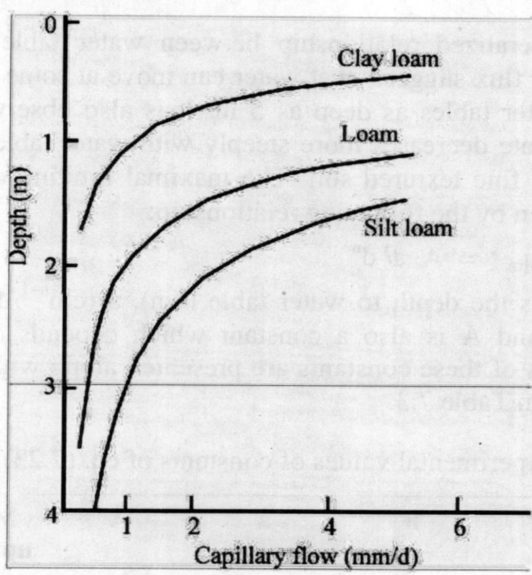

Fig. 7.7. Capillary flow velocity and depth of water table in 3 texture soil groups

Permissible Maximum Flux

Water is the main carrier of salts in the soil. As the soil solution reaches the site of evaporation at or near the soil surface, the water is evaporated and the salts are left behind. In cropped lands, the zone of salt build-up also depends upon the depth and the distribution of active roots, which may transpire a part of the water even before it reaches the soil surface. Salinity in the profile increases with decreasing depth to groundwater table but increases with increasing salinity of the groundwater at the same depth to water table. For favorable water and salt balance, understanding and knowledge of the maximum flux, q_{max} is quite important. For good quality groundwater, conventionally 1 mm/d is taken as the limiting value but for poor quality groundwater, the value is reduced to minimize soil salinization. The effect of groundwater salinity (0.6-3.7 dS/m) was observed to be more pronounced at shallow than at deeper depths. The studies have also shown that correlation between the concentration of the salts at the soil surface and the groundwater is significant only when water table is at or above 1.2 m.

The generalized relationship between water table depth and upward water flux suggests that water can move at some appreciable rate from water tables as deep as 5 m. It is also observed that the evaporation rate decreases more steeply with water table depth in a coarse than a fine textured soil. The maximal limiting upward flux (cm/d) is given by the following relationship:

$$q_{max} = A.\ a/\ d^n \qquad (7.23)$$

Here d is the depth to water table (cm), a (cm^{n+1}/d) and n are coefficients and A is also a constant which depends on n. Some typical values of these constants are presented along with maximum upward flux in Table 7.3.

Table 7.3. Experimental values of constants of eq. (7.23)

Soil texture	n	a (cm^{n+1}/d)	A	Maximum upward flux[#] (cm/d)
Sand	4	$1.7*10^8$	1.52	2.58
Fine sandy loam	3	$3.2*10^5$	1.76	0.56
Loam	2	$1.7*10^3$	2.46	0.42
Clay	2	$1.1*10^3$	2.46	0.27

Source: Gardner and Fireman (1958); [#] For d=100 cm

Soil Salinization due to Capillary Movement

The total quantity of water brought into the root zone and the salt concentration of the groundwater determine the extent of soil salinization during the fallow period. The amount of salts brought into the root zone is calculated by the following equation:

$$S = C*m \qquad (7.24)$$

Here S is the amount of salt added, kg/ha, C salt concentration of groundwater, g/L or kg/m^3 and m is amount of capillary flow entering the targeted zone. The change in soil EC_e in terms of water entering the root zone in depth units and EC of groundwater (EC_g), can be given by the equation:

$$\Delta EC_e = D_g/D_s * \rho_w/\rho_b * EC_g/SP * 100 \qquad (7.25)$$

Here ΔEC_e is the change in salt concentration (EC) of the soil depth, Ds, cm, D_g is the amount of water that enters the given depth of the soil, cm, SP is the water content in mass percentage at saturation, ρ_b and ρ_w are the bulk density of soil, g/cm^3 and density of water respectively usually taken as $1 g/cm^3$.

If salt concentration of groundwater is known in g/L or kg/m^3 such that one can calculate the salt added to the root zone in kg/ha then the above equation for the change in EC_e of the soil due to added salts is written as:

$$\Delta EC_e = (\Delta S/0.7)*1/w_e* \rho_w/\rho_b*1/D \quad (7.26)$$

Such that ΔS is the amount of salt added through capillary rise in kg/ha. The factor 0.7 is used to convert the salt in terms of EC. It may be noted that this factor varies from 0.64 to 0.80 and therefore, an average value of 0.7 is used in these calculations. For accuracy one may change this value accordingly.

Example 7.1

Determine the change in soil salinity in 30 cm soil depth if during the period of 3 months 15 cm of water gets evaporated from the soil surface. EC of groundwater is 5 dS/m. The bulk density of the soil is 1.5 g/cm^3 and saturation percent is 60. Assume any other appropriate value required for calculations. Show that change is nearly same for eq. (7.25) and eq. (7.26).

We have all the values for use in the eq. (7.25).

$$\Delta EC_e = D_g/D_s* \rho_w / \rho_b * EC_g/SP *100$$

Assuming ρ_w as 1.0 g/cm^3, we have:

$$\Delta EC_e = 15/30 * 1.0/1.5 * 5/60 *100 = 2.8 \text{ dS/m.}$$

To use eq. (7.24), we have EC of the water as 5.0 dS/m or the water contains approximately 3.5 g/L (kg/m^3) of salts.

The amount of water entering the soil profile is 15 cm or 1500 m^3.

Thus, salt added to the profile are 1500*3.5 = 5250 kg/ha. Assuming ρ_w as 1.0 g/cm^3 and using eq. (7.26) we have:

$$\Delta EC_e = (5250/0.64)*1/60* 1/1.5*1/30 = 2.77 \text{ dS/m.}$$

The values are nearly same.

Salinization Due to Saline Water Irrigations

Every irrigation event, howsoever small the salt content may be, adds a certain amount of salt into the root zone. In irrigated lands particularly when limited quantities of water is applied, the salts accumulate in the root zone if equivalent amounts are not removed through appropriate means. For example; if 10 cm of water having EC = 0.5 dS/m is applied to irrigate an area, it will add about 320 kg/ha of salts. If the EC is 1 dS/m, the addition will be 640 kg/ha.

Several factors control the extent of salinity development in the soil profile, five being the most important. These are (a) quality of irrigation water, (b) number of irrigations, (c) soil texture (d) leaching fraction and (e) rainfall.

In irrigated lands where there is no drainage and there are no other sources of salts, the salt build-up is determined from the following relation:

$$\Delta EC_e = D_i/D_s * \rho_w/\rho_b * EC_i * 100/SP \qquad (7.27)$$

Although eq. (7.25) has limited application to assess soil salinity development in the soil profiles as some water always percolates below the zone of interest, nonetheless, a very useful application of this equation can be made in the preparation of saline soils for conducting pot/lysimeter experiments. The amount of saline water required to salinize a soil to a pre-decided level is given as:

$$D_i = D_s * \rho_b/\rho_w * \Delta EC_e / EC_i * SP/100 \qquad (7.28)$$

Example 7.2

Determine the change in EC of the top 30 cm soil profile if 50 cm depth of saline water of 2 dS/m is applied and the whole water remains within this depth without any deep percolation. The bulk density of soil may be taken as 1.3 g/cc and the density of water as 1 g/cc. The soil saturation percentage is 42. What would be the depth of water applied if the change in salinity is not to exceed 4.0 dS/m?

Incorporating all the known values in eq. (7.27)

$$\Delta EC_e = D_i/D_s * \rho_w/\rho_b * EC_i * 100/SP$$

$$\Delta EC_e = 50/30 * 1/1.3 * 2.0 * 100/40 = 6.4 \text{ dS/m}$$

Incorporating all the known values in eq. (7.28)

$$D_i = D_s * \rho_b/\rho_w * \Delta EC_e / EC_i * SP/100$$

$$D_i = 30 * 1.3/1.0 * 4.0/2.0 * 40/100 = 31.2 \text{ cm}$$

■■■■

CHAPTER 8

QUALITY OF IRRIGATION WATER

INTRODUCTION

The quality of water is decided on the basis of the purpose for which it is used. Water that may not be fit for domestic use may well be fit for irrigation. Similarly, all irrigation waters contain dissolved salts in different amounts, yet all of these are not qualified as saline. For a geologist, all waters which have TDS less than 2000 ppm (EC < 3.0 dS/m) are fresh. For an agriculturist, conventionally, water containing total dissolved salts exceeding 1000 ppm (EC > 1.5 dS/m), has been qualified as saline. Currently, even in this sector this limit has been raised to 2.0 dS/m to designate water as saline. With increasing concerns on water quality, it ie realized that agriculture sector, which consumes the largest amount of water is capable to use relatively poorer quality water that other sectors can't use or even discard. This chapter describes the basic concept used in categorization of water quality for irrigation. It also gives an overview of the anticipated problems from the use of such water in agriculture.

CHARACTERISTICS OF WATER

The characteristics of irrigation water that are most important in determining its quality, depending on climatic conditions, irrigation practices, soil-water retention characteristics, crop tolerance, depth to water table and agronomic practices, etc. are the following:

- **Salinity hazard** (Total concentration of soluble salts); Electrical conductivity (EC)
- **Sodicity hazard** (Relative proportion of sodium to other cations); Sodium Adsorption Ratio (SAR) or Sodium to

Calcium Activity Ratio (SCAR), adjusted SAR (Now obsolete)

- **Alkalinity hazard** (Bicarbonate concentration as related to the concentration of calcium plus magnesium), or calcium alone; Residual Sodium Carbonate (RSC) or Residual Sodium Bicarbonate (RSBC)
- **Specific ion toxicity hazard** (Ionic composition): Major constituents (Sodium, Magnesium, Chloride, Bicarbonate, Carbonate, Silica, Nitrate) and minor constituents (Boron, Lithium, Fluorine and other micro-toxicants)

In addition to above problems permeability hazard is commonly encountered with continuous use of sodic/alkali water. Moreover, besides the use of individual parameters to describe a problem, combinations of two or more parameters as given below are also of practical importance in anticipating the likely problems:

- EC and SAR
- EC, SAR, RSC
- SAR and RSC

Since EC and SAR have been described previously in Chapter 2, herein only the criteria used in classifying the irrigation water on the basis of EC and SAR and the concept of RSC is included.

EC

The total concentration of soluble salts is the single most important criterion which has been used conventionally to determine the quality of irrigation water. A salinity problem related to water quality occurs if the total concentration of salts as measured by EC is high enough for salts to accumulate in the crop root zone to the extent that yield are adversely affected. Ideally, it may be inferred that EC of irrigation water should be as low as possible but the water which is completely free of the soluble salts is never the best for irrigation. The waters having EC less than 0.2 dS/m have no fertility value and are well known to create permeability problem in the soil. The irrigation water should, however, have EC preferably less than 2.0 dS/m so that the irrigated soils do not ever become saline and there is full choice to grow the crops. *Per se*, salinity of irrigation water has no direct effect on crops (except in sprinkler irrigation) but with excessive accumulation of soluble salts in the root zone, the crop has extra difficulty in extracting enough water from the root zone, resulting in what is described as physiological drought. The classification criterion of irrigation waters on the basis of EC is given in Table 8.1.

Table 8.1. Classification of waters based on EC

Water class	EC (dS/m)	Type of water
Non saline	<0.7	Drinking/irrigation water
Slightly saline	0.7-2.0	Irrigation water
Moderately saline	2.0-10.0	Drainage/groundwater
Highly saline	10.0-25.0	Drainage/groundwater
Very highly saline	25.0-45.0	Very saline groundwater
Brine	>45.0	Sea water

Example 8.1

Prepare a saline water of EC 3 dS/m using various salts.

Using the relation: me/l = 10 * EC, we have to prepare a solution of 30 me/l

If Na_2SO_4 (Equivalent weight 71.0) is used then dissolve 2.13g Na_2SO_4 per liter (30* 71/1000)

If $CaCl_2$ (Equivalent weight 55.5) is used then dissolve 1.667 g $CaCl_2$ per liter (30* 55.5/1000)

If $CaCl_2.\ 2H_2O$ (Equivalent weight 73.5) is used then dissolve 2.20 g $CaCl_2\ 2H_2O$ per liter (30*73.4/1000)

Example 8.2

Prepare a saline water of EC 8 dS/m using sodium chloride.

Since me/l = 10 * EC is not applicable, EC being more than 5, we follow the other conversions

Salt present in saline water = 8*800 = 6400 mg/l

If we divide this value with equivalent weight of the salt to be added (Annexure I), say sodium chloride, we have

6400/58.5 =109.40 me/l

Dissolution of 58.5 mg of sodium chloride in 1 liter gives 1000 me/l.

Therefore,

NaCl to be dissolved in 1 liter = (58.5/1000)* 109.4 = 6.40 g.

Compare with the approximate relation (me/l = 10 * EC), we need to dissolve only 4.68 g of NaCl (58.5*8/1000).

Example 8.3

To prepare a salt water of say 4 dS/m having Na:Ca ratio as 50:50 with a single anion chloride.

Since the relation, me/l = 10 * EC can be used. Therefore me/l = 40. For Na and Ca to be in 50:50 ratio, we have to add 20 me/l of Na and 20 me/l of calcium salt. We prepare the following table.

Salt	Concentration (me/l)	Quantity to be added per liter
NaCl	20	58.5*20/1000= 1.17 g
$CaCl_2$	20	55.5*20/1000= 1.11 g
or $CaCl_2.H_2O$	20	73.5*20/1000= 1.47 g
Total	40	

Therefore, either 1.17 g of NaCl and 1.11 g $CaCl_2$ or 1.17 g of NaCl and 1.47 g of $CaCl_2.H_2O$ should be dissolved in 1 liter of water.

Harmful Effects of Saline Water

Some of the visual symptoms irrigated with saline water are: plants look stunted; leaves are smaller but thicker and have often-dark green colour as compared to plants growing in a salt free soil irrigated with good quality water. The decline in crop growth is reflected in lower crop yields. It happens because of the following reasons.

- First and the most important is that the water availability to the plants decreases with increasing amount of salts in the soil water. Although the soil may contain sufficient moisture, it will not be available to the plants. Since the osmotic pressure of the soil solution increases with increasing salt concentration, plants are unable to extract water as readily as they can from a relatively non-saline soil solution. It is why excessive salts in root zone can induce early wilting and the effects are almost similar to those of drought.
- Secondly, at high concentration, absorption of an individual ion (e.g. Na^+, Cl^-, SO_4^{--}, Mg^{2+}, B) may prove toxic to the plants and can induce physiological disorders.
- Thirdly, preferential absorption of one ion may also retard the absorption of other essential plant nutrients necessary for the normal growth of plants. For example, high concentrations of chloride may depress the uptake of phosphorus.
- Finally, some elements like lithium, boron etc present in irrigation water may prove toxic to plants even though they may be present only in very less quantities.

It is believed that the adverse effects are usually due to combined effects of these factors although one might dominate the others (Gupta and Gupta, 1997).

Classification of Water on the Basis of EC

On the basis of electrical conductivity, the irrigation waters may be classified into four salinity classes (Gupta, 1979a, Gupta, 1984):

(a) C-1: Low salinity water (EC 0.2-1.5 dS/m) can be used for irrigation with most crops on most soils with little likelihood that soil salinity will develop. Large areas of agricultural lands in India and abroad have been irrigated continuously for many years with such waters without any adverse effects. These waters do not have any special leaching requirement. The waters are relatively high in Ca and low in chlorides. There is seldom any problem of high SAR/SCAR or of RSC/RSBC or toxic elements for the waters falling in this category. The waters having EC <0.2 dS/m may create a severe permeability problem in the soil because infiltration rate is adversely affected to an extent that the crop is not adequately supplied with water. In the current classification, the upper value has been raised from 1.5 dS/m to 2.0 dS/m.

(b) C-2: Medium salinity waters (EC 1.5-3.0 dS/m) can be used if a moderate amount of leaching occurs under the current irrigation practices. Most of the crops except sensitive ones comprising some horticultural and leguminous plants can be grown on all soils except for very heavy textured soils or soils with impeded drainage. These waters do not have any significant problem of SAR/SCAR and toxic elements but may have some problem due to RSC/RSBC.

(c) C-3: High salinity waters (EC 3.0-5.0 dS/m) can be used on soils provided with good drainage such that leaching fraction is not less than 0.3. Most of semi-tolerant and tolerant crops can be grown under good management. These waters may have a moderate problem of high SAR/SCAR and RSC/RSBC but seldom a significant problem of toxic elements. Interestingly, there is evidence to show that optimum yields of many tolerant crops are obtained under irrigation with waters belonging to this EC group.

(d) C-4: Very High salinity waters (EC 5.0-10.0 dS/m) are suitable for irrigation on soils provided with very good drainage so that leaching fraction is always greater than 0.3 and EC_e does not exceed EC_{iw}. Only salt tolerant crops can be grown under good management. Higher the total and seasonal rainfall during the crop growth period, better the usability. These waters may have a significant problem of medium to high SAR/SCAR but no problem due to RSC/RSBC. However, high SAR as high as 30 $(mmol/l)^{1/2}$ disqualify these waters for satisfactory use for prolonged periods. Summer fallowing is sometimes necessary to ensure natural desalinization. Application of some amendments may be necessary to take care of sodium and/or toxic elements hazards.

SODIUM/ALKALI HAZARD

pH

The pH, a measure of the acidity or alkalinity of water, is of interest as an indicator but is seldom of any real importance by itself. The main use of pH is quick evaluation of the possibility that the water may be abnormal. The normal range for irrigation water is 6.5 to 8.4. Within this range the crops do well. Irrigation waters having pH outside this range may still be satisfactory but other problems of nutrition or toxicity become suspect. For example; high pH above 8.5 is often associated with high bicarbonate (HCO_3^-) and carbonate (CO_3^{2-}) concentrations, which might turn the water alkali in nature. Thus, abnormal value of pH serves as a warning that the water needs further evaluation for possible corrective measure.

Sodium Adsorption Ratio, SAR

Apart from the total salinity, the next important consideration is whether the use of irrigation water of given quality will cause alkalinity in the soil. Water which might be considered suitable for irrigation on the basis of EC may not be suitable if sodium predominates. For this purpose, ratio of the monovalent and the divalent cations in the water is used to assess the sodium hazard of water. As the potassium is comparatively less in waters than the sodium, it is neglected in the calculations. Sodium adsorption ratio is calculated either by eq. (1.2) or as per the following equation:

$$SAR = Na/\sqrt{(Ca + Mg)} \qquad (8.1)$$

Here concentration of Na, Ca and Mg are taken in mmol/l. The units of SAR are mostly based on this definition although units based on the basis of eq. (1.2) are also reported.

Concentrations of the ions in water samples are typically reported in mg/l. To convert Na, Ca, and Mg from mg/l to me/l, one needs to divide the concentration by 22.9, 20, and 12.15 respectively. To convert me/l to mmol/l, divide by the charge or the valency of the ion.

If the SAR of the water is more than 15 $(mmol/l)^{1/2}$, the water with its continuous use for irrigation may pose problems by affecting soil permeability. Since in heavy textured soil SAR of even 10 might create problems, some classification of water on the basis of SAR treat such water of medium hazard (Table 8.2).

Table 8.2. Classification of water based on SAR values

SAR	Sodium hazard of water	Remarks
1-9	Low	Cautious use on sodium sensitive crops
10-17	Medium	Amendments (such as gypsum) and leaching needed
18-25	High	Generally unsuitable for continuous use
>26	Very high	Generally unsuitable for use

High SAR waters cause permeability problem in the soil wherein the rate of water infiltration into and through the soil is reduced to such an extent that the crop is not adequately supplied with water and consequently the yield is reduced. The adverse effect of high SAR depends upon total electrolyte concentration, bicarbonate and carbonate concentrations, presence of silica in irrigation water and clay mineral of the soil. Higher the total electrolyte concentration, lower the bicarbonate, carbonate and silica content, lesser the adverse effect on the permeability. For stable permeability, the irrigation waters having SAR 10, 20, 30 should have total electrolyte concentration of about 10, 20 and 30 me/l (EC 1, 2, and 3 dS/m) respectively.

An unusual aspect of the SAR is that Ca^{++} and Mg^{++} have been lumped together. According to one group of scientists, Mg acts like Ca in influencing the physico-chemical properties of the soil, owing to divalent nature of the cation. Others, however, believe that Mg should not be coupled with Ca in view of the varying affinity for adsorption on the exchange complex of the soil. There seems to be no theoretical basis for combining the concentrations of these two ions, except for the fact that ion valency is considered to be most sensitive parameter for predicting ion-exchange relations as it governs the ions size. Garrels and Christ (1965) have shown that the Debye Huckel distance of the closest approach for magnesium is approximately one-third larger than for calcium. Hence, over the range of solution salt concentration common to irrigation agriculture (2 to 200 me/l or 0.001 - 0.1 molar) the activity coefficient for magnesium should be 2 to 20% higher than for calcium. Many soils and clays also tend to prefer Ca^{++} over Mg^{++} in ion-exchange, with a relative increase in preference of about 20% being common. These two effects have opposing trends and may account for at least a part of the widespread success when using the SAR for ion-exchange predictions. Another important factor is that many waters and soil solutions contain two to five times more Ca^{++} than Mg^{++}. Moreover, unlike Indian waters, Mg is only a minor constituent of the vast majority of irrigation waters in the USA. More important and adequate, justifications for the widespread usage of the SAR are its relative simplicity and the fact that it generally works. However, to resolve sme of the issues raised, following alternatives have been suggested.

Adj. SAR

Ayers and Westcot (1976) proposed the concept of adj. SAR, which is calculated by the following relation:

$$\text{Adj. SAR} = \text{SAR}\,(1 + 8.4 - pH_c) \tag{8.2}$$

The pH_c is a theoretical parameter which is the calculated pH of the irrigation water in contact with lime and in equilibrium with soil CO_2. Since the concept has become obsolete (Ayers and Westcot, 1985), detailed discussion on this parameter is being avoided.

Sodium to Calcium Activity Ratio (SCAR)

The application of SAR to the group of waters which have EC higher than 5 dS/m and Mg/Ca ratio more than 1 is questionable. Therefore, Gupta (1984) proposed that SCAR meaning sodium to calcium activity ratio may be used instead of SAR.

$$\text{SCAR} = \text{Na} / \sqrt{\text{Ca}} \tag{8.3}$$

Later, it was also suggested by Keren (2000). Since Mg is neglected in this equation, it overestimates the problem when Mg < Ca, gives the same value for Mg = Ca and again overestimates the problem if Mg > Ca. Overestimation of the problem in the latter case seems justified but needs careful handling when Mg < Ca.

Classification of Water Based on SAR/SCAR

The classification of irrigation waters with respect to sodium hazard on the basis of SAR/SCAR is established primarily on the increase of exchangeable sodium and its effect on the physical condition of the soil. On the basis of SAR/SCAR, the irrigation waters may be classified in four classes:

(a) S-1: Low sodium waters (SAR/SCAR<10) can be used for irrigation on almost all soils with little possibility of the development of harmful levels of exchangeable sodium. It can be used to grow all crops except some stone-fruit tree or woody trees which are sensitive to sodium. These waters do not require additional leaching and/or application of amendments.

(b) S-2: Medium sodium waters (SAR/SCAR 10-18) can be used for crops which are semi-tolerant or tolerant to sodium on almost all soils so that LF is around 0.3. These waters can be used successfully in soils containing gypsum or calcium carbonate, EC should not be less than 1 to 2 dS/m lest permeability problem develops.

(c) S-3: High sodium waters (SAR/SCAR 18-26) can be used only for crops which are tolerant to sodium on soils provided with good drainage such that LF is always greater than 0.3. If soils contain gypsum or calcium carbonate and if rainfall is appreciable and effective, these waters can be used quite successfully. EC should not be less than 2 to 3 dS/m lest permeability problem develops.

(d) S-4: Very high sodium waters (SAR/SCAR>26) are not suitable for irrigation but may be used in cycle or in conjunction with low sodium waters or with amendments such as gypsum.

Alkalinity Hazard

Residual Sodium Carbonate

Eaton (1950) proposed the concept of residual sodium carbonate (RSC). He observed that when water containing CO_3+HCO_3 higher than Ca+Mg is used for irrigation, it may lead to the development of alkali soils (saline or non-saline) especially if water is used so sparingly that little leaching occurs. RSC is defined as:

$$RSC = (CO_3 + HCO_3) - (Ca + Mg) \qquad (8.4)$$

All ions are expressed in me/l resulting in units of RSC as me/l. In water containing high concentrations of bicarbonate ions, there is tendency for calcium and to some extent for magnesium to precipitate as carbonates as the soil solution becomes more concentrated. On the other hand, if sodium is not too greatly in excess of Ca+Mg and enough of such water is used such that the drainage water continues to contain a substantial proportion of Ca and Mg, there may be no ill effects. The following reaction takes place:

$$Ca^{++} + 2HCO_3 = CaCO_3 + H_2O + CO_2 \uparrow \qquad (8.5)$$

This reaction does not go to completion under ordinary circumstances, but in so far as it does proceed, the concentrations of calcium and magnesium are reduced and the relative proportion of sodium is increased causing thereby a decrease in soil salinity and increase in ESP.

The residual bicarbonate reacts with sodium and forms sodium bicarbonate which upon evaporation is converted into sodium carbonate according to the following reactions:

$$Na^{+} + HCO_3 = NaHCO_3 \qquad (8.6)$$

$$2NaHCO_3 = Na_2CO_3 + H_2O + CO_2 \uparrow \qquad (8.7)$$

The sodium carbonate, thus formed reacts with normal soil and replaces exchangeable calcium with sodium on exchange complex (EX).

$$Ca^{++}(Ex) + Na_2CO_3 = 2Na^{+}(Ex) + CaCO_3 \qquad (8.8)$$

Formation of indurated layers of $CaCO_3$ has often been associated with the use of irrigation waters containing bicarbonate ions higher than calcium ions. Several layers of carbonate deposition become indurated or hardened, and essentially are impermeable, except for cracks and other discontinuities. Indurated layers normally are considered to be a product of hundreds or even thousands of years of carbonate accumulation, so that their formation during short-term agricultural operations is probably unimportant.

Richard (1954) recommended that both RSC and SAR be used to evaluate the sodicity hazard. Subsequent investigators showed that general reliability of the RSC concept for evaluating the sodium hazard of bicarbonate water is questionable. In the Indian context, RSC is still being widely used as a means of managing alkali water. If the RSC of the irrigation water is 2.5 me/l or less, the water is considered to be of good quality. In that event it is not necessary to apply gypsum or any other amendment. For every additional 1 me/l RSC to be neutralized, agriculture grade gypsum (70% purity) @ 12 kg/ha should be added for each cm of alkali water applied. The quantity of gypsum to be added is, thus determined by the RSC of irrigation water and quantity of water required for irrigation during the growing season or on yearly basis as per the following equation:

Gypsum to be added (t/ha)= 12 x (RSC-2.5)*IW/1000 (8.9)

Here RSC is the RSC of the irrigation water, IW is the depth of irrigation water applied during the season/year, cm. Division by 1000 is to convert kg/ha in t/ha. The beneficial effects of soil application of gypsum on rice and wheat yield are depicted in Table 8.3 (Sharma *et al.*, 1975).

Table 8.3. Effect of gypsum application on grain yield of rice and wheat in farmer's fields

Wheat 1973-74			**Rice 1974**		
Likely RSC after gypsum application*	**Gypsum (t/ha)**	**Wheat yield (t/ha)**	**Likely RSC after gypsum application**	**Gypsum (t/ha)**	**Rice yield (t/ha)**
12.5	Nil	0.95	10.0	Nil	0.63
7.5	2.0	2.10	7.5	2	0.99
5.0	3.0	2.63	3.0	4	1.62
2.5	4.0	2.63	2.5	6	2.02

*Assuming that 50 cm of water is applied, gypsum applied would neutralize the RSC to the level indicated.

Besides several other amendments such as acid forming substances (notably pyrites), distillery spent wash (DSW), phosphogypsum (PG) and sulphitation press mud (SPM) can be used as soil application or through gypsum beds or mixing in water as in the case of DSW. The benefits of acid forming amendments would depend upon the calcareous nature of the soils.

Example 8.4

The water of a tube well has a RSC of 10.5 me/l. The water is to be used to irrigate wheat crop, which needs about 5 irrigations of 6 cm each. Calculate the gypsum requirement to treat the soil to avoid adverse effects.

RSC to be neutralized = 10.5-2.5 = 8.0

Number of irrigation = 5 (6 cm each)

Applying eq. (9.8), we get

Gypsum to be added (t/ha) = (12 x 8 x 5 x 6)/1000 = 2.88 t/ha

Recent developments have shown that if the RSC is more than 2.5 me/l, Adj. R_{Na} comes into play to assess precipitation of calcium from the water and/or dissolution of native calcium of the soils. This concept has been described manner in a subsequent section in this chapter. In several textbooks, Adj. SAR comes into play is also mentioned but as already pointed out Adj. SAR concept is obsolete.

Residual Sodium Bi-Carbonate (RSBC)

Eaton (1950) suggested that RSC should be determined as per eq. (8.4), yet RSC was actually calculated as (HCO_3) - (Ca +Mg). In subsequent pot culture experiment conducted by Wilcox *et al.* (1954) to substantiate RSC concept, RSC was determined as (HCO_3-Ca). Similarly, in green house and lysimeter experiments, Bower (1961) and Pratt *et al.* (1960), considered only HCO_3 and Ca. Likewise, in many experiments conducted in India only $NaHCO_3$ salt was added in very low salinity waters to prepare waters of different RSC (Gupta, 1984; Anonymous, 1983; Bajwa *et al.*, 1983; Manchanda *et al.*, 1985). Since carbonate ions do not occur very frequently in appreciable concentrations and because bicarbonate ions do not precipitate magnesium ions, Gupta (1984a) suggested that the alkalinity hazard should be determined through the index called "Residual

Sodium Bi-carbonate (RSBC)" to be calculated as below, all ions expressed as me/l.

$$RSBC = HCO_3\text{-}Ca \qquad (8.9)$$

Studies have also shown that the absolute concentration of bicarbonate ions in amounts greater than 10.0 me/l is likely to affect plant growth. It happens because (i) it may affect the mineral nutrition, (ii) it tends to reduce the iron availability in many plants thereby causing chlorosis, and (iii) it may create severe problems if used for irrigation through overhead sprinklers.

Classification of Waters on the Basis of RSC/RSBC

On the basis of RSC/RSBC, the irrigation water may be classified in four classes:

(a) A-1: Low-alkalinity water (RSC/RSBC <2.5 me/l) can be used for irrigation on almost all soils for all crops. These waters may not create any permeability problem unless drainage is impeded, rainfall is unduly low and evaporation is very high.

(b) A-2: Medium-alkalinity waters (RSC/RSBC 2.5-5.0 me/l) can be used for irrigation on almost all soils with little danger of the development of harmful levels of alkali for growing all crops except those which are specifically sensitive to carbonate or bicarbonates. Optimum yields of several alkali tolerant crops are obtained when RSBC is in this range.

(c) A-3: High -alkalinity waters (RSC/RSBC 5.0-10.0 me/l) can be used for irrigation on soils provided with good drainage such that LF is not less than 0.3, for growing semi-tolerant and tolerant crops to sodium. EC should be <3.0 dS/m and SAR <10. Rainfall should be appreciable and effective (400 mm) and evaporation must be low (<2000 mm) for the prolonged successful utilization of such waters. If SAR is >10, use of gypsum may be required.

(d) A-4: Very high-alkalinity waters (RSC/RSBC >10.0 me/l) are directly not suitable for irrigation but may be used in cycle or in conjunction with low alkalinity waters or with the use of amendments such as gypsum. High alkalinity with preponderance of carbonates associated with high EC and SAR/SCAR, prohibits the use of these waters for irrigation.

Adjusted R_{Na}

To assess the dissolution and precipitation of calcium, Suarez (1981) proposed the following equation and designated it as adjusted R_{Na}. It is given as follows:

$$\text{Adj. } R_{Na} = Na / [(Ca_x + Mg)/2]^{1/2} \quad (8.10)$$

Here, Na and Mg concentration of irrigation water is expressed as me/l. Ca_x is the value of Ca in the water that takes care of EC_{iw} and HCO_3/Ca ratio of the water and the estimated partial pressure of CO_2 in the surface few mm of the soil (Table 8.3). The Ca_x represents me/l of Ca expected to remain in the soil solution at equilibrium. It assumes a soil source of calcium from lime ($CaCO_3$) or silicates, no precipitation of magnesium, and partial pressure of CO_2 near the soil surface (pCO_2) as .0007 atmospheres. Ca_x, HCO_3, Ca are in me/l and EC_{iw} is in dS/m (Suarez, 1981; Rhoades, 1983).

Table 8.3. Calculation of Ca_x as affected by EC_{iw} and HCO_3/Ca ratio of the water for use in eq. (8.10)

HCO_3/Ca	Salinity of applied water (EC_{iw}) (dS/m)											
	0.1	**0.2**	**0.3**	**0.5**	**0.7**	**1.0**	**1.5**	**2.0**	**3.0**	**4.0**	**6.0**	**8.0**
.05	13.20	13.61	13.92	14.40	14.79	15.26	15.91	16.43	17.28	17.97	19.07	19.94
.10	8.31	8.57	8.77	9.07	9.31	9.62	10.02	10.35	10.89	11.32	12.01	12.56
.15	6.34	6.54	6.69	6.92	7.11	7.34	7.65	7.90	8.31	8.64	9.17	9.58
.20	5.24	5.40	5.52	5.71	5.87	6.06	6.31	6.52	6.86	7.13	7.57	7.91
.25	4.51	4.65	4.76	4.92	5.06	5.22	5.44	5.62	5.91	6.15	6.52	6.82
.30	4.00	4.12	4.21	4.36	4.48	4.62	4.82	4.98	5.24	5.44	5.77	6.04
.35	3.61	3.72	3.80	3.94	4.04	4.17	4.35	4.49	4.72	4.91	5.21	5.45
.40	3.30	3.40	3.48	3.60	3.60	3.82	3.98	4.11	4.32	4.49	4.77	4.98
.45	3.05	3.14	3.22	3.33	3.42	3.53	3.68	3.80	4.00	4.15	4.41	4.61
.50	2.84	2.93	3.00	3.10	3.19	3.29	3.43	3.54	3.72	3.87	4.11	4.30
.75	2.17	2.24	2.29	2.37	3.43	2.51	2.62	2.70	2.84	2.95	3.14	3.28
1.00	1.79	1.85	1.89	1.96	2.01	2.09	2.16	2.23	2.35	2.44	2.59	2.71

HCO$_3$/ Ca	Salinity of applied water (EC_{iw}) (dS/m)											
	0.1	**0.2**	**0.3**	**0.5**	**0.7**	**1.0**	**1.5**	**2.0**	**3.0**	**4.0**	**6.0**	**8.0**
1.25	1.54	1.59	1.63	1.68	1.73	1.78	1.86	1.92	2.02	2.10	2.23	2.33
1.50	1.37	1.41	1.44	1.49	1.53	1.58	1.65	1.70	1.79	1.86	1.97	2.07
1.75	1.23	1.27	1.30	1.35	1.38	1.43	1.49	1.54	1.62	1.68	1.78	1.86
2.00	1.31	1.16	1.19	1.23	1.26	1.31	1.36	1.40	1.48	1.54	1.63	1.70
2.25	1.04	1.08	1.10	1.14	1.17	1.21	1.26	1.30	1.37	1.42	1.51	1.58
2.50	0.97	1.00	1.02	1.06	1.09	1.12	1.17	1.21	1.27	1.32	1.40	1.47
3.00	0.85	0.89	0.91	0.94	0.96	1.00	1.04	1.07	1.13	1.17	1.24	1.30
3.50	0.72	0.80	0.82	0.85	0.87	0.90	0.94	0.97	1.02	1.06	1.12	1.17
4.00	0.71	0.73	0.75	0.78	0.80	0.82	0.86	0.8	0.93	0.97	1.03	1.07
4.50	0.66	0.68	0.69	0.72	0.74	0.76	0.79	0.82	0.86	0.90	0.95	0.99
5.00	0.61	0.63	0.65	0.67	0.69	0.71	0.74	0.76	0.80	0.83	0.88	0.93
7.00	0.49	0.50	0.52	0.53	0.55	0.57	0.59	0.61	0.64	0.67	0.71	0.74
10.00	0.39	0.40	0.41	0.42	0.43	0.45	0.47	0.48	0.51	0.53	0.56	0.58
20.00	0.24	0.25	0.26	0.26	0.27	0.28	0.29	0.30	0.32	0.33	0.35	0.37
30.00	0.18	0.19	0.20	0.20	0.21	0.21	0.22	0.23	0.24	0.25	0.27	0.28

[1]Adapted from Suarez (1981)

It can be stated with certainty that both RSC and SAR influence the physical properties of the soil in an integrated manner. Whereas in low salinity waters, the effect of RSC is more prominent, in high salinity waters, it is because .of high SAR. However, it will be desirable to determine RSBC in waters having EC less than 3 dS/m, RSC and SAR for waters having EC between 3 and 5 dS/m and SCAR for waters having EC greater than 5 dS/m. RSC/RSBC and SAR/SCAR both should not be high. RSBC up to 10.0 me/l is permissible provided that SAR is less than 10 and SAR up to 20 and 30 is permissible provided that RSC is less than 5 and 10 me/l, respectively.

Harmful Effects of Alkali Waters

Continuous application of alkali water adversely affects the crop growth. It is because of the following factors that come into play either singly or more often in a combination.

- The increased exchangeable sodium percentages (ESP) resulting from long-term use leads to breakdown of soil structure due to swelling and dispersion of clay particles.

Fine textured soils remain dispersed and puddled when wet and hard when dry. Thus, less air and water enter the soil resulting in poor environment for root growth. Internal drainage is restricted resulting in temporary water logging. These soils do not easily attain proper soil moisture conditions for cultivation.

- A thin crust is often formed at the soil surface, which acts as a barrier to water penetration and seedling emergence.
- Compacted topsoil and sub-soil creates problems for root penetration and cultivation.
- Adverse crop growth may also be due to direct toxic effects of sodium in many crop plants.
- Increase in soil pH reduces the availability of a number of plant nutrients like nitrogen, zinc, iron, etc. It also results in decreased availability of calcium, magnesium and potassium.

Example 8.5

Determine the quality parameters of irrigation water if water analysis report of a given water sample is as follows:

Ca = 2.54 me/l Mg= 1.34 me/l

Na = 6.87 me/l EC = 1.1 dS/m

CO_3 = 0.38 me/l HCO_3 =4.21 me/l

To find RSC, SAR, and adj. R_{Na}, we proceed as follows.

Using eq. (8.4) RSC is given as:

$$(0.38+4.21) - (2.54+1.34) = 4.59 - 3.88 = 0.71 \text{ me/l}$$

Using eq. (1.2) SAR is given as:

$$6.87 / \sqrt{(2.54+1.34)/2} = 4.93.$$

Using eq. (8.10) and Table 8.3, Adj. R_{Na} is given as:

EC is 1.1 dS/m, HCO_3/Ca ratio 4.21/2.54 = 1.66

Ca_x value from Table 8.3 corresponding to the nearest values of HCO_3/Ca and EC_{iw} is 1.43. Therefore

$$6.87/\sqrt{(1.43+1.34)/2} = 6.87/1.18 = 5.82$$

PROBLEMS OF INFILTRATION/SOIL STRUCTURE

For extremely low salinity water, even low SAR may cause water movement problems. It is difficult to specify universally applicable critical values of SAR and EC of irrigation water to maintain appropriate infiltration rates. Therefore, a diagram has been devised by which the permeability hazard is assessed by ascertaining whether the SAR-EC combination lies to the left (problem likely) or to the right (no problem likely) of the threshold line (Fig. 8.1; Rhoades, 1977). Values of EC that must be used with a given SAR of water to avoid adverse impact on infiltration rate and soil structure are also given in Table 8.4.

Table 8.4. Irrigation water quality criteria (EC, dS/m) for various SAR values of water

SAR	EC for various degrees of restriction on use		
	No	**Slight to moderate**	**Severe**
< 3	<0.7	0.7--0.2	< 0.2
3-6	>1.2	1.2-0.2	< 0.3
6-12	>1.9	1.9-0.5	<0.5
12-20	>2.9	2.9- 0.5	<0.5
20-40	>5	5 - 2.9	<2.9

Source: Ayers and Westcot (1985)

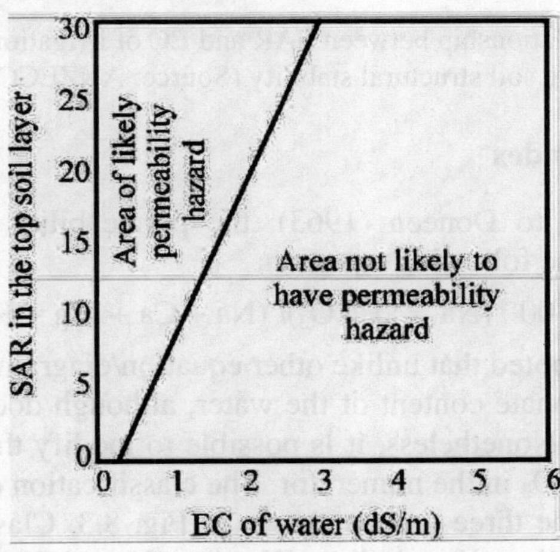

Fig. 8.1. Relative rate of infiltration as affected by salinity and SAR

Similarly, to predict soil structural stability, a diagram showing relationship between SAR and EC has been proposed (Fig. 8.2). Water quality that falls to the right of the dashed line is unlikely to cause soil structural problems. Water quality that falls to the left of the solid line is likely to induce soil structure degradation calling for corrective management (e.g. application of gypsum or some suitable amendment). Water that falls between the lines is of marginal quality and needs to be applied with caution depending upon the soil properties and rainfall.

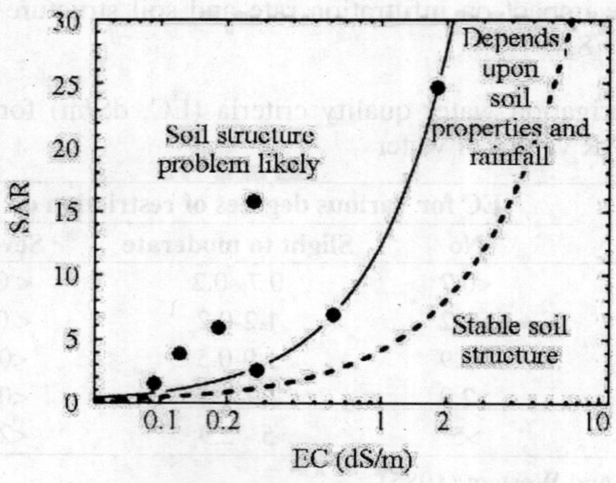

Fig. 8.2. Relationship between SAR and EC of irrigation water for predicting soil structural stability (Source: ANZECC, 2000)

Permeability Index

According to Doneen (1963), the permeability Index (PI) is calculated by the following equation:

$$PI = 100 *(Na + \sqrt{HCO_3})/ (Na + Ca + Mg + K) \qquad (8.11)$$

It may be noted that unlike other equation/diagrams, it takes into account bicarbonate content of the water, although does not account for carbonates. Nonetheless, it is possible to modify the equation by adding K and CO_3 in the numerator. The classification of the water is then made in the three groups shown in Fig. 8.3. Class 1 shows no problem while class II and Class III mean 25 and 75% reduction in permeability.

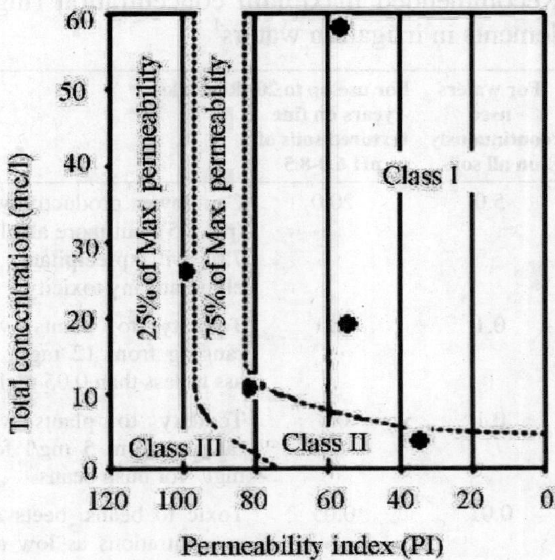

Fig. 8.3. Classification of irrigation water on the basis of permeability index

TOXICITY HAZARD

The toxic waters have variable EC, SAR and RSC but have excess of an individual ion beyond permissible limits that causes toxicity to plants resulting in reduced crop growth/yield. A toxicity problem is different than the problems due to salinity or alkalinity as the toxicity occurs within the crop as a result of the uptake and accumulation of certain constituents from the irrigation water. The constituents of concern are: chloride, sulphate, sodium, nitrate, boron, fluoride, or heavy metals such as selenium, cadmium, lead and arsenic etc. Presence of silica (Silicon) can also affect productivity. High concentration of these ions may also cause imbalance in the uptake of various ions or even reduce the uptake of plant nutrients. Moreover, these ions may affect the micro-irrigation systems or cause foliar injury to plants. The adverse effects may occur even at low salinity or sodicity. Toxicity problem accompanying salinity or sodicity complicates the issue with simultaneous development of salinity and infiltration problems. The permissible limits of trace elements in irrigation water are given in Table 8.5.

Table 8.5. Recommended maximum concentration (mg/l) of trace elements in irrigation waters[1]

Element	For waters used continuously on all soils	For use up to 20 years on fine textured soils of pH 6.0-8.5	Remarks
Aluminium (Al)	5.0	20.0	Can lower productivity in acid soils (pH 5.5), but more alkaline soils at pH 7.0 will precipitate the ion and eliminate any toxicity
Arsenic (As)	0.1	2.0	Toxicity to plants varies widely, ranging from 12 mg/l, for Sudan grass to less than 0.05 mgl^{-1} for rice
Beryllium (Be)	0.1	0.5	Toxicity to plants varies widely, ranging from 5 mg/l for kale to 0.5 mgl^{-1} for bush beans
Cadmium (Cd)	0.01	0.05	Toxic to beans, beets and turnips at concentrations as low as 0.1 mg/l in nutrient solutions. Conservative limits recommended due to its potential for accumulation in plants and soils to concentrations that may be harmful to humans.
Cobalt (Co)	0.05	5.0	Toxic to tomato plants at 0.1 mg/l in nutrient solution. Tends to be inactivated by neutral and alkaline soils
Chromium (Cr)	0.10	1.0	Not generally recognized as an essential growth element. Conservative limits recommended due to lack of knowledge of its toxicity to plants
Copper (Cu)	0.2	5.0	Toxic to a number of plants at 0.1 mg/l in nutrient solutions
Fluorine (F)	1.0	15.0	Inactivated by neutral and alkaline soils
Iron (Fe)	5.0	20.0	Not toxic to plants in aerated soils, but can contribute to soil acidification and loss of availability of essential phosphorus and molybdenum. Overhead sprinkling may result in unsightly deposits on plants, equipment and buildings.
Lead (Pd)	5.0	10.0	Can inhibit plant cell growth at very high concentrations

Element	For waters used continuously on all soils	For use up to 20 years on fine textured soils of pH 6.0-8.5	Remarks
Lithium (Li)	2.5	2.5	Tolerated by most crops up to 5 mg/l, mobile in soil. Toxic to citrus at low concentrations (0.075 mg/l). Acts similarly to boron.
Manganese (Mn)	0.2	10.0	Toxic to a number of crops at a few-tenths (Mn) to a few mg/l, but usually only in acid soils.
Molybdenum (Mo)	0.01	0.05	Not toxic to plants at normal (Mo) concentration in soil and water. Can be toxic to livestock if forage is grown in soils with high concentrations of available molybdenum.
Nickel (Ni)	0.2	2.0	Toxic to a number of plants at 0.5 mg/l to 1.0 mg/l; reduced toxicity at neutral or alkaline pH.
Selenium (Se)	0.02	0.02	Toxic to plants at concentrations as low as 0.025 mg/l and toxic to livestock if forage is grown in soils with relatively high levels of added selenium. An essential element to animals but in very low concentrations.
Vanadium (V)	0.1	1.0	Toxic to many plants at relatively low concentrations
Zinc (Zn)	2.0	10.0	Toxic to many plants at widely varying concentrations; reduced toxicity at pH 6.0 and in fine textured or organic soils.

[1] The maximum concentration is based on a water application rate which is consistent with good irrigation practices (10,000 m^3/ha per year). If the water application rate greatly exceeds this, the maximum concentrations should be adjusted downward accordingly. No adjustment should be made for application rates less than 10,000 m^3/ha per year. The values given are for water used on a continuous basis at one site.

Several classifications for toxic ions are now available and to discuss each of these classifications is beyond the scope of this book. As an illustration, a classification for the most commonly found element boron in saline/alkali soils is given (Gupta, 1979a). As boron is adsorbed by the soil, more so by heavy textured soils. Permissible limits of B are higher in heavy textured soils.

B1. Normal water (B < 3 ppm): All irrigation waters which have B less than 3 ppm can be used for most of the crops on all soils without any injurious effect on the crop yields. However, toxicity symptoms, *viz.* chlorosis, etc. might be seen in some sensitive plants.

B2. Low boron waters (B = 1-3 ppm): These waters can be used for irrigation on deep black or clayey as well as medium textured soils without any problem for all crops but only up to 1 ppm for semi tolerant crops and up to 3 ppm for tolerant crops, respectively, on light textured soils.

B3. Medium boron waters (B = 3.5 ppm): Such waters can be used for irrigation on deep black or heavy textured soils for most of the crops. In the case of medium textured soils, water containing B up to 3 and 5 ppm could be used without problem for semi tolerant and tolerant crops, respectively. The use of such waters may create problems in light textured soils.

B4. Boron waters (B = 5-10 ppm): Can be used for irrigation for semi tolerant and tolerant crops, on deep black or heavy textured soils. There may be a problem in light textured soils. Presence of native $CaCO_3$ in soils may alleviate the toxic effect of boron.

B5. High boron waters (B >10 ppm): The waters are not suitable for irrigation under ordinary conditions but may be used occasionally under very special circumstances. The waters which have low salinity and low SAR may be used successfully, particularly in soils which contain good reserves of $CaCO_3$ and the soils are well drained.

CHARACTERIZATION AND CLASSIFICATION OF WATERS

Several classification criteria are available in the literature. In these classifications, it is presumed that the water will be used under average conditions with respect to several factors which influence its quality in relation to its use. Whether the salts dissolved in irrigation water will accumulate in the soil to injurious level and affect plant growth depends on one or more of the several parameters discussed in Chapter 2. Large deviations from the average for one or more of the variables may make it unsafe to use what, under average conditions, would be a good water; or may make it safe to use, even if under average conditions, the quality is doubtful. One of the classification proposed by Gupta as ABC classification has been discussed under EC, SAR/SCAR and toxic element boron (Gupta, 1972). Two other commonly used classifications are given in the following sections.

USSL Diagram

The United States Salinity Laboratory (Richards, 1954) published a semi-logarithmic diagram, which rated waters with respect to EC and SAR. Four classes each of salinity and SAR were devised (Table 8.5, Fig. 8.4). Each salinity class was associated with all the four sodium classes and vice versa, thus giving a total of

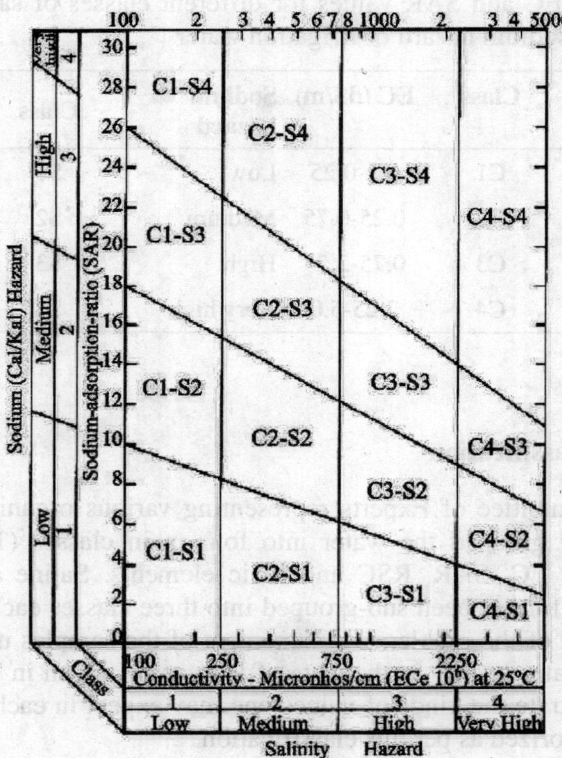

Fig. 8.4. USSL diagram to categorize water based on EC and SAR

sixteen classes for the purpose of evaluating the quality of waters for irrigation. It was concluded that low salinity water (C1) can be used for irrigation with most crops on most soils, with little likelihood that soil salinity will develop, and that very high salinity water (C4) was not suitable for irrigation under ordinary conditions, but might be used occasionally under very special circumstances. The soil must be permeable, drainage must be adequate, irrigation water must be applied in excess to provide considerable leaching and high salt tolerant crops should be grown. In the Indian context, classification of waters on this basis has been questioned by many authors. In view of this criticism and several problems encountered in the use of this diagram, it has emerged that it should not be used any more for the evaluation of the quality of irrigation waters (Gupta, 1972a).

Table 8.6. EC and SAR values for different classes of salinity and sodium hazard of irrigation water

Salinity hazard	Class	EC (dS/m)	Sodium hazard	Class	SAR
Low	C1	0.1-0.25	Low	S1	<10
Medium	C2	0.25-0.75	Medium	S2	10-18
High	C3	0.75-2.25	High	S3	18-26
Very high	C4	2.25-5.0	Very high	S4	26-31

AICRP Classification

A Committee of Experts representing various organizations in the country grouped the water into four main classes (Table 8.7) considering EC, SAR, RSC and toxic elements. Saline and alkali water have further been sub-grouped into three classes each to grade the severity of the problems. A statement of the samples under each group of water quality in the state of Haryana is given in Table 8.8, which illustrate the kinds of values one may expect in each group of water categorized as per this classification.

Table 8.7. Grouping of poor quality water

Water quality	EC_{iw} (dS/m)	SAR_{iw} (mmol/ l)$^{1/2}$	RSC (me/ l)
A. Good	<2	< 10	< 2.5
B. Saline			
i. Marginally saline	2-4	< 10	< 2.5
ii. Saline	> 4	< 10	< 2.5
iii. High-SAR saline	> 4	> 10	< 2.5
C. Alkali water			
i. Marginally alkali	< 4	< 10	2.5 – 4.0
ii. Alkali	< 4	< 10	> 4.0
iii. High-SAR alkali	variable	>10	> 4.0
D. Toxic Water	The toxic water has variable salinity, SAR and RSC but has excess of specific ions such as nitrate, boron, fluoride, chloride, sodium, silica or heavy metals such as selenium, cadmium, lead and arsenic etc.		

Source: Minhas and Gupta (1992), Toxic waters added by Gupta (2011)

Table 8.8. Chemical constituents of various kinds on water categorized as per Table 8.6 in the state of Haryana

Parameter	Good	Marginally saline	Saline	High SAR saline	Marginally alkali	Alkali	High SAR alkali
EC (dS/m)	0.61	2.56	6.52	14.04	1.04	1.83	2.53
pH	8.60	8.30	8.3	8.55	8.05	8.20	8.50
Na (me/l)	2.20	7.00	30.8	103.60	6.10	12.20	21.70
Ca (me/l)	2.10	3.00	8.3	22.64	1.50	2.70	2.10
Mg (me/l)	1.70	9.70	19.7	12.60	3.30	3.10	1.70
Cl (me/l)	2.20	6.00	33.00	102.00	3.60	3.80	6.80
SO_4 (me/l)	0.90	3.10	25.40	36.75	0.50	1.70	9.80
CO_3+HCO_3 (me/l)	4.20	13.20	6.00	2.40	7.60	13.00	10.60
SAR (me/l)$^{1/2}$	1.60	2.89	8.24	24.80	3.90	7.20	15.72
RSC (me/l)	0.40	1.50	Nil	Nil	2.80	7.20	6.80
Mg/Ca ratio	0.81	3.23	2.37	0.56	2.20	1.15	0.81
Cl/SO_4 ratio	2.44	1.94	1.30	2.78	7.20	2.24	0.69

MANAGEMENT TECHNOLOGIES

A great deal of work has been done in India and abroad to develop technologies for using saline/alkali waters in agriculture to get sustainable yields on a long-term basis. These can be classified in five groups namely crop management, soil management, water management, chemical management and rainwater management (Table 8.9). Thus, one can adopt several permutations and combinations of these technologies including their sub-groups to achieve similar results. You may notice that the technologies for both kinds of water are similar. However, in practice major differences are observed as given in remarks column. For example a cropping sequence or variety which may work well for saline water may not work well for alkali/sodic water. It is not intended to dwell upon these technologies in detail as most of these are discussed in other chapters on management of saline and alkali soils.

Table 8.9. Reclamation and management package for saline and alkali water

Group/sub-group	Technology	
	Saline or alkali/sodic water	**Remarks**
Crop management	Selection of crops and cropping sequences	The crops and cropping sequences may be different for saline and alkali/sodic water
	Exploitation of varietal differences	Varieties may be different for saline and alkali/sodic waters
	Improved agronomic practices such as 25% higher seed rate/plant population	-
Soil management	Land forming/seeding	Benefits may not be realized in alkali/sodic water
Irrigation water management	Shallow depth-high frequency irrigation	True for both the waters
	Pre/post-sowing irrigation with fresh water	As above
	Switchover to improved irrigation techniques	As above

Group/sub-group	Technology	
	Saline or alkali/sodic water	**Remarks**
	Leaching for salt balance	Limited benefits in alkali/sodic water
Chemical management	Application of additional nutrients, organic manures	Organic manure without amendment may cause problems with sodic water
	Application of chemical amendments (only for high RSC/SAR water to improve infiltration rate)	Application of amendments to neutralize RSC of water is necessary
Rainwater management	*In-situ* rainwater conservation	True for both kinds of water
	Rainwater harvesting and reuse	
	Fallowing	

■■■■

CHAPTER 9

CHEMICAL AMENDMENTS

INTRODUCTION

A chemical amendment is a material that is added to the soil to improve its physico- chemical properties. It aids in plant growth in degraded lands. These amendments respond on the basis of chemical, biological, and physical actions that are initiated upon their application. Theoretically, even fertilizers, that contain useful amounts of one or more plant nutrients and help in better plant growth, can be called amendments. But in practice, these are not categorized as amendments. The chemical amendments are commonly used to reclaim alkali lands. These amendments help to replace the exchangeable sodium by calcium and neutralize the pH of the soil. The soil ESP, time span within which one wants to reclaim the land and the presence of *kankar* (impure $CaCO_3$) in the soil help to decide the kind of amendment to be used. These factors also determine the effectiveness of a chemical amendment. Besides, the success of a chemical reclamation program depends to a great extent on the quality of irrigation water and the depth to water table. The good quality irrigation water and deeper depth to water table helps to achieve complete, early and sustainable reclamation. This chapter discusses various kinds of amendments, their availability, methods of application and their response in the land reclamation for increasing crop productivity.

KINDS OF AMENDMENTS

The chemical amendments are broadly classified in three categories, namely

- Soluble calcium salts, viz. calcium chloride and gypsum which directly supply soluble Ca for the replacement of exchangeable Na.
- Acidifying materials, viz. sulphuric acid and elemental sulphur which due to their acidulating action and/or increasing the partial pressure of CO_2 supply Ca indirectly by dissolving calcite natively found in alkali soils. The Ca so mobilized is used to replace Na from the exchange complex.
- Organic materials viz. rice husk, FYM and green manure etc. Compost and other organic materials improve soil physical structure by two distinct mechanisms – by stimulating microbial activity, which helps in soil aggregation, and at high application rates, by directly decreasing soil density. These also help to improve soil structure as well as dissolve the native $CaCO_3$ albeit weakly. The reclamation efficiency is usually low with these kinds of amendments.

Besides these many industrial wastes are used as amendments that work on similar principles depending upon their constituents. Amongst them pyrites, distillery spent wash, dairy waste, phosphogypsum and press mud have been widely used.

ROLE OF CALCIUM CARBONATE IN NATURAL RECLAMATION

Most alkali soils contain *kankar* (impure $CaCO_3$) in sufficiently good amount in the root zone depth. It is a slowly soluble compound with a solubility of about 13.1 mg/l. Its solubility increases with decreasing soil pH (Table 9.1) and increasing concentration of CO_2 in the soil environment. In the presence of CO_2, $CaCO_3$ gets converted into calcium bicarbonate, which is more soluble, its solubility being in the range of 60 to 140 mg/l. Thus, any process that aids in increasing the CO_2 concentration in the soil and/or lowers the soil pH, will initiate natural reclamation process of the alkali soil especially of the low and medium degrees. The most practical means to reduce the pH and permanently increase the CO_2 concentration is to improve the microbiological properties. It can be achieved through

Table 9.1. Solubility of $CaCO_3$ (mg/l) as affected by the CO_2 content and pH of the soil

Content of CO_2 in air (% by volume)	pH	Solubility of $CaCO_3$
0.00	10.23	13.1
0.03	8.48	63.4
0.30	7.81	133.4
1.00	7.47	202.9
10.00	6.80	470.0
100.00	6.13	1098.6

Source: Kovda *et al.* (1973)

the decomposition of organic substances ploughed into the soil in the form of green or farmyard manure. The application of small amounts of gypsum or other acidifying materials and the conservation of rain water during the monsoon season can also help to speed up the process of the natural reclamation. Following processes come into play.

- Impure $CaCO_3$ present in alkali soils releases Ca albeit slowly but it helps in reclaiming alkali soils
- The irrigation water especially from canals contains Ca + Mg, eight to ten times higher than Na. It results in constant replacement of Na with Ca on the exchange complex particularly after the application of gypsum
- The use of nitrogenous fertilizers, particularly, $(NH_4)_2 SO_4$ helps to lower down the soil pH. This also exerts a marked influence in releasing more calcium from $CaCO_3$
- Consequent upon the start of the process of reclamation through cultivation of crops, the plant roots evolve CO_2 which help in the conversion of $CaCO_3$ into $Ca(HCO_3)_2$. As already stated its solubility is more than the former
- The rain water being acidic in nature due to presence of CO_2, the natural reclamation of the soil to some extent also takes place provided rainfall is appreciable and effective

Gupta and Abichandani (1970) reported that the ESP of a highly calcareous alkali soil was reduced from 59.5 to 18.0 and from 34.9 to 11.0 as a result of leaching caused by the rainfall alone (350 to 450 mm rainfall) in Western Rajasthan. The pH_2 values of these soils were less than 9.0. Besides the remarkable changes in ESP, EC_e in the top 0-20 cm decreased from 11.8 to 2.6 and 22.2 to 2.1 dS/m in two soils, respectively. Surface dressing of such soils with small amounts of gypsum in summer before rainy season expedites reclamation.

Besides what is stated, the presence of $CaCO_3$ is also essential pre-requisite if acidifying materials are used as amendments in alkali soils. This issue will be discussed in greater details in relevant section of this chapter. On the other hand, because of its low solubility at high pH and because of its natural availability in most alkali soils, $CaCO_3$ is not applied as an amendment to reclaim alkali soils.

GYPSUM

Gypsum by far is the most commonly used amendment, being used in India for more than a century. Leather (1897) reported for the first time that, the addition of small amounts of gypsum to reduce Na_2CO_2 to 0.1%, in the fine size, applied at the surface of alkali soils, was beneficial. Gypsum at this time was not used for field reclamation programs mainly because of poor economics. Three main commercial forms of calcium sulphate are (i) anhydrite form, also known as land plaster ($CaSO_4$); (ii) semi hydrate form, commonly called plaster of paris ($CaSO_4$, 1/2 H_2O); and (iii) dihydrate form, generally known as gypsum ($CaSO_4$, $2H_2O$). The last one, commonly known as mineral gypsum, is applied in land reclamation. The other kind of gypsum is derived as a byproduct of industry producing tartaric acid, formic acid, oxalic acid, citric acid, common salt and phosphoric acid. Since nearly 90% of it is derived from phosphoric industry, it is commonly referred as phosphogypsum (PG). For every ton of P_2O_5 produced, 5.5 tons of phosphogypsum containing 25% water are produced. More than 10.0 lakh tons of phosphogypsum is available from industries in India (Rohtagi *et al.* 1977). Relative composition of two types of gypsum is reported in Table 9.2.

Table 9.2. Chemical analysis (%) of mined and byproduct phosphogypsum

Constituents	Mined gypsum (70-80% purity)*	Phospho-gypsum§	BIS limits for Phospho-gypsum**
$CaSO_4.2H_2O$	77.56	85.0	85.0 min
SiO_2	12.37	0.45- 4.0	0.40 max
R_2O_3	1.59	0.05-0.3	0.15 max
Magnesia as MgO	0.81	-	0.10 max
Lime as CaO	28.56	32.5-44.0	-
Loss on ignition	20.44	-	-
Chloride as NaCl (% by mass)	0.016	-	0.10 max
Water soluble P_2O_5	-	0.2-1.2	0.40 max
F (% by mass)	-	0.5-1.5	0.40 max
SO_3	-	45-46	0.30 max
pH of 10% aqueous suspension of gypsum			5.0 min

*Mined by Fertilizer Corporation of India
§Phosphogypsum from Morocco rock phosphate. Range shows the variation due to method to produce phosphoric acid. [Source: 85§;**16]; max =maximum; min=minimum

Gypsum reacts with sodium on the exchange complex as per the following reaction:

$$\underset{\text{Alkali soil}}{2\,Na - soil} + 2CaSO_4 \rightarrow \underset{\text{Normal soil}}{Ca - soil} + Na_2SO_4 \quad (9.1)$$

There are other kinds of gypsum as well. Fluoro-gypsum is produced as a by-product during the manufacture of hydrofluoric acid. Marine gypsum is obtained during the process of recovering common salt by solar evaporation of sea water. Salt manufacturers in the coastal regions of Gujarat, Maharashtra and Tamil Nadu recover marine gypsum as a by-product of salt industry. Currently, a large amount of flue gas desulfurization gypsum commonly referred as FGD gypsum is being used as an amendment in Ohio, USA. It is produced by removal of sulphur dioxide (SO_2) from flue gas streams produced as a result of burning of coal containing high sulphur (S).

Quality of Gypsum

Gypsum occurs extensively in natural deposits as a white coloured secondary mineral. In India, rich deposits of gypsum are found in Rajasthan. The resources of gypsum in India are estimated to be about 100.4 x 10^7 tons, 90% of which are located in the mines of Rajasthan only. Limited quantities of gypsum are also mined in Tamil Nadu, Gujarat and Himachal Pradesh.

The mines operated by the Fertilizer Corporation of India Ltd. (FCI) produce three grades of gypsum having different purity levels: (i) Grade 1 85-90% $CaSO_4$ - $2H_2O$, (ii) Grade II 80-85% $CaSO_4$ - $2H_2O$, and (iii) Grade III 70-80% $CaSO_4$ $2H_2O$. The chemical analysis of these three grades of gypsum is given in Table 9.3.

Table 9.3. Chemical analysis of gypsum mined by the FCI

Constituents (%)	Gypsum Grade		
	I	II	III
$CaSO_4$ $2H_2O$	88.67	83.32	77.56
SIO_2	4.84	9.75	12.37
Mixed oxides (R_2O_3)	0.68	0.85	1.59
MgO	1.12	0.88	0.81
CaO	30.59	28.08	28.56
Loss on ignition	21.17	20.40	20.44
Chlorides as NaCl	0.008	0.014	0.016

According to the revised specifications of Indian Standard Institution (ISI now BIS), vide IS : 1290 - 1965, the mined gypsum is categorize in five grades, viz. (i) Grade I for use in surgical plaster industry; (ii) Grade II for ammonium sulphate industry; (iii) Grade III for pottery industry; (iv) Grade IV for cement industry, and (v) Grade V for soil reclamation. The specifications for different grades of gypsum are given in Table 9.4. The gypsum of grade III conforms to the ISI specifications for grade V gypsum, suitable for the reclamation of alkali soils. The gypsum used in soil reclamation should not contain more than 1.0% of Na_2O.

Table 9.4. ISI specifications for gypsum of different grades

Constituents (% by weight)	Grades				
	I	II	III	IV	V
Free water, maximum	1.0	-	1.0	-	-
CO_2, maximum	1.0	-	3.0	-	-
Silica and insolubles, maximum	1.0	6.5	6.5	-	-
Fe_2O_2 and Al_2O_3 maximum	1.0	1.5	1.0	-	-
MgO maximum	0.5	1.0	1.5	3.0	-
$CaSO_4$ minimum	96.0	86.0	83.8	80-85	70-80
Chlorides (as NaCl) maximum	0.1	0.01	0.01	0.5	-

Fineness

Gypsum at the mine site is quite cheap. Its transportation to the field sites and powdering adds to its cost. It is true that solubility of gypsum increases with increasing fineness because the contact area with water increases with increasing fineness of the particles. On the other hand, the cost increases with increasing fineness and thus, becomes a limiting factor. Therefore, a compromise has to be arrived at between the fineness viz-a-viz the solubility and cost. Conventionally, gypsum of about 2 mm (10 mesh) size has been used for the reclamation of alkali soils (Kovda *et al.* 1973). As per the ISI specifications, the whole of the gypsum must pass through 10 mesh and at least 50% of it must pass through 100 mesh (0.149 mm) sieve. This standard achieves nearly 30 mesh (0.595 mm) fineness, which adds about 15% of the cost to the total cost of gypsum. It can be reduced if one adheres to the standard of 10 mesh fineness.

Notwithstanding what is stated in the previous paragraph, experiments are in progress to produce nano-particles (size range of 1 - 100 nano-meters, nm, 1 nm = 10^{-9} m) of gypsum using various techniques so as to improve the gypsum solubility and to examine its role in degree of land reclamation, time taken to reclaim the lands and crop productivity of the reclaimed lands. Cost effectiveness and the simplicity of the process will determine the application of nano-particles in land reclamation.

Solubility

The rate of reaction of gypsum during reclamation is limited by its moderate solubility, being soluble in water only to the extent of about 0.25%. Scientists in the past believed that 90 to 120 cm of irrigation water will be required to dissolve 10 to 12 t/ha of high grade gypsum (Richards *et al.* 1954). Now the scientists are veering to the view that such a proposition is true only in normal soils. On the other hand, in alkali soils, the solubility of gypsum depends upon pH, ESP and type of salts present in the soil. In alkali soils, the solubility of gypsum is nearly ten times more than its normal solubility (Oster and Halvorson, 1978). It is ascribed to the fact that Ca in the soil solution is constantly withdrawn as it is being used to replace the Na on the exchange complex. Thus, the exchange complex acts as a sink for Ca ions. According to Hira and Singh (1980), the amount of water required to dissolve the added gypsum to reclaim an alkali soil does not increase with the amount of gypsum applied, provided it equals the gypsum requirement. In the presence of mixture of salts with dissimilar ions commonly found in saline soils, the solubility of gypsum increases with increasing concentration of soil solution in the presence of NaCl but reduces with increasing concentration in the presence of $CaCl_2$ (Table 9.5). The actual solubility increase is due to the effect of Na^+ and Cl^- in solution so that the concentrations of the ions must increase to maintain the constancy of the solubility product.

Redly *et al.* (1980) studied the solubility of $CaSO_4$ containing chemical amendments, hemihydrates and mechanically activated anhydrites in the saturation extracts of alkali soils differing in the concentration and composition of water soluble salts and the degree of sodium saturation. The equilibrium concentration of Ca^{++} ions in the soil extracts saturated with these materials exceeded 10-50 times the Ca^{++} ion concentration in the original solution (Table 9.6). It increased with the increase in the ionic strength for solutions poor in sulphates. On the other hand, the relatively stable Ca^{++} concentration was observed in the studied concentration range with the increasing ionic strength for solutions where sulphates dominated. The thermodynamic solubility product of $CaSO_4$ calculated from the reduced activities of Ca^{++} and SO_4 ions is practically independent of the ionic strength of the solution but it depends on the kind of the material, the average values being as follows:

Analytical $CaSO_4$ 1/2 H_2O	3.03 x 10^{-5}
Fine grained $CaSO_4$	4.23 x 10^{-5}
Coarse grained $CaSO_4$	4.78 x 10^{-5}

Table 9.5. Solubility of gypsum in water in the presence of NaCl and $CaCl_2$ (in g per 100 g of water)

Chlorides (me/l)	Solubility of $CaSO_4$
NaCl	
0	0.204
17.2	0.787
20.0	0.823
24.4	0.820
29.3	0.614
35.8	0.709
$CaCl_2$	
0	0.204
5.2	0.103
9.9	0.086
18.9	0.077
40.8	0.035

Source : Kovda *et al*. (1973)

The pH value and the ESP of the soils are usually lower after saturation with $CaSO_4$. The amount of Ca^{++} ions from the dissolution of $CaSO_4$ resulting in the change of a unit value of ESP (in the range of 5-40) in 0.15 and 0.26 g/kg soil for samples with low and high SO_4^- content, respectively.

The solubility of gypsum is increased if it is applied in conjunction with farm yard manure or green manures (Maharudrappa *et al.*, 1972, Mendiratta *et al.* 1972). Dargan *et al.* (1976 a, b) observed that the use of FYM in conjunction with gypsum resulted in higher yields of the first crop of *berseem*, sugarcane and rice grown in separate fields (Table 9.7). There was a high order of synergy when FYM and gypsum were applied together.

Table 9.6. The average values of Ca^{++} and SO_4 concentration (me/l) in the saturation extracts of soil profiles

pH	Soil characteristics EC_e (dS/m)	ESP	Treatment*	Ca^{++}	SO_4^{--}
5.9-9.6	10.7-3.0	30-75	0	3.4	84.8
			1	29.8	107.1
			2	32.6	115.7
			3	38.7	119.8
7.4-9.5	1.8-0.4	8-29	0	1.8	5.8
			1	31.4	42.5
			2	36.6	43.5
			3	43.5	49.7
6.6-8.7	1.8-9.9	2-32	0	3.0	44.0
			2	46.7	87.5
			3	48.8	92.4
7.5-8.9	0.9-1.3	3-12	0	1.2	5.2
			2	66.5	74.5
			3	67.3	72.7

*Treatment : 0 = Original soil, 1 = Analytical $CaSO_4$ hemihydrate, 2 = Fine grained $CaSO_4$ 3 = Coarse grained $CaSO_4$

Table 9.7. Effect of FYM and gypsum on the yield of crops (t/ha)

Treatment	*Berseem* fodder	Sugarcane (stripped cane)	Rice
Control	0.15	-	2.45
FYM 25 t/ha	0.83	0.49	3.59
FYM 50 t/ha	1.74	0.47	3.74
Gypsum (Gyp) 25% GR	-	14.32	4.23
FYM 25 t/ha + Gyp25% GR	-	21.37	5.29
FYM 50 t/ha + Gyp25% GR	-	35.77	5.46
Gypsum 50% GR	9.49	28.65	5.70
FYM 25 t/ha + Gyp 50% GR	29.48	37.46	5.76
FYM 50 t/ha + Gyp 50% GR	31.89	39.11	6.89
C.D. at $p = 0.05$	4.77	9.29	0.89
Soil pH_2 at start	10.5	10.6	10.6

Van Alphen and Romero (1971) observed that up to 2% gypsum in the soil favours crop growth, between 2 and 25% has little or no adverse effect if present in a powdery form, but more than 25% can cause substantial reductions in the crop yield. The yield reductions were attributed partly to imbalanced ion ratios its particular to K/Ca and Mg/Ca ratios. It was also noted that gypsum is readily redistributed in soil and frequently forms cemented and indurated layers. With the gypsum percentage between 14 and 80% these layers form a mechanical impediment to root growth and adversely affect water retention and transmission.

QUANTITY OF GYPSUM

The quantity of gypsum to be applied obviously depends upon the initial exchangeable sodium content of the soil and the level to which it has to be reduced. It is presumed that the irrigation water is of good quality with RSC <2.5 me/l. The level to which the exchangeable sodium content of the soil needs to be reduced depends upon the soil texture and the first crop to be grown at the beginning of reclamation. For a soil fully saturated with the exchangeable sodium initially (ESP = 95 to 100), it may suffice to reduce the ESP to about 50 if a tolerant crop like transplanted rice is grown. Likewise, ESP needs to be reduced to 30 if semi-tolerant crops like wheat, or to 10 if sensitive crops like gram, is to be cultivated. Similarly, for the same crop, coarse textured soils can support optimum growth at relatively higher ESP than fine textured soils. Bhumbla and Abrol (1975) reported that satisfactory yields of rice, wheat and barley can be obtained by adding gypsum @ 50% of gypsum requirement as determined in the laboratory by Schoonover's method for 0-15 cm layer.

The quantity of gypsum also depends upon the depth of the soils to which it has to be applied. It is usually not practical or even necessary, to remove all of the exchangeable sodium from the entire soil profile by the application of gypsum. A common practice is therefore, to apply sufficient gypsum to remove sodium from the top 10-20 cm of the soil. This improves the physical condition of the surface soil very quickly to permit cultivation.

Theoretically, the Gypsum Requirement (GR) can be calculated by the following relationship:

$$GR\ (me/100\,g) = \frac{CEC\ (initial\ ESP - final\ ESP)}{100} \quad (9.2)$$

Since 1 me of gypsum/100 g of soil is equivalent to 860 ppm of gypsum, and since 1 ha of soil to a depth of 20 cm may be taken to weigh 3,100,000 kg (Bulk density 1.55 g/cm^3) or approximately 3x 10^6 kg (bulk density is 1.5 g/cm^3), the amount of gypsum theoretically required to treat this soil depth will be

$$GR\,/\,ha - 20\,cm = GR\ (me/100\,g) \times 860 \times 10^{-6} \times 3.10^6$$

$$= GR(me/100\,g) \times 2580\ (kg) \quad (9.3)$$

Since soil weight of the 15 cm is usualy taken as 2,250,000 kg (bulk density 1.5 g/cm^3) the amount of gypsum theoretically required to treat this depth will be

GR/ha-15 cm = GR (me/100 g) x 1935 (kg) (9.4)

These calculations are based on 100% pure gypsum. If the percentage purity of gypsum is less than 100%, the quantity of gypsum has to be increased proportionately by dividing the calculated GR by percent purity.

Laboratory Method

Schoonover method is commonly used to work out the gypsum requirement (Schoonover, 1952). In this method, saturated gypsum solution of known calcium concentration is prepared by adding 5 g of chemically pure $CaSO_4 . 2H_2O$ to one liter of distilled water. It is vigorously shaken for about 10 min on a mechanical shaker or 1 hr by hand and then filtered through Whatman filter paper no. 1.

Take 5 g of air dry soil in 250 ml conical flask. Add 100 ml of the saturated gypsum solution. Firmly put a stopper and shake for 5 min on a mechanical shaker or 50 min by hand. Filter the contents through Whatman filter paper no.1. Transfer 5 ml of aliquot into 100 ml porcelain dish. Determine the calcium and magnesium concentration. Run a blank using 5 ml of the saturated gypsum solution prepared before.

GR (me/100 g soil) = 2 [(Ca conc. in blank, me/l)
– (Ca conc. in the filtrate, me/l)] (9.5)

GR/ha-15 cm (t/ha) = GR (me/100 g soil) * 1.72 (9.6)

Abrol *et al.* (1975) proposed an amendment to this method. They argued that the gypsum requirement as calculated by Schoonover method includes the gypsum required to neutralize the soluble sodium carbonate, if present in alkali soils. It is therefore, prudent to leach the sample of the soil (5 g) with 20 ml of 60% ethanol. It will leach down the soluble sodium carbonate and therefore, will not add to the gypsum requirement of the soil. Rest of the procedure is same as discussed in the foregoing paragraph.

It may be noted that if the gypsum is applied first before the soil is leached, GR calculated with the original soil sample using Schoonover method will be more realistic. The values of GR will be high in this case because it has to account for both the neutralization of free Na_2CO_3 and replacement of exchangeable sodium. It is therefore, recommended that the soil should be leached first and then the gypsum should be determined as per Schoonover method with soil samples taken after leaching or as per Abrol *et al.* (1975) method using the original sample (Table 9.8, Arora *et al.*, 2015).

Table 9.8. Effect of flushing on gypsum requirement

Location	Soil pH_2	Soil EC_2 (dS/ m)	Gypsum requirement (me/100g)	
			Without leaching	After leaching
Koithar	10.1	8.18	5.70	1.40
Kausambhi	9.2	0.74	3.70	2.90
Rasoolpur	8.9	2.43	1.60	1.10
Lalganj	10.0	1.28	5.80	3.40
Sandila	10.3	2.50	5.30	3.70
Dhaura	10.1	4.04	4.50	2.20
Vishnupur	9.8	2.25	5.80	4.80

pH versus GR graph

The absolute quantity of exchangeable sodium percentage of a alkali soil will continue to be the basis to accurately determine the quantity of gypsum required for reclamation. In practice, pH of the saturated soil paste (pH_s) or pH of 1:2 soil water suspension (pH_2) may also be used. It has been established that pH_s, pH_2 and the ESP of the alluvial soils of the Indo-Gangetic plains are fairly correlated with each other. This method is especially useful when skilled staff and laboratory facilities to determine the exchangeable sodium and gypsum requirement are not available. Abrol *et al.* (1973) prepared a graph between pH_2 and GR of the soils for different soil textures (Fig. 9.1). This graph is a quick and fairly good method to estimate GR of alkali soils of the Indo-Gangetic plains. The GR calculated with this graph need not be reduced to half as in the case of laboratory determined GR. It may however, be noted that this graph is not universal as pH_s and ESP of black cotton soils (Vertisols) are not well related. Therefore, use of such a graph is not advisable for such soils.

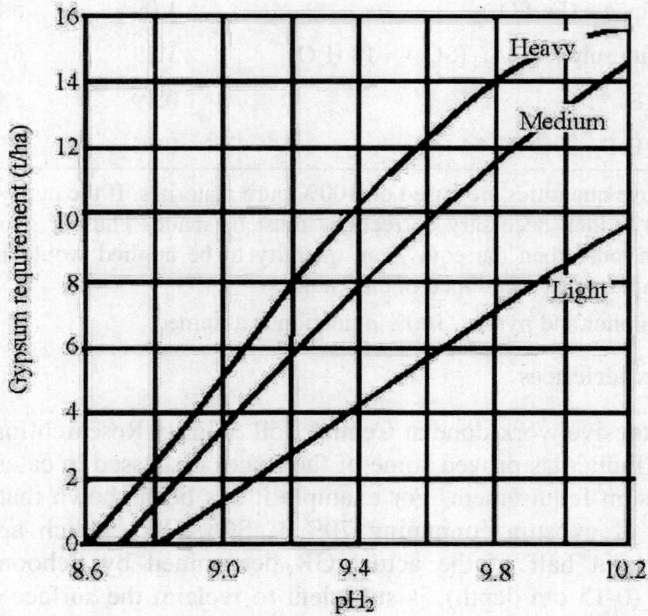

Fig. 9.1. pH_2 versus GR graph for alluvial soils of Indo-Gangetic plains

Once the GR is known, equivalent quantities of other amendments are calculated using Table 9.9.

Table 9.9. Commonly used amendments and their equivalent amendment values

Amendments (Chemical formula)	**Tons of amendment material equivalent to**	
	1 ton of pure gypsum	**1 ton of pure sulphur**
Gypsum ($CaSO_4 . 2 H_2O$)	1.00	5.38
Calcium chloride ($CaCl_2 . 2 H_2O$)	0.86	--
Calcium nitrate ($Ca (NO_3) . 2 H_2O$)	1.06	--
Press mud (lime–sulphur, 9% Ca+24% S)	0.78	4.17
Sulphuric acid (H_2SO_4)	0.61	3.20
Iron sulphate ($FeSO_4 . 7 H_2O$)	1.62	
Ferric sulphate ($Fe_2SO_4)_3$.	1.09	5.85
Aluminium sulphate ($Al_2 (SO_4)_3 . 18 H_2O$)	1.29	6.34
Sulphur (S) **	0.19	1.00
Pyrites ((FeS_2, 30% S) **	0.63	

*The above quantities are based on 100% pure materials. If the purity is less than 100%, then necessary corrections must be made. Thus, if gypsum is only 80% pure then the equivalent quantity to be applied would be 1.25 times of quantities mentioned in the Table.

** For sulphur and pyrites, 100% oxidation is assumed.

Field Experiences

Extensive work done at Central Soil Salinity Research Institute, Karnal, India, has proved some of the issues discussed in calculating the gypsum requirement. For example it has been shown that 10 to 15 t/ha of gypsum containing 70% $CaSO_4.2H_2O$, which approximates about half of the actual GR determined by Schoonover's method (0-15 cm depth), is sufficient to reclaim the surface 15 cm soil if one is to start with rice as the first crop (Table 9.10). This is keeping in view the high tolerance of transplanted rice to soil ESP and its shallow root system. Moreover, its roots capacity to facilitate the soil water to dissolve soil $CaCO_3$ also helps to reduce the soil exchangeable sodium. Over time, ESP of the surface as well as deeper layers gets reduced so as to allow cultivation of other crops. To grow sensitive crops like pulses in the initial years, one needs to add high amounts of amendments for reclamation.

Table 9.10. Yield (t/ha) of crops as affected by gypsum treatment of an alkali soil

Gypsum (t/ha)	*Ses-bania** 1970	Wheat 1970-71	Rice 1971	Wheat 1971-72	Rice 1972	Wheat 1972-73	Mash 1973	Lentil 1973-74	Pearl millet 1974	Gram 1974-75
0.0	0.44	0.0	4.39	1.52	7.18	1.35	0.05	0.28	1.06	0.12
7.5	3.99	1.89	6.21	2.79	7.23	1.96	0.25	0.82	1.33	0.47
15.0	4.35	3.49	6.39	3.45	7.13	2.43	0.32	1.16	1.96	0.69
22.5	4.70	4.16	7.08	3.72	7.17	2.26	0.38	1.36	1.81	0.90
30.0	4.50	3.79	6.66	3.98	7.03	2.74	0.42	1.40	1.94	0.93
LSD at P = 0.05	0.51	0.64	0.81	0.70	NS	0.50	0.12	0.35	0.52	0.44
Normal yield	-	4.5	7.0	4.5	7.0	4.5	1.0	2.0	3.0	2.0

* Green matter yield; Initial soil pH = 10.3; ESP = 94 in surface 15 cm soil;
Source: Abrol and Bhumbla (1979)

The gypsum calculated @ 50% GR (0-15 cm soil depth), must be uniformly applied to the whole field (Fig. 9.2) and mixed in the shallow layer of 10 cm depth (Sharma *et al.*, 2016). It has been found to be superior to mixing in the 20 or 30 cm depth (Table 9.11).

Fig. 9.2. Application of gypsum in the fields

Table 9.11. Effect of depth of mixing gypsum on the yield of crops (t/ha)*

Gypsum (t/ha)	Depth of mixing (cm)	Barley 1970-71	Rice 1971	Wheat 1971-72
0	10	0.00	5.18	2.04
	20	0.00	5.69	1.77
	30	0.00	4.62	0.93
13.5	10	2.64	7.02	3.28
	20	2.46	6.05	3.16
	30	0.53	5.64	2.06
27.0	10	2.83	6.71	3.67
	20	3.22	6.28	3.67
	30	2.30	5.98	3.53
CD at p = 0.5 for gypsum and depth		0.35	0.73	0.41

* Initial soil pH_2 = 10.5 Source : Khosla *et al.* (1973)

Single application of gypsum at the rate of 50% GR (6.5 t/ha), before the first crop of rice has proved as effective as the higher and repeated doses in subsequent years on a highly alkali soil (Table 9.12). The pH came down from 10.7 in the first year before the rice crop to 9.5, 8.6 and 8.4 after the wheat crop in the first, second and third year, respectively. It shows that if reclamation has to proceed with the rice as the first crop, application of gypsum @ 25% GR would be enough, if average yields for 3 or more years are taken as the basis. However, the yields may be low in the first year especially that of wheat if gypsum at the rate of 25% GR is applied as compared to 50% GR.

Crop Tolerance and Gypsum

The soils seldom require gypsum application if pH_2 is less than 8.5. Yadav and Sharma (1980) reported that even gram (*Cicer arietinum*), a sensitive pulse, can be grown well if pH_2 is ≤ 8.5 in 0-30 cm soil depth. If the pH_2 is above 8.5, some amount of gypsum is necessary to lower the ESP to a level that can be tolerated by the crop. A thumb rule for this purpose is to apply 1.0 t/ha gypsum (70-80% purity) for every 0.1 unit of pH_2 higher than the tolerance limit of the crop. Thus, gypsum requirement depends upon the degree of ESP (pH) tolerance of the crop. The sensitive, semi-tolerant and tolerant crops can be cultivated fairly well if pH_2 is less than 8.5, 9.3 and 9.8 respectively.

Table 9.12. Effect of repeat gypsum doses applied over three years on crop yields (t/ha)

Gypsum (t/ha)			Mean grain yield	
Ist year	**2nd year**	**3rd year**	**Rice**	**Wheat**
6.5	-	-	7.01	4.33
6.5	6.5	-	6.40	3.98
6.5	6.5	6.5	7.46	4.16
13.0	-	-	6.64	4.51
13.0	6.5	-	7.04	4.35
13.0	6.5	6.5	7.24	4.50
19.5	-	-	6.92	4.62
19.5	6.5	-	7.31	4.72
19.5	6.5	6.5	6.96	4.57
26.0	-	-	7.20	4.69
26.0	6.5	-	7.24	4.89
26.0	6.5	6.5	7.14	4.17

Source: Gaul and Dargan (1978)

PHOSPHOGYPSUM

In addition to the mineral gypsum, large quantities of byproduct gypsum are available from different industries, which are produced when rock phosphate is treated with sulphuric acid as per the following reaction:

Apatite (rock phosphate) Phosphogypsum

$$Ca_{10}(PO_4)_6F_2 + 10H_2SO_4 + 20H_2O \rightarrow 10\,CaSO_4 \cdot 2H_2O + 6H_3PO_4 + 2HF \qquad (9.7)$$

The detailed composition of phosphogypsum available from different industries is given in Table 9.13. The phosphogypsum is reported to contain about 90% $CaSO_4$ and 1 to 2% fluorine.

Table 9.13. Chemical composition of phosphogypsum

Characteristics	Albright Morarji and Pandit Ltd. Ambernath	FCI Trombay	Coromandel Fertiliser Ltd., Visakha-patnam	FCI Haldia	SPIC Madras
Combined water	19.5	20.2	-	-	-
Total P_2O_5	1.0	0.15-1.0	0.8-1.0	-	-
CaO	31.1	30-32	30.8	-	32.
Mixed oxides	0.9	-	0.45	-	0.5
SiO_2 and other insolubles	1.8	-	4.6	0.7	0.9
F	1.52	-	0.6	0.35	1.1
Na_2O	0.56	-	1.3	-	-
K_2O	0.6	-	-	-	-
Equivalent $CaSO_4$ $2H_2O$	-	95-98	90.2	70.2	-

Source: Rohtagi *et al.* (1977)

Mehta and Yadav (1977) reported that phosphogypsum is a promising chemical amendment and can be used effectively for the reclamation of alkali soils and good crop production on them. A number of demonstration on the use of PG to reclaim alkali lands have been conducted by Hindustan Copper Limited, New Delhi during 1974-77 in Punjab, Haryana and Uttar Pradesh in association with research organizations (FCI-FAI, 1977). Shrotiya and Mishra (1977) reported high yields of rice and wheat even in soils with a pH as high as 10 or more. In these demonstrations phosphogypsum @ 10 t/ha was applied and the cultural practices advocated by CSSRI, Karnal were implemented (Table 9.14).

Table 9.14. Performance of phosphogypsum in alkali land reclamation

State	Soil pH before amendment application	Average yield in demonstrations (t/ha)	
		Rice	Wheat
Haryana	9.5-10.6*	4.70	2.84
Punjab	9.0-10.0^	4.75	2.98

*range of 8 demonstrations, ^ range of 9 demonstrations

Current researches in India have proved that PG is a good amendment to reclaim alkali lands and alkali water. The PG treatment resulted in greater reduction in pH and ESP of both the surface and the lower layers (Table 9.15). It is attributed to higher solubility of PG over mined gypsum, which allows soluble Ca to permeate to lower depths. Application of phosphogypsum also gave comparatively higher yield over equivalent dose of mined gypsum. Phosphogypsum @ 10 t/ha (equivalent to 50% GR) resulted in slightly higher yield of 4.5 t/ha over the treatment 50% mineral gypsum (4.22 t/ha) and 25% mineral gypsum+25% PG (4.32 t/ha) in the first year of reclamation.

Table 9.15. Effect of phosphogypsum on soil ESP after first and third year of rice-wheat cropping

Treatments	2006-07		2008-09	
	0-15 cm	15-30 cm	0-15 cm	15-30 cm
50 GRm	26.5	31.6	24.6	28.5
50 GRpg	20.4	27.3	18.3	25.7
50 GR_{PG}-KFC	18.3	25.2	17.9	23.4
25 GRm+ 25 GRpg	17.8	23.5	16.7	21.5
25 GRpg	19.4	25.6	17.9	24.3
12.5 GRm+ 12.5 GRpg	25.2	30.8	23.5	27.5
LSD (p=0.05)	1.5	2	1.5	2

50GRm = 50% gypsum requirement through mineral gypsum;
50 GRpg = 50% gypsum requirement through phosphogypsum;
25 and 12.5 means 25% and 12.5% of GR

Two major ecological problems associated with phosphogypsum relate to its high fluorine content and presence of certain radioactive elements in the amendment. Singh *et al.* (1980) conducted detailed studied on the phosphogypsum to evaluate its effect on the soil properties and crop growth. The possibility that phosphogypsum may cause adverse effect to crop growth because it contains high concentrations of fluorine has been discounted. It has been explained that the adverse effect of fluorine is nullified because of the phosphorus contained in this amendment. Recent studies also revealed that one time application of phosphogypsum for land reclamation and/or small doses for water reclamation are unlikely to result in phytotoxic level of fluoride. As a matter of abundant caution, the rice and wheat cultivated on reclaimed lands with PG in fluoride affected areas should be consumed with care for first two years of reclamation especially in areas having high fluoride groundwater.

Fears of radioactive elements have also been discounted. The fate of radium-226 in Florida phosphogypsum was investigated by Mays and Mortvedt (1986). They applied phosphogypsum containing 25 pCi.1g 226Ra at rates up to 112 Mg/ha to the surface of silt loam soil and grew successive crops of corn (*Zea mays* L.), wheat (*Triticum aestivum* L.) and soybean (*Glycine max* L.). Application of phosphogypsum even at the 112 Mg/ha rate had no effect on the radioactivity levels in the grains of corn, wheat, or soybeans. Nayak *et al*. (2011) also reported that radioactive nuclei present in the phosphogypsum used in their study have no adverse effect on crop yield or its quality.

PYRITES

Pyrites, FeS_2 is a mineral containing iron and sulphur. It occurs extensively as natural deposits as black coloured sediment. In India, rich deposits are located in the Vindhya hills at Amjhore in Rohtas district in Bihar. The pyrite bed has a thickness of about 0.6 m and is overlaid by 5-6 m thick pyrites ferrous shales, covering an area of 48 km^2. The total resource in India is estimated to be 1500 million tons of low grade (10% S), and 350 million tons of high grade (40% S), available exclusively in Bihar.

Quality

Pyrites Phosphates and Chemicals Ltd. is intensively exploring and exploiting this deposit in Amjhore block. In the present set-up of mining, 0.6 m of pyrites (40% S) and 1 m of pyritesferous shale (10% S) is mined simultaneously. The runoff mine thus obtained contains about 22% sulphur. High grade pyrites is supplied to the Fertilizer Corporation of India, Sindri, for the manufacture of sulphuric acid. The low grade pyrites available for agricultural use contains 22% `S'. The residual left over product is treated as rejects containing about 10-12% `S' (Jaggi, 1978). The pyrites is highly pyrophoric in nature and immediately reacts with air and water to form sulphuric acid and iron sulphate. Because of this nature of the pyrites, the mine water becomes acidic with a pH around 3. The chemical composition of the pyrites of the two grades is given in Table 9.16.

Table 9.16. Chemical composition of pyrites

Constituent	(Expressed as percentage)	
	Pyrites (powder)	**Pyrites (shales)**
Sulphur	22-24	11.7
Iron	20-22	11.7
Al_2O_2	6-8	11.3
SiO_2	35-40	46.9
CaO	0.1	4.3
MgO	0.5-0.6	0.9
Cu	0.05	-
Mn	0.01	-
C	-	3.2
Moisture	-	14.7

Source: Jaggi (1978)

Chemical Reaction

The low quality pyrites having different percentage of sulphur (15-30%) classified as agricultural grade is normally used in the reclamation of alkali soils. Pyrites *per se* do not react with the soil directly and therefore, it´s mode of reaction in alkali soils must be clearly distinguished from that of elemental sulphur. The elemental sulphur has to be oxidized to sulphuric acid by the sulphur oxidizing bacteria like *Thiobacillus thiooxidans,* etc. Sulphur oxidation is therefore, quite slow in alkali soils. On the contrary, the sulphur present in the pyrites is in a highly reduced state. It readily reacts with the oxygen of the air and forms sulphur compounds of a higher oxidation state. Both the products formed directly or indirectly react with soil as illustrated by the following reactions.

$$2FeS_2 + 2HOH + 7O_2 \rightarrow 2FeSO_4 + 2H_2SO_4 \quad (9.8)$$

$$CaCO_3 + 2H_2SO_4 \rightarrow CaSO_4 + CO_2 + H_2O \quad (9.9)$$

$$2\ Na\text{–}soil + CaSO_4 \rightarrow Ca\text{–}soil + Na_2SO_4 \downarrow \quad (9.10)$$

Equation (9.8) shows that the effectiveness of pyrites in alkali soils varies with the date of mining of the material and storage conditions. The oxidation of pyrites begins as soon as the mined material is exposed to the air. The reaction is slow but wasteful as some of the S escapes to the atmosphere as follows:

$$FeS_2 + 3O_2 \rightarrow FeSO_4 + SO_2 \uparrow \quad (9.11)$$

Singh *et al.* (1978) observed that more than 36% of the sulphur is extracted in water in the first two days from the pyrites containing 30% sulphur. Later on, the oxidation is slow. This aspect has to be kept in mind while judging the effectiveness of sulphuric acid produced due to the oxidation of pyrites in the reclamation of alkali soils.

The rate of reaction depends upon several factors, viz. moisture content, availability of oxygen, temperature and presence of toxic constituents around the granules. For efficient oxidation, a warm, well aerated moist soil is required. The biological oxidation of sulphur in the amendment to some extent is carried out by *Thiobacillus ferrooxidans* but only at a specific pH range normally obtainable in rice crop fields (Singh *et al.,* 1978). In buffer suspensions, it has been found that the oxidation of pyrites is maximum at pH 5.0. At pH 2.5 it is low and at pH 8.0 it is negligible (Venkat Krishnan, 1979).

Method of Application

Not much information is available as to what is the appropriate method of application of pyrites for its optimum efficacy. Singh *et al.* (1978) observed that the surface application of pyrites lowered pH and ESP of the soil to a greater extent as compared with its mixing in the surface 10 cm of the soil. The efficiency of the amendment also decreased when soil water content was lowered from full saturation to one fourth saturation.

Venkat Krishnan (1979) reported that mixing of the pyrites in the whole soil column was inferior to surface placement. Mixing the surface pyrites with the rest of the soil after 12 and 18 days of surface application performed comparably and resulted in better and more uniform improvement of the soil than retaining the pyrites on the surface unmixed. Mixing of the pyrites after 6 days of application led to poor improvement. Higher amounts of leachates were collected from columns where pyrites were surface placed. Pyrites showed signs of appreciable oxidation forming a rusty coloration after about 15 days of retention on the surface. Some guidelines on pyrites application are:

- The pyrites must be properly crushed before use. Its soluble S content should be more than 5%
- Apply pyrites only when the field moisture content is around field capacity or above
- For proper oxidation, it should be applied as thinly as possible. Besides, sufficient time should be allowed to oxidize the material before the leaching process is initiated
- The soils should be leached with good quality water for 7-15 days preferably in intermittent leaching mode before transplanting the first rice crop.

Pyrites containing 6-8% soluble sulphur has a pH of 2-2.5 when dissolved in water (1:100). As such, it is used as a field test to assess the quality of the pyrites for land reclamation. The test procedure is as follows:

Dissolve 1-g pyrite in 100 ml distilled water and check the pH using a litmus paper or a pH meter. If the pH is more than 2.5, the pyrites needs to be further stored. It may be periodically moistened by sprinkling water over the bags. The material should be stored until the test reveals a pH of 2.5 or less.

Practical Problem

In the initial years when pyrites disposal was a problem, large scale use of this amendment was made and it helped to reclaim as much as 25,000 ha of alkali soils in Uttar Pradesh and 10,000 ha in Bihar. Lately because of other applications, its cost has increased and it is no more used in land reclamation in India.

SULPHUR AND SULPHURIC ACID

Mineral sulphur is converted to H_2SO_4 through the process of biological oxidation by *Thiobacillus thiooxidens* as per the following reaction:

$$2\ S + 3\ O_2 \rightarrow 2\ SO_3 \text{ (microbiological oxidation)} \quad (9.12)$$

$$SO_3 + H_2O = H_2SO_4 \quad (9.13)$$

Oxidation in this manner being a very slow process at high pH commonly encountered in alkali soils, the relative efficiency of mineral S as an amendment is low. Further reaction will be as per eq. (9.14) and eq. (9.15) in the case of sulphuric acid in this section.

The farmers in the Western States of the USA (California, Arizona, etc.) and erstwhile USSR (Araraat Valley, Armenia) have been using concentrated sulphuric acid as a means of reclaiming the calcareous alkali soils under irrigated farming. The reclamation is based upon the following chemical reaction:

$$H_2SO_4 + CaCO3 \rightarrow CaSO_4 + CO_2 + H_2O \quad (9.14)$$

$$2Na - clay + CaSO_4 \rightarrow Ca - clay + Na_2SO_4 \quad (9.15)$$

but if the soil is non-calcareous,

$$2Na - clay + H_2SO_4 \rightarrow 2H - clay + Na_2SO_4 \quad (9.16)$$

Although the initial cost of the sulphuric acid is apparently high but on equivalent basis, only half of the quantity as that of gypsum is required to reclaim the soil. As such, its transport cost is much lower than gypsum. Moreover, if the time taken to reclaim the soil is considered, it may turn out to be less costly than gypsum, since expenses begin to defray earlier than gypsum (Kovda *et al.,* 1973). Since the reclamation is complete, there is no restriction in the choice of crops. Even the most sensitive crops comprising horticultural plants can be established well because the soil profile is reclaimed through a deeper depth.

Overstreet *et al.* (1951) compared the effectiveness of gypsum, sulphur and sulphuric acid on a highly alkali soil (ESP 60-100). After seven months of experimentation, including leaching of the soils, sulphuric acid reduced soil ESP to 10, more efficaciously than gypsum. Large scale reclamation programs on alkali soils are being executed in Armenia by the application of 1% sulphuric acid solution, heavy leaching (15000-30,000 m^3/ha) sometimes combined with rice cultivation and an extensive deep drainage (horizontal drains combined with tube wells). Since the alkali horizons of these soils are very thick (as much as 60-100 cm), the sulphuric acid is applied layer by layer, with multi-tier ploughs or using a special ripper. The standard application of sulphuric acid is 30 t/ha (expressed as 80% acid), and the crop (cotton, fruits and vines) yields for the first two years is sufficient to pay for the reclamation cost.

In the Indian context, Shah (1975) used waste sulphuric acid (70% strength) in the field and achieved remarkable improvement in the alkali soils. However, farmers being more dependent on subsidies, and due to the non-availability of adequate machinery for transport and application of H_2SO_4 , probably use of sulphuric acid is not gaining acceptance.

OTHER AMENDMENTS

Although sulphur and H_2SO_4 have not become popular in India, some of the acidic waste or inert materials available from several industries are being increasingly tested/used for land reclamation.

Distillery Spent Wash

Distillery spent wash (DSW), a byproduct of the molasses based distilleries, is a nutrient rich wastewater. It has high biological oxygen and chemical oxygen demand (BOD and COD). It has 0.10-0.15% N, 0.01-0.03% P_2O_5, 0.80-1.20% K_2O and appreciable quantities of Ca, Mg, S and other micronutrients (Table 9.17). It is estimated that for each liter of alcohol, about 15 liters of DSW is produced. In Tamil Nadu, under the AICRP on Management of Salt Affected Soils and Use of Saline Water in Agriculture, distillery spent wash (DSW) has been used to reclaim alkali lands. Around 0.5 million l/ha of DSW needs to be applied for reclamation (Fig. 9.3; Haroon and Bose, 2004). Tests on this amendment in Madhya Pradesh, under the AICRP on Management of Salt Affected Soils and Use of Saline Water in Agriculture, have also proved the efficacy of this amendment in soybean-wheat crop rotation. Singh *et al.* (1980) reported that the addition of spent wash was very effective in increasing the water intake rate of a highly alkali soil. ESP was reduced from 100 to 2 in the top 15 cm of the soil in the column. The application of spent wash followed by irrigation rather than the dilution of spent wash at the time of its application was effective.

Table 9.17. Chemical Characteristics of pre-treated distillery spent wash (undiluted)

Parameters	Values
pH	7.7
Electrical conductivity (dS/m)	26.4
Total solids	47200
Total dissolved solids	37100
Total suspended solids	10240
Settle able solids	9880
BOD	16100
Carbonate	Nil
Bicarbonate	12200
Total phosphorus	40.5
Total potassium	7500
Calcium	900
Magnesium	1244
Sulphur	70
Sodium	520
Chlorides	6204
Iron	7.5
Manganese	980
Zinc	1.5
Copper	0.25
Cadmium	0.005
Lead	0.16
Chromium	0.05
Nickel	0.09
Ammoniacal nitrogen	750.8
Carbohydrates (%)	22.8

Units: All values are in mg/l, except pH or otherwise specified; Source: Chandraju *et al.* (2013)

Fig. 9.3. Application of DSW for reclamation of alkali land

Press Mud

Press mud produced during clarification of sugarcane juice is about 3.6% to 4% of sugarcane crushed i.e. 36 to 40 kg of press mud is produced for each ton of cane crushed. Press mud is a soft, spongy, amorphous and dark brown material containing sugar, fiber and coagulated colloids including cane wax, albuminoids, inorganic salt, sand and soil particles (Table 9.18). It consists of 80% water and 0.9%-1.5% sugar organic matter, nitrogen, phosphorus, potassium, calcium, coagulated colloids and other materials in varying amounts. Sugar mills either use the carbonation (calcium carbonate-sulphur) or the sulphitation process (Sulphur) to purify the sugarcane juice. Press mud coming out of the sugar mills with the carbonation process contains about 70% calcium carbonate and 11% organic matter. Since the solubility of calcium carbonate is quite low, the press mud available through this process cannot be used as an effective amendment. However, it is applied along with farmyard manure as an amendment on mild alkali soils or to improve soil physical properties. The press mud, produced in sugar mills using sulphitation process, contains 9.3% calcium sulphate and 35.8% organic matter. Sugar mills in India produce about 12 million tons of press mud (filter cake) as a waste from double sulphitation process. Since the percent calcium sulphate is too less, the quantities required to reclaim barren lands is quite high. Nonetheless, it has a good potential to reclaim mild alkali lands as well as to ameliorate soils where alkalinity build-up is due to irrigation with alkali water. Evidences have been generated to prove its usefulness as an amendment.

Table 9.18. Values of some selected physico-chemical parameters of press mud (PM) (The values are mean ± SE of 10 observations each)

Parameters	Press mud (PM)
Water holding capacity (%)	78.2± 1.323
pH	7.66± 0.047
Electrical conductivity (dS/m)	1.77± 0.092
Sodicity (me/100 g)	2.89± 0.198
Carbon content (%)	29.67± 1.058
Organic matter (%)	51.2± 1.827
Available phosphorus (%)	1.80± 0.157
Sulphate (%)	2.297± 0.123
Nitrate nitrogen (%)	0.271± 0.044
Total nitrogen (%)	1.13± 0.095
Total calcium (%)	0.204± 0.015
Total magnesium (%)	0.088± 0.011
Total sodium (%)	0.005± 0.001
Total potassium (%)	0.037± 0.003
C:N	26.25
C:P	16.48

Source: Joshi and Sharma (2010)

Addition of 10-20 t/ha press mud can help to reduce the gypsum requirement of soils by about half because of the addition of calcium sulphate and organic matter. For example; Isaac *et al.* (2010) reported that the highest yield were obtained when gypsum was applied at 50% of GR, and it was at par with gypsum applied at 25%

GR when press mud @ 10 t/ha was applied along with. It was also observed that phosphogypsum @25% GR+press mud can help in reducing soil pH from 10.1 to 9.05. Maximum grain yield of paddy was observed in phosphogypsum@25% GR+press mud treatment compared to gypsum @50%GR and gypsum@25%GR+ press mud. Also, maximum wheat yield was obtained in treatment where gypsum @ 25% GR+ press mud+ bio-inoculant was applied which was at par with treatment phosphogypsum @ 25% GR + press mud. It is also a good substitute as a fertilizer, both in the unprocessed or processed form (Yaduvanshi and Yadav, 1990). Processes used to improve its fertilizer value include composting, treatment with micro-organisms and mixing with distillery effluents (Nasir, 2006).

So far disposal of press mud has been a major issue. Industries therefore, sold it free. Now, it is being used as a fuel in brick kilns. Such a use of the material is a huge loss as it results in wastage of millions of tons of nutrients besides causing environmental degradation. But it has dried up the supply of press mud for agriculture because it is now a priced product besides the transportation cost. Dry press mud can be exploited for generation of energy as it contains higher percentage of combustibles. It is also a good source of biogas as it contains about 5–15% of sugar.

Fly Ash

Fly ash (FA), a by-product of coal based thermal power plants, besides many elements contains soluble phosphorus, magnesium, sodium and calcium. Around 10 million tons of fly ash is generated by various thermal power plants in India. Depending upon the origin of coal used in a particular power plant, fly ash can be alkali or acidic in nature. Few experiments have established its usefulness in alkali soils (Table 9.19). Guidelines for its application as an amendment are yet to be developed. Integrated use of acidic FA and gypsum decreased SAR and improved nutrient availability, saturated hydraulic conductivity (HC) and soil water retention in a alkali soil (pH=9.07, EC= 3.87dS/m, ESP=26) leading to higher rice and wheat yields than control indicating the possibility of partial gypsum substitution (up to 40% of GR) with FA (Kumar and Singh, 2003). At 10 and 20 Mg/ha doses, FA considerably lowered soil pH, soluble salts and Na^+ saturation. Soil pH reduced from 10.0 to 8.3 and Na saturation decreased from 64.5% to 24.0% after reclamation. Appreciable increase in rice and wheat yields was seen in the ameliorated soils (Tiwari *et al.*, 1992). Although FA can improve soil fertility, texture and water-holding capacity, it may also raise the pH and salinity of the treated soils.

Table 9.19. Effect of fly ash and pyrites on rice and wheat yield in alkali soils

Amendment	Quantity (t/ ha)	Yield (t/ha)			
		1979-80		1980-81	
		Rice	Wheat	Rice	Wheat
Control	0	1.02	0.57	1.98	0.91
Fly ash	10	2.85	2.40	4.03	0.75
Pyrite	10	3.03	2.87	4.33	2.79

Source: Tiwari *et al.* (1989)

Recent studies have shown that results of fly ash are inconsistent over locations and the kind of fly ash used in land reclamation. Thus, while its utility for land reclamation still needs to be established, field application of fly ash in a limited manner can be a good way to dispose of this waste (Chhabra and Lal, 2001-02) without impacting the soil resource or the crop yields (Table 9.20).

Table 9.20. Effect of fly ash on the yield of rice in a reclaimed alkali soil

Fly ash (t/ha)	Yield (t ha)	
	Grain	Straw
0.0	5.83	5.00
2.5	5.78	4.86
5.0	5.84	4.57
7.5	5.94	4.76
10.0	5.82	4.76
20.0	5.94	5.42
30.0	5.83	5.03
40.0	5.83	5.06
50.0	5.69	4.75
CD : 5%	Non-significant (NS)	NS

Dairy Waste and Polymers

Besides what has been stated, the dairy waste available after the testing of milk for its fat content using H_2SO_4 is a very useful material for the reclamation of alkali soils. Several polymers and nano-scale materials have been used to improve the soil physical properties of alkali lands. However, their long-term effects have so far not been evaluated.

ORGANIC AMENDMENTS

Most organic amendments like straw, rice husk (rice hulls), groundnut and safflower hulls, farm yard manure (FYM), compost, poultry dropping, green manure, tree leaves, saw dust etc. on their decomposition produce high partial pressure of CO_2 and organic acids, which help in increasing the electrolyte concentration and lowering the soil pH. Organic materials also feed soil microbes that produce polysaccharides and long chain aliphatic compounds capable of binding and stabilizing soil aggregates (Lado *et al.*, 2004), which tremendously helps to maintain soil structure. Such processes also increase the solubility of *kankar* to lower the ESP. Since alkali soils are generally deficient in organic matter, addition of crop residues and other organic materials to the soil improves and maintains the soil structure, supplies needed plant nutrients and hastens the soil reclamation process.

To be very effective alone, relatively large quantities (20 to 400 t/ha representing 1/5 to 1/3 of the soil volume) of organic amendments have to be used. Moreover, C:N ratio of the organic amendment needs to be considered while choosing these for field application. Materials with high C:N ratio such as saw dust, rice husk and barley straw decompose slowly and are less effective as compared to *Sesbania* with a low C:N ratio. Organic amendments with high C:N ratio may cause N deficiency which should be addressed through appropriate nutrient management strategies.

Besides, the prevailing field conditions during decomposition impact the performance of organic amendments. Singh (1974) observed that straw of barley and *Sesbania* were less efficient in terms of ESP reduction and improvement in soil physical properties when decomposed under aerobic conditions as compared to when these were decomposed under saturated (anaerobic) conditions (Table 9.21). Thus, efficiencies of organic amendments can be improved by maintaining continuous ponding during the course of decomposition. Organic materials together with inorganic amendments are more cost effective as they hasten the reclamation process and increase the yield. Therefore, their integrated use should be encouraged (Dargan *et al.*, 1976a).

Table 9.21. Effect of organic materials on soil physical properties

Treatment	pH_s	ESP	Modulus of rupture (bars)	Hydraulic conductivity (cm/h)	Na in leachate (me/l)
Barley straw aerated	9.1	81.5	12.9	0.03	16.6
Sesbania straw aerated	8.8	68.5	11.8	0.04	26.3
Barley straw saturated	8.4	56.1	11.3	1.72	30.2
Sesbania straw saturated	8.3	47.2	7.5	1.29	43.0
Original soil	9.5	95.0	-	-	-

Source: Singh (1974)

In recent studies, incorporation of municipal solid waste compost (MSWC) to soils hastened the dissolution of precipitated $CaCO_3$ leading to increased soluble Ca^{2+} content and the consequent displacement of the harmful Na^+ ions from the soil exchange complex. However, many times such additions cause unintended side effects in plants due to the presence of heavy metals. It calls for complete laboratory analysis of the amendment before actual field application to ensure that their use is safe for soil and plant health.

■■■■

CHAPTER 10

INTRODUCTION TO INNOVATIVE METHODS OF RECLAMATION

INTRODUCTION

Amelioration of alkali soils has been mainly achieved through the application of chemical amendments. Owing to the gradual increase in amendment cost in some parts of the world, this strategy has become cost-intensive, particularly for the subsistence farmers in the developing countries. There can be many other settings where it may not be possible to reclaim the soils using the amendments. In many cases, resource constraints of the farming community may inhibit the adoption of this technology. It is in this context that scientists are working on new innovative, cost-effective and ecologically sound sustainable reclamation and management techniques. This chapter introduces the readers with some of these technologies.

PHYTOREMEDIATION

Remediate or remediation means to solve a problem and bio-remediation in the context of reclaiming degraded soils/water means to use biological materials to remediate or improve the quality of contaminated soil or water. More specifically bio-remediation is further categorized as phytoremediation in which salt tolerant crops/plants species are used to reclaim salt affected soils/water and bio-remediation refers specifically to the use of organisms to remediate the soil/water. Phytoremediation has also been defined as the use of plants to remove pollutants from the environment and to render them harmless (Hasanuzzaman *et al.*, 2014). Phytoremed-iation for alkali soils works through plant root action that helps dissolve native soil calcite ($CaCO_3$) of low solubility to supply Ca^{2+} for an effective Na^+–Ca^{2+} exchange without the application of an amendment. The phyto-remediation is a clean and environmentally safe technology. Phytoremediation technology has many advantages (Qadir *et al.*, 2010). Among them following are the major ones:

- Less investments as it either obviates or minimizes the need of chemical amendments
- Financial or other benefits from crops grown during amelioration
- Improves physical condition of the soil
- Higher organic matter and availability of plant-nutrients
- Amelioration is more uniform and to deeper depths than chemical amelioration
- No environmental issues rather carbon sequestration is achieved if plants are used as ameliorants

The phyto-reclamation approach is not a substitute for chemical amendment model, but may be adopted as an alternate approach by resource poor farmers who may not be able to afford the investments on chemical amendments.

The technology makes use of:

- Grasses
- Salt tolerant crops
- Forest plantations
- Halophytic plants

Highly Saline/Alkali soils

The alternate options depending upon the kind of crops/plants used are briefly described in the following paragraphs.

Grasses are, in general, more tolerant of alkali conditions than most field crops. Karnal grass (*Diplachne fusca*), Rhodes grass (*Chloris gayana*). Para grass (*Brachiaria mutica*) and Bermuda grass (*Cynodon dactylon*) all are highly tolerant and can be successfully grown in alkali soils (Kumar and Abrol, 1983, 1986). In heavy textured Verisols highly salt tolerant and high biomass producing grass species besides above mentioned species include *Aeluropus lagopoides*, *Sporobolus helvolus*, and *Brachiaria ramose*. Karnal grass grows extremely well in soils of very high ESP (80 to 90) even without application of any amendment (Kumar *et al.*, 1980). Rhodes grass is the next best. Karnal grass and para grass are also highly tolerant to ponded water conditions, a condition that is commonly encountered in alkali soils during reclamation. There is a continuous decrease in soil pH resulting in improved soil physical properties with time. Within 2-3 years, one can switchover from grasses to rice-whet rotation. Besides reclamation, cultivation of grasses provides forage, which can be fed to the animals.

The second approach comprises of growing CSR 10 variety of rice during *kharif* and sugar beet during the *rabi* season (rice-sugar beet cropping sequence) for first three years. Within this period, the soil pH comes down from 10.0 to 9.3. In the fourth year and later any other high yielding variety of rice in *kharif* and wheat (KRL 1-4) or raya CS 52 can be grown successfully.

The third approach comprises of growing agro-forestry systems on alkali or even saline lands. Agroforestry systems have the potential to make use of marginal and degraded lands as well as initiate land reclamation through the soil improving effects of trees. Trees species *Prosopis juliflora*, *Acacia nilotica*, *Prosopis chinensis*, *Acacia auriculiformis*, *Azadirachta indica*, *Casurina equisetifoia*, *Dalbergia sissoo*, *Alianthus excelsa*, *Prosopis cineraria*, and *Acacia tortilis* are used for phytoremediation of the saline and alkali soils. The degree to which the salt affected soils are ameliorated by afforestation and agroforestry practices depends upon the selection of tree and grass species, nature of growth, quality and quantity of litter, ability of tree species to fix nitrogen, planting density, age of plantation and management practices. The *Prosopis*-based silvi-pastoral system have proved more effective in reducing soil pH, displacing Na^+ from the exchange complex, increasing organic carbon and available N, P and K (Singh *et al.*, 2014, Table 10.1). A combination of *Prosopis juliflora* and Karnal grass improves the soil condition to such an extent that after some years, less tolerant but more palatable fodder species such as *Berseem* (*Trifolium alexandricum*), *senji* (*Melilotus parviflora*) and *shaftal* (*Trifolium resupinatum*) can be grown under trees (Singh *et al.*, 1991; Singh, 1995). After the completion of one cycle of trees, one can even go for rice-wheat cropping.

Table 10.1. Effect of *P. juliflora* - *Leptochloa fusca* silvopastoral system on alkali soil properties after 6 years

Soil property	**Original (year 0)**	***P. juliflora* only**	***P. juliflora* + *L. fusca***
pH	10.3	9.3	8.9
EC (dS/m)	2.2	0.46	0.36
Organic carbon (%)	0.18	0.43	0.58
Available nitrogen (kg/ha)	79	133	165

Halophytes, plants that survive to reproduce in environments where the salt concentration is around 200 mM NaCl or more, constitute about 1% of the world's flora. Many of these plant species grow naturally on the coastal saline areas such as salt marshes and salt deserts. Under those extreme conditions of soil or water salinity where no crop of agricultural interest can be grown, it is possible to cultivate dedicated halophyte plantations for forage production, soil rehabilitation, bioenergy generation, landscaping, carbon sequestration and a number of other useful purposes. Among conventional crops, beetroot (*Beta vulgaris*) and date palm (*Phoenix dactylifera*) having halophytic ancestors are well known for their food value. Intertidal soils of India are dominated with *Salicornia brachiata, Aeluropus lagopoides*, and *Suaeda maritima* halophytes. *Salvadora persica*, a facultative halophyte appears to be a potentially valuable oilseed crop for saline and alkali soils, since the seed contains 40–45% of oil rich in industrially important lauric (C12) and myrestic (C14) acids. Halophytes also represent an important genetic reservoir that can be exploited for improving crops for marginal land.

Moderately Alkali Soils (pH 9.2 to 9.8)

Transplanted rice is tolerant to exchangeable sodium (ES), some of varieties being highly tolerant. The ability of rice to withstand, stagnating water on the field throughout the growing season is an added advantage, since the high pH of alkali soils is reduced under continuous flooding. Ponnamperuma and his colleagues observed that pH decreased from 8.8 to 7, twelve weeks after flooding. This was ascribed to evolution of large quantities of carbon dioxide from bacterial action and its accumulation because of restricted diffusion of gases in flooded soils (Ponnamperuma, 1965; Ponnamperuma *et al.*, 1966). The technology comprises of growing rice (variety CSR 10) in the *kharif* season as the first crop without application of soil amendment (Fig. 10.1). As a result, the soil pH will come down to a level so that wheat KRL 1–4 or Indian mustard varieties such as Pusa bold, Kranti, Varuna or CS 52 can be grown during the *rabi* season.

Fig. 10.1. Alkali tolerant variety of rice (CSR-10) released by CSSRI, Karnal

Bio-remediation

Microbes, the most abundant entity in the universe, are increasingly being used in various bio-remediation processes. Halo-tolerant or alkali tolerant microbes can be used to reclaim alkali soils as these are able to dissolve the sodium and other cation complexes by producing organic acids (Paikray and Singh, 2012). A microbial technology package for the reclamation of alkali soils is disclosed in patent number WO 2012046908A1. The microbial culture used is the mixed enriched culture isolated from alkali soil (pH>10) of Kakori block, Lucknow, Uttar Pradesh selected for its ability to grow and survive on high stress of sodium ion concentration and pH. The microbial composition comprises of bacteria strains such as *Pseudomonas fluorescens* - MTCC 5525, *Bacillus* sp. -MTCC 5536, *Micrococcus* sp. - MTCC 5534 bacterial strain, *Kocuria* sp. - MTCC 5535, *Kocuria carniphila* - MTCC 5537 and *Trichoderma herzianum* - MTCC 5530 fungal strain. The Microbial consortia of 10^9 colony forming unit (cfu) along with *jaggery*/molasses/sugar and Farm Yard

Manure (FYM)/ organic manure/agricultural waste are applied to the affected soil. The method however, needs testing under field conditions for its effectiveness to reclaim alkali lands.

Bio-inoculation of seeds of wheat with plant growth promoting halophilic bacteria resulted in increase of 18.1 and 24.2% in grain and straw yield, respectively under alkali conditions. Also, soil pH of alkali soil reduced from 9.4 to 8.6 when consortia of halophilic bacteria were applied. Soil biochemical properties improved and there was increase of microbial biomass C up to 137 μg/g as compared to 82 μg/g in control (Arora *et al.*, 2016). In greenhouse pot experiments, saline water irrigated wheat and maize plants exhibited better growth and yield when seeds were inoculated with halophilic strains. It was observed that there was 10-12% increase in yield attributes and yield of wheat at 6% NaCl as compared to 2% NaCl. Plants inoculated with a consortium of halophilic bacteria also showed growth at 10% NaCl, whereas inoculation with a single isolate did not promote plant growth at this salt concentration. The maximum fresh weight, dry weight, shoot length and root length of plants were noted in consortium 5% NaCl treated pots.

Many halophilic bacterial strains viz. *Bacillus subtilis, B. aerius, B. aquimaris, B. megaterium, Brevibacterium halotolerans, Virgibacillus salarius, Paecilomyces variotii, Pseudomonas indica, Pseudomonas stutzeri, Sinorhizobium fredii*, isolated from coastal saline soils are found to have plant growth promoting traits as well as nutrient mobilization to mitigate salt stress. Halophilic endophytic bacteria *Acinetobacter baumannii, Bacillus cereus, Bacillus firmus, Bacillus aerius, Pseudomonas fluorescens* and *Bacillus subtilis* isolated from leaves of halophytes were positive for ammonia production while phosphate solubilization was positive for *A. baumannii* and *P. fluorescens*. Halophilic strain CSSRO2 (*Planococcus maritimus*) was more efficient in reducing sodium concentration from 112,230 ppm in supernatant to 100,190 ppm at 24 hr while strain CSSRY1 (*Nesterenkonia alba*) reduced Na concentration to 92,730 ppm at 48 hr in halophilic broth with 15% NaCl. These halophilic strains also showed high potential for removal of sodium ions from soil. CSSRY1 efficiently removed sodium at higher (6%, 8%, 10% NaCl) salt concentration in comparison of CSSRO2 and association of both organisms (CSSRY1 and CSSRO2). This was also confirmed by reduction of electrical conductivity or total dissolved salts (TDS). It is hypothesized that once the sodium ion concentration is reduced in rhizosphere, plants are able to resume nutrient and water uptake. The plant growth promoting effects of these bacterial isolates in salt affected soils primarily appear to be due to higher biosynthesis of auxin and thus enhanced root growth. Another very likely mechanism is the alleviation of salinity stress via plant growth promoting rhizobacteria that express ACC deaminase activity. This enzyme removes ethylene from the rhizosphere. Also, the halophilic/halo-tolerant bacteria remove excess sodium from the surrounding soils and thus create favourable conditions for plant growth.

A growth promoting technology for alkali lands comprising of a microbial consortium of *Bacillus pumilus*, *Baciillus thurigensis* and *Trichoderma harzianum* in a dynamic media has been prepared by CSSRI, RRS, Lucknow. It has been patented and commercialized (Damodaran, 2014). The four types of bio-formulations namely Halo-Azo, Halo-PSB, Halo-Rhizo and Halo-Mix are effective in improving crop growth and yield as well as improve the soil fertility and health.

It is now emerging that phytoremediation or bio-remediation can be developed as an effective amelioration strategy for calcareous alkali and saline alkali soils having comparable performance to chemical amendments albeit the time scale that may be large, say 2-3 years before the land can be put to normal cropping.

ENHANCING PLANTS' ADAPTATION MECHANISMS

Under saline conditions, salt-sensitive glycophytes have to cope up with osmotic and ionic stresses. In the former rise in salt levels around the roots inhibits water uptake, cell expansion and lateral root development. In the latter case when ions such as Na and Cl accumulate in excess in plants, particularly in leaves, it leads to a decrease in the activity of primary metabolism, including photo-synthesis, with an increase in chlorosis and cell death. The ability of plants to tolerate salts is determined by the plant adaptation to abiotic salt stress that involves complex physiological traits, metabolic pathways, and molecular or gene networks. These facilitate retention and/or acquisition of water, protect chloroplast functions and maintain ion homeostasis. Physiological and molecular genetic studies have led to increasing knowledge of the protective mechanisms that plants use to cope with the detrimental effects of salinity. Osmoregulation and/or osmoprotection and ion homeostasis through compartmentalization are part of the main adaptive mechanisms found in halophytes subjected to salinity. Thus, exogenous application of some osmolytes (Proline, Glycinebetaine, mannitol, ascorbic acid or thiamine or sodium salicylate) has been used to perform osmoprotective roles under abiotic stress and has been suggested as an alternative approach to improve crop productivity under abiotic stress conditions (Ashraf and Foolad, 2007). An improvement in growth after applying exogenous osmolytes has been observed in tobacco, wheat, barley, rice, soybean and common beans (Borojevic *et al.*, 1980; Agboma *et al.*, 1997; Lutts, 2000). Different proteomics and transcriptomics studies have also indicated the major contribution of antioxidant defense related genes or proteins for salt tolerance. However, the results are not uniform in some cases so that sometimes some antioxidant enzymes were related, while others are not related to salt tolerance. There is need to carry forward these studies to arrive at doable technologies to impart salt tolerance and hasten the varietal development programs (Zhawar and Kaur, 2018).

ELECTRICAL METHOD

Investigations have shown that the treatment with electrical current may stimulate the reclamation of alkali soils, although it cannot replace conventional procedures involving controlled leaching. A remarkable result was obtained by the U.S. Bureau of Reclamation on a field trial in Montana where saline, boggy land that produced no crop for 19 years had been improved to such an extent that a heavy growth of grasses and sweet clover covered the area 50 days later. In this work, after the current had been applied for a short time, the sodium content of the drainage water from the area increased from 1500 ppm to 37,000 ppm. The expenses for electrical current amounted to 6 dollars per hectare (Kovda *et al.,* 1973). The penetration of the chemical amendments at high densities of the electrical current (0.1 and 0.5 mA/cm^2) and complete reclamation of the alkali soil was achieved, the treatments proving to be the most efficacious. The results of these investigations have been used in field experiments also.

Vadjunina and El-Sawaby El-koumi (1972) conducted a laboratory experiment on alkali soil monoliths (15 x 15 x 50 cm) to examine the effect of electrical reclamation as compared to gypsum and acid treatment. The treatments were:

(i) Control
(ii) Gypsum @ 30 t/ha
(iii) H_2SO_4 @ 10 t/ha
(iv) H_2SO_4 @ 20 t/ha
(v) Electrical current (E) (i) for 3000 hours — 0.01 mA/cm^2, 0.10 mA/cm^2
(vi) Electrical current (i) for 1000 hours — 0.50 mA/cm^2

The soils were leached after the above treatments. The changes in the soil pH, ESP and infiltration rate are shown in Table 10.2. The reclamative effect of electric current application at a density of 0.01 mA/cm^2 was found to be equal to the application of gypsum or H_2SO_4. The remarkable improvements in pH, ESP and the infiltration rate took place at 0-30 cm soil depth.

Table 10.2. The effect of electrical and other treatments to the soil on pH, ESP and infiltration rate

Treatment	pH	ESP	pH	ESP	Average infiltration rate (mm/d)
	0-10 cm		10-30 cm		
Original soil	10	72.8	10.09	89.9	
Control	9.74	49.2	9.95	82.6	4.4
Gypsum	8.90	19.2	9.72	62.2	13.0
H_2SO_4 (10 t/ha)	9.46	31.1	9.88	79.3	10.0
H_2SO_4 (20 t/ha)	7.56	2.9	9.11	39.4	13.0
E (0.01 mA/cm^2)	8.68	10.3	9.85	78.5	14.4
E (0-10 mA/cm^2)	8.20	1.7	8.26	1.2	57.0
E (0.5 mA/cm^2)	7.98	1.0	8.09	1.2	86.4

THERMAL METHOD

Investigations have shown that the heating of an alkali soil to a high temperature may bring in significant changes in soil properties. Gupta and Bandyopadhya (1972) achieved remarkable improvement in the alkali soil which was heated to a high temperature in the muffle furnace. On heating an alkali soil containing illite, possibly vermiculite, at 500ºC for 2 hours, the EC_e dropped from 13.4 to 4.8 dS/m, ESP reduced from 91.5 to 36.7, boron decreased from 9 to 1 ppm and hydraulic conductivity (K) increased from 0.2 to 1.0 cm/hr (Table 10.3).

Table 10.3. Effect of heating on properties of an alkali soil

Treat-ment	ECe (dS/m)	Soluble ions (me/l)					CEC (me/100g)	ESP	B (ppm)	K (cm/hr)
		Na	Ca + Mg	CO_3	HCO_3	Cl				
Control	13.4	108	1.5	54	25	49	8.3	92	9	0.2
75ºC for 24 hrs	12.0	102	1.5	44	25	45	8.3	91	-	-
100ºC for 24 hrs	11.5	97	1.5	35	19	45	8.3	92	-	-
200ºC for 6 hrs	10.4	88	1.5	16	9	45	8.3	91	-	-
400ºC for 3 hrs	5.8	52	1.5	2	4	35	7.6	68	-	-
500ºC for 2 hrs	4.8	30	1.5	0	3	35	6.8	37	1	1.0

The mechanism of the remarkable changes that take place due to heating is related with the clay mineral of the soil. Gupta and Bandyopadhya (1972) explained that on heating at high temperatures of 400-500ºC, part of the sodium ions in the exchange complex of the soil penetrates into the vacant octahedral positions in the dioctahedral lattice of the illite, possibly vermiculite clay mineral.

After entering the vacant octahedral positions, either the sodium exchange is rendered difficult or Na is fixed in a non-exchangeable form. However, the nature of the clay mineral did not undergo any change even at 500°C and illite, possibly vermiculite dominated. The collapse of the interlayer space of the crystal lattice, follows the loss of swelling potential of the soil and increase in hydraulic conductivity (K) from 0.2 to 1.0 cm/hr is therefore, quite obvious. Further, it seems that part of sodium carbonate is lost due to volatilization. Boron occurs in the soil in organic form as well as inorganic form. Since the organic matter is destroyed beyond 200°C, some decrease in boron is quite probable. Partly, boron may be reduced due to volatilization too.

Preliminary studies on germination and early seedling growth of wheat were further carried out in the saturation extract of original soil and that of one heated at 500°C for 2 hr. At the end of four days after the seed sowing, the germination percentage was recorded and after 12 days, the length of the seedlings measured (Table 10.4.). Figs. 10.2 and 10.3 show the growth of seedlings in the extracts of the two soils, and in the heated soil *in-situ* in a petri-dish, respectively.

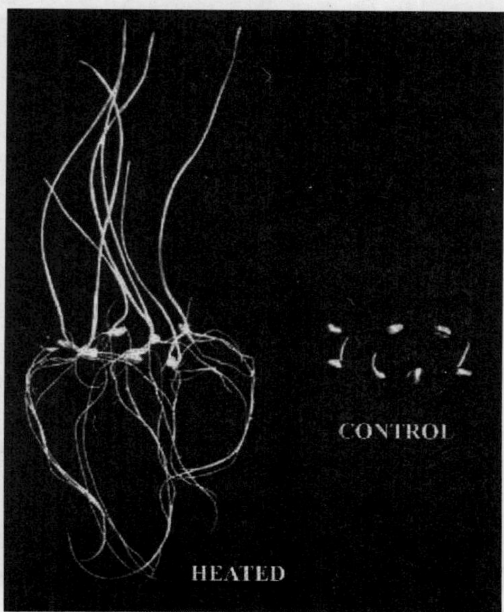

Fig. 10.2 Effect of heating in the reclamation of a salt affected soil

Table 10.4. Germination percentage and seedling growth of wheat

Soil	Germination percentage	Length of seedlings (cm)
Original soil	0.0	-
Heated soil	100.0	14.3

It was further examined whether the type of clay minerals influences the reduction of ESP. It was observed that the illite mineral is more amenable to changes in ESP on heating to a high temperature followed by bentonite. The kaolinite mineral is not affected at all (Table 10.5).

Fig. 10.3. Seedling growth of wheat in heated soil *in-situ*

Table 10.5. Effect of heating on ESP of different clay minerals

Clay mineral	ESP	
	Original	**After heating at 500°C for 2 hr**
Illite	63.0	24.5
Bentonite	80.1	63.8
Kaolinite	79.6	79.6

The experiments conducted on laboratory scale produced encouraging and promising results. It is suggested that if a slow moving down draft furnace with its flame directed down towards the ground on alkali lands is moved in the field, the soils could be reclaimed instantaneously. Such type of down draft furnaces are manufactured in Australia. One such machine was gifted to the Central Road Research Institute, CSIR, New Delhi, for heating the black cotton soils around Bhopal. The resulting hard substance, a kind of clinker, was used as a sub base material for making roads. The machine heats the soil to about 15 cm to a temperature of about 700°C. The results could be translated into the field for practical purposes by the use of this type of machine. Alternatively, the soils could be heated in the field *in-situ* by electricity which may prove to be prohibitive because of the cost involved. Harnessing of solar energy in the near future seems to be a more rational method to implement this principle in practice.

■■■■

CHAPTER 11

IMPROVED AGRONOMIC AND CULTURAL PRACTICES

INTRODUCTION

The term agronomy is derived from Greek words "Agro" meaning field and "Nomo" meaning to manage. Thus, agronomy is a branch of agricultural science which deals with principles and practices of soil, water and crop management. It essentially deals with all aspects of soil, crop and water management to increase productively of crops in relation to environment in which crops are cultivated. Cultural practices on the other hand are those measures that we take to ensure maintenance of the soil productivity and help sustaining the yield stability of crops. Growers design and implement culture practices which include land selection, tillage, crop and variety selection, crop rotation, fertilization, integrated pest management (insect, disease and weed control), transplanting, seedbed preparation, irrigation, mulching, windbreak management, etc. Proper planning and implementation is required to make all these cultural practices to work favourably. This chapter discusses some of the improved agronomic and cultural practices to sustain yield on saline/alkali soils.

CROPS AND CROPPING PATTERNS

Different crops respond differently when exposed to salinity or alkalinity stresses. While some crops are quite sensitive, others are tolerant at a given level of salinity or alkalinity, the latter showing little loss in crop yield as compared to the former. Therefore, the selection of the first crop is a subject of practical importance and depends upon the degree of soil salinity or alkalinity after the basic reclamation measures are implemented. Since the complete reclamation is not attainable from the very beginning, the crops which are tolerant to salinity or alkalinity conditions are preferred in the early years of reclamation. Later on, one can switchover to sensitive crops depending upon the degree of reclamation achieved at a particular time. Crops, such as rice, that contribute to the process of reclamation should be preferred in alkali soils. Basu and Tagore (1943) tried several crops on moderately alkali soils of the Deccan canal area and found that from amelioration point of view the order of crops is: *Berseem* > rice > *dhaincha* > cotton > *rabi* sorghum > tobacco > wheat. Rice crop also suits to saline soils where it should be included in the medium-term cropping pattern to leach down the salts.

The cultivation of different crops at different times rather than growing the same crop continuously on the same land is a very old practice. The yearly sequence of crops in India, the cropping pattern, follows *kharif* season from July to October and *rabi* season from October to March. The crops grown between March to June are called *zaid.* The crops are grown solo or mixed (mixed-cropping) or in a definite sequence (rotational cropping). Cropping systems has to be evolved based on climate, soil and water availability for efficient use of available natural resources. The objective is to ensure enough food for the family, fodder for cattle and to generate sufficient cash income to meet domestic and cultivation expenses. Health of the soil resource adds another dimension in the decision to adopt a given cropping pattern. The legume crops are added in the sequence to take advantage of N added to the soil by the bacteria. The fixed N can be utilized by non-legume crop during the following season. In soils having physical and chemical problems, crop rotations need to be decided on the basis of their tolerance to a given problem as well as their role in managing the given problems.

Salt Tolerance and Crop Selection

As discussed in the previous sections, crops differ in their tolerance to water logging, soil salinity and alkalinity. Adverse impacts of surface water stagnation and the need for agricultural land drainage therefore, to a great extent is influenced by the kind of crops grown. A rice crop can withstand surface stagnation much more than maize crop, later being sensitive to surface stagnation. Studies have shown that the yield reduction varies from 5.3 to 23.2% for each day of submergence in the case of wheat, pigeon pea, cowpea, pearl millet, maize and groundnut. It appears that submergence exceeding 1–2 days will result in more than 10% reduction in yield of dry food crops. Even the stage of growth at which stagnation occurs may influence the adverse impacts. As an example, for the maize crop, seedling stage is the most sensitive followed by the silking stage. Tolerance of crops to shallow groundwater is shown in Table 11.1 and the depth to water table at which optimum crop growth can be obtained is given in Table 11.2

Table 11.1. Crops tolerance to shallow groundwater table (Groundwater at 50 cm)

Tolerance level	Crops
High tolerance	Sugarcane, potatoes, rice, willow, plum, broad beans strawberries, some grasses
Medium tolerance	Sugar beet, wheat, oats, citrus, bananas, apple, barley, peas, cotton pears, blackberries, onion
Sensitive	Maize, tobacco, peaches, cherries, olives, peas, beans, date palm

Table 11.2. Optimum depth to water table for various crops

Crop	Depth to water table (m)
Rice	≤ 0.30
Sugarcane	0.60
Wheat	0.60
Cowpea	0.75
Pea	0.75
Barley	0.90
Gram	0.90
Safflower	1.00
Maize	1.20
Cotton	1.22
Pearl millet	1.25
Soybean	1.25

Crop tolerance to soil ESP reported in Table 11.3 shows that transplanted rice is the most tolerant while some of the pulses the most sensitive crops. Tolerance of crops to soil salinity (Table 11.4) shows that barley is the most tolerant while some of the pulses and vegetable crops the most sensitive crops.

Table 11.3. Tolerance of crops to soil ESP

Range of ESP	Crops
10 – 15	Safflower, mash, peas, lentil, pigeon pea, *urd* bean
15 –20	Bengal gram, soybean, maize
20 – 25	Groundnut, cowpea, onion, pearl millet
25 – 30	Linseed, garlic, guar, lemon grass, pamarosa, sugarcane, cotton
30 – 50	Wheat, *raya*, sorghum, sunflower, *berseem*, *shaftal*, *senji*, blue panic, gutton panic
50 – 60	Barley, *Sesbania*, para grass, Rhodes grass, matricaria
60 - 70	Rice, sugar beet, Karnal grass

Note: Relative crop yields are only 50% of the maximum in the indicated ESP range

The recommended cropping sequences for alkali soils are rice-wheat, rice-rice or rice-shafal (fodder). Results of various experiments that compared different cropping systems such as rice-wheat, rice-*berseem*, rice-mustard, sorghum-wheat, sorghum-mustard and sorghum-*berseem* at Karnal indicated that the rice based cropping patterns are more promising and profitable than sorghum based sequences. Dargan (1979) concluded that two rotations, viz. rice-wheat-*dhaincha* (green manure) and rice-*berseem* are more efficacious and remunerative in the early phase of reclamation in highly alkali soils (pH_2 > 10.5). Rice-wheat- *dhaincha* (green manure) is a better crop rotation, since reclamation process is slow under rice-wheat rotation alone. When these crops are cultivated successfully for about 3 years, the pH_2 of the surface soil (0-40 cm) comes down to 8.5 and then it is possible to grow even semi-tolerant crops like *raya* (*Brassica juncea*), pearl millet, sorghum, sugarcane etc. However, Singh *et al.* (1980) showed that *raya* can be raised successfully even as the first crop on highly alkali soils, if reclamation has to commence from the *rabi* season. Satyanarayana (1953) stated that the temporary substitution of a second crop of rice in place of pulses between January and June pushes down the salts enabling the cultivation of pulse crop in the second season. The two years crop rotation is rice-rice in the first year and rice-pulse in the second year.

Table 11.4. Crop groups based on response to salt stress

Sensitive Group		**Resistant Group**	
Highly sensitive	**Medium sensitive**	**Medium tolerant**	**Highly tolerant**
Lentil	Radish	Spinach	Barley
Mash	Cow pea	Sugarcane	Rice (transplanted)
Chickpea	Broad bean	Indian mustard	Cotton
Beans	Vetch	Rice (direct sowing)	Sugar beet
Peas	Cabbage	Wheat	Turnip
Carrot	Cauliflower	Pearl millet	Tobacco
Onion	Cucumber	Oats	Safflower
Lemon	Gourds	Alfalfa	*Taramira*
Orange	Tomato	Blue panic grass	Karnal grass
Grape	Sweet potato	Para grass	Date palm
Peach	Sorghum	Rhodes grass	*Ber*
Plum	Minor millets	Sudan grass	Mesquite
Pear	Maize	Guava	*Casuarina*
Apple	Clover, *berseem*	Pomegranate	*Tamarix*
		Acacia	*Salvadora*

In the mono-cropped coastal belt (saline soils) most area is under rice-fallow, except for some northern parts in Gujarat state. Rice crop is taken during the *kharif* season and the land remains fallow during the rest of the year. Recent work done at Canning (West Bengal) shows that barley, linseed, mustard, sunflower and sugar beet are some of the promising crops for the *rabi* season which do fairly well in the EC_e range of 5 to 10 dS/m. Chilli is also quite salt tolerant crop which performs well under limited irrigated water (Yadav *et al.*, 1979).

The cropping pattern for the reclamation and post reclamation phases is shown in Table 11.5 and 11.6 respectively. Crops other than rice in different parts of India can be selected on the basis of climatic conditions and eating habits in that particular region. Sharma *et al.* (1983) studied the effect of four cropping patterns in an alkali soil (pH_2 10.0). Grain yield of subsequent rice crop was higher after rice-Persian clover (*shaftal*) and rice-*berseem* rotations than after rice-wheat or rice-wheat- *dhaincha* (green manuring) rotations (Table 11.7). It was more profitable to raise Persian clover or *berseem* for fodder in winter than wheat on farmer's fields.

Table 11.5. Cropping patterns for reclamation phase of alkali soils

Season	Year of reclamation		
	First	**Second**	**Third**
Summer (April – May)	Leaching/ *Sesbania* for green manuring	*Sesbania* green manuring	*Sesbania* green manuring/ Fallow/ Cow pea (fodder)/ Maize (fodder), *Mung* bean
Kharif	Rice	Rice	Rice
Rabi	Rice/Wheat/ sugar beet	Wheat/ Barley/ *Raya/Shaftal Senji/* Sugar beet	Wheat/Barley/*Raya/ Berseem/Senji/*Sugar beet

Table 11.6. Cropping sequences for alkali soils (Post reclamation phase)

Soil pH_2 (0-15 cm layer)	Years after reclamation	Crops	
		Kharif	***Rabi***
9.2 to 9.3	3 to 4	Cotton, sorghum, pearl millet, rape seed	Mustard
8.9 to 9.0	5 to 8	Sugarcane, groundnut, *arhar*, soybean	*Berseem*, sunflower, safflower, gram, pea, lentil
8.5 to 8.8	After 8 to 10 years	Any kind of crops including vegetables, flowers etc.	

Table 11.7. Average yield (t/ha) of crops under different crop rotations on farmer's fields

Rotation	***Rabi* yield**		***Kharif* yield**		**Profit (Rs./ha)**
	1979-80	**1980-81**	**1980**	**1981**	
Rice-wheat	1.7	1.9	5.7	6.0	7,528
Rice-wheat- *dhaincha*	1.7	2.1	6.1	6.7	8,403
Rice-*berseem*	18.0	52.8	6.8	7.2	8,810
Rice-Persian clover (*shaftal*)	24.0	55.8	7.0	7.1	9,930
CD, P = 0.05	--	---	0.7	0.6	--

For saline soils the cropping pattern for north India are pearl millet-barley, pearl millet-wheat, pearl millet-mustard, sorghum-wheat or barley, sorghum-mustard, cluster bean-wheat or barley and cotton-wheat or barley. Pearl millet and sorghum based cropping sequences are more remunerative in saline soils. Cotton based cropping sequences are not beneficial since the yield of the winter crops that follow cotton are usually low. In areas with water scarcity, mustard may replace wheat in the cropping sequence since its water requirement is low compared to wheat. Irrespective of the cropping pattern, it must be ensured that at no time land is left fallow over a prolonged period. It will reverse the reclamation process and resalinize/resodify the soils.

It may be noted that all crops discussed in this section may not perform equally well under various degrees of alkali/salinity stresses (Chandra, 1979). In alkali soils with pH_2 up to 9.0, wheat performs better than barley, in the range 9.0-9.3, both are at par but in the higher range (9.3-9.5), barley performs better than wheat. In saline soils with EC_e in the range of 5 to 10 dS/m, wheat is better than barley whereas in a higher range (EC_e 10-15 dS/m), barley performs better than wheat (Chandra, 1980).

Besides the crops, even crop varieties differ in their tolerance to salts (Table 11.8). CSR 3 and Kharchia 65 are ranked as the most tolerant genotypes in rice and wheat crops, respectively. Although the salt tolerant varieties may be region specific, varieties developed at CSSRI, Karnal have wider application and yield higher than conventional high yielding varieties on stressed soils. Once the land is reclaimed, it is recommended that any high yielding variety can be used for cultivation.

Table 11.8. Salt tolerance of various varieties and cultivars

Crop	Varieties/cultivars	Adaptability	
		Alkali (pH_2)	Salinity (EC_e, dS/m)
Rice	CSR 10, CSR 11, CSR 12, CSR 13	9.8-10.2	6-11
	CSR 19, CSR 23, CSR 27, CSR 30, CSR 36, CSR 1, CSR 2, CSR 3, CSR 4	9.4-9.8	6-11
	CST-7-1, SR 26B, Sumati	9.4-9.8	4
Wheat	Kharhia 65, KRl 1-4, WH 157	< 9.3	6-9
	Raj. 3077, KRL 19, KRL 210 and KRL 213	<9.3	6-10
Barley	DL 200, Ratna, BH 97, DL 348	8.8-9.3	6-10
Indian mustard	Pusa bold, Varuna, Kranti, CS 52, CS 54, CS 56, CS 60, CSTR 330-1, CST 609-B 10	8.8-9.3	6-9
Gram	Karnal Chana 1	<9.0	<6.0
Sugar beet	Ramonskaaya 06, Maribo, Resistapoly	9.5-10.0	<6.5
Sugar-cane	Co 453, CO 1341, others (see Chapter 13)	<9.0	10

Crops not only differ in their tolerance to salts but also in their water requirement, optimum growing season, rooting depth, moisture extraction pattern and cultural requirements. Thus, in the absence of proper water and soil management practices, salinity of the soil may be affected differentially under various crop rotations. Cropping sequences such as rice, *berseem* and those requiring frequent irrigations reduce salinity/alkalinity effectively, where drainage is adequate. Therefore, knowledge of the expected salt balance of the root zone under various crop rotations will be extremely useful in planning the best cropping sequences during and after reclamation.

CULTURAL PRACTICES

In saline or alkali soils, poor germination and crop stand and tillering are few problems that can be counterbalanced by several improved cultural practices. Some of these practices are discussed in the following paragraphs.

Avoid Scrapping of the Surface Soil

Many farmers believe that saline soils can be reclaimed by scrapping top few cm of the soil surface where most soluble salts accumulate. However, it may be noted that it is not a viable cure. Besides high cost, the relief provided if any, is temporary and short-lived. In fact soil salinization of these lands is rather quick because now the water table is nearer to the soil surface than before scraping. It is recommended that this practice may not be adopted as a means of land reclamation.

Pre-Sowing Irrigation

Salts often accumulate in the top few cm of the soil during intervening periods between the two crops especially under shallow water table situations or in arid climates. Plants, on the other hand, are most susceptible to salinity at germination, becoming more salt-tolerant as they mature. Therefore, it is necessary to manage the salts at this stage so that seeds germinate fully leading to a better crop stand and yields. Commonly two approaches can be adopted:

A heavy pre-sowing irrigation may be given before seeding so that the accumulated salts are leached from the root zone. It should improve germination and early seedling growth. Wherever available, the pre-sowing irrigation should be given with the fresh quality canal/tube well water. If the groundwater is saline efforts should be made to sow the seeds in the conserved moisture rather than apply saline water at this stage.

Land Forming and Seeding Practices

Under limited water situations, practices that move the salts away from the germinating seed should be adopted. Some of the land forming and seeding techniques that help the seed to avoid points of salt accumulation are illustrated in Fig. 11.1 and Fig. 11.2. For example, in a broad bed furrow system, the salts accumulate at the center of the bed. Therefore, seeding near the edges of the bed allows the plants to grow in a relatively salt free environment.

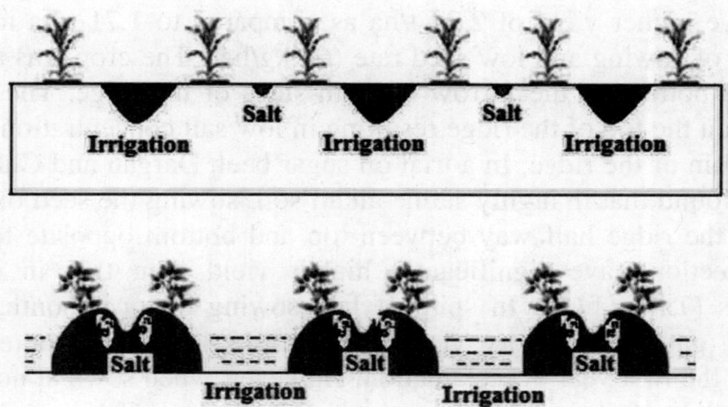

Fig. 11.1. Salt accumulation in double row planting (top) and furrow top with a notch (bottom) when each furrow is irrigated

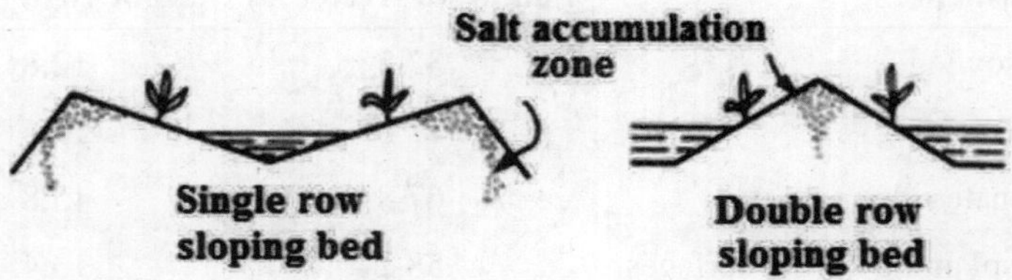

Fig. 11.2. Planting on the sides of slanted ridges to minimize salt stress on growing plants

In a furrow irrigation system, salts accumulate at the top of the bed while the furrow remains relatively salt free. Crops that are tolerant to short-term flooding can be cultivated in furrows to avoid excessive exposure to salts. Plantations in saline soils have been found to perform well when planted in furrows. In drip irrigation, salts move away and accumulate at the edges of the wetting front. Therefore, seeding near the emitters can help to avoid excessive salt exposure in drip irrigation especially when saline water is used for irrigation.

Nalamwar and Dastane (1966) reported that sowing of barley by ridge and furrow method using a higher seed rate (90 kg/ha) in a saline soil (EC_e 4.7-5.2 dS/m) with a shallow water table (69-110 cm) gave higher yield of 2.21 t/ha as compared to 1.71 t/ha in flat method of sowing and low seed rate (60 kg/ha). The crop was sown near the bottom of the furrow on both sides of the ridge. The salts moved on the top of the ridge resulting in low salt concentration near the bottom of the ridge. In a trial on sugar beet, Dargan and Chhillar (1973) found that in highly saline-alkali soil, sowing the seed on one side of the ridge half way between top and bottom opposite to the sun direction gave significantly higher yield than that in other methods (Table 11.9). In spite of late sowing by one month, this method, gave satisfactory yield of about 19.4 t/ha of marketable roots in the first year of reclamation. However, when sown at normal time and by following this method, Dargan and Chhillar (1975) obtained quite high yields of 44.5 t/ha in the first year of reclamation and 48.3 t/ha in the second year of reclamation.

Table 11.9. Effect of sowing method on sugar beet yield*

Treatment	Plants survived (%)	Root yield (t/ha)
Flat sowing	37.5	12.8
Ridge sowing	71.6	19.4
Alternate ridge sowing	67.8	15.8
Transplanting on flat surfaces	58.2	13.4
Transplanting on ridges	84.3	13.7

C.D. at 5%; *pH_2 = 10.5; EC_e = 6.36 dS/m

Dargan *et al.* (1973) found that the trench method of sowing in sugarcane, gave significantly higher yield than flat sowing (Table 11.10), in highly saline alkali soil having pH_2 10.6 and EC_e 10.5 dS/m. In trench sowing, ridges were made 65 cm apart and trenches were dug in furrows where setts were sown after applying gypsum, FYM and fertilizers. Higher yields in trenches were due to less concentration of salts which moved to the top of adjoining ridges.

Table 11.10. Yield of stripped cane (t/ha)

Amendments (t/ha)			Flat sowing	Trench sowing
Gypsum	+	FYM		
9	+	25	15.3	26.5
12	+	30	28.0	43.5
18	+	25	28.6	40.6
18	+	50	35.3	43.4

Puddling

Puddling before rice transplanting is a common cultural practice on normal soils. It minimizes deep percolation losses and weed infestation and creates conditions conducive to proper transplanting of rice seedlings. In alkali soils, seedlings (variety Jhona 349) transplanted under unpuddled conditions gave significantly higher grain yield as against the puddled conditions (Fig. 11.3; Dargan, 1979). The beneficial effects are attributed to:

- Infiltration rate in the initial years of reclamation is low and some leaching is essentially required to leach down the salts
- Weed problem is not as acute as in normal soils
- The physical conditions remain superior in non-puddled than puddled fields

Beneficial effects of puddling are observed only after 3-4 years of cropping when the pH_2 of the soil comes down to about 8.5 (Singh, 1985, 1988).

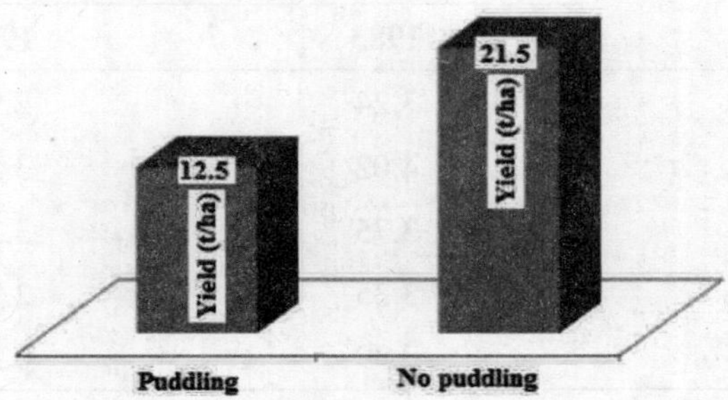

Fig. 11.3. Effect of puddling on rice yield in an alkali soil

Seed Rate and Plant Population

High mortality of seeds and seedlings in salt affected soils results in poor crop stand. To compensate for poor germination to improve crop stand, seed rate should be increased by about 25% and in transplanted crops, closer plant to plant and row to row spacing should be adopted. Nalamwar and Dastane (1966) found that increased seed rate in barley increased the grain yield by 18.3% under saline and shallow water table conditions in a saline soil (EC_e 5 dS/m) (Table 11.11). Similarly, the rice variety, CSR1 transplanted at a spacing of 15 x 15cm resulted in higher yield (Singh, 1988) than other spacing (Table 11.12). It is therefore, recommended that rice should be transplanted at a row to row and plant to plant spacing of 15 cm.

Table 11.11. Seed rate and grain yield of barley (t/ha)

Seed rate (kg/ha)	Grain yield	
	1964	1965
60	2.0.6	1.72
75	2.32	1.95
90	2.32	2.14
CD at 5%	0.058	0.083

Table 11.12. Effect of spacing on the yield of rice (CSR 1) in a alkali soil (Initial pH_2 9.8)

Spacing (cm)	Grain yield (t/ha)	
	1983	1984
15 x 10	3.24	2.97
15 x 15	4.02	3.50
20 x 10	3.75	3.36
20 x 15	3.35	2.97
CD at 0 = 0.05	0.39	0.35

Dargan *et al.* (1974) reported that the grain yield of rice increased significantly in a highly alkali soils with the increase in number of plants per hill (Fig. 11.4). In the initial years of reclamation at least 4 to 6 seedlings per hill should be transplanted. After taking 2-3 rice crops, the number of seedlings may be reduced to 2-3 per hill. In coastal saline soils also, 4-5 seedlings or even more per hill with a spacing combination of 15 x 15 cm proved beneficial in terms of crop yield. Number of seedlings and the spacing between the plants also depends upon the time of transplanting, more seedlings and lesser space being useful in delayed transplanting (Bhattacharyya, 1977).

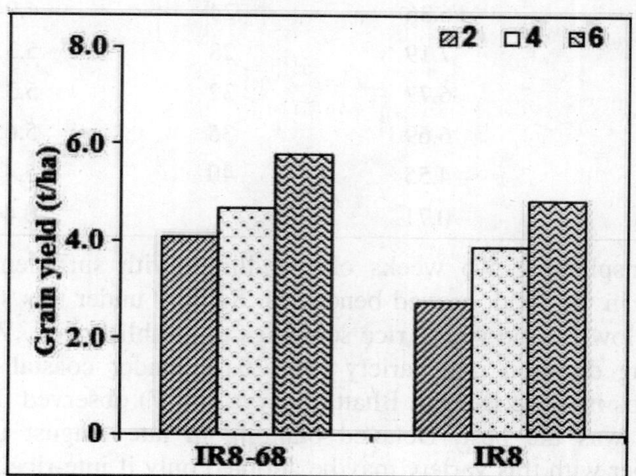

Fig. 11.4. Grain yield of two rice varieties as affected by number of seedlings per hill

Age of Seedlings and Date of Transplanting

The young seedlings of rice are relatively more sensitive to alkali conditions; somewhat older seedlings are therefore, preferred for transplanting. Seedlings 35 to 40 days old performed better in alkali soils instead of the usually recommended 30 days old seedlings for normal soils. Still better results are obtained when rice seedlings are grown in normal soils as it provides healthy seedlings for transplanting. Similarly, to establish tree species, planting older seedlings grown in normal soil is beneficial. In a trial on age of seedlings and date of transplanting of rice crop, Dargan and Gaul (1974b) found that older seedlings of 5-6 weeks age gave higher yield than younger seedlings. Transplanting done on 30th June in case of variety IR 8-68 gave significantly higher yield than on 12th and 18th July (Table 11.13).

Table 11.13. Effect of date of transplanting and age of seedlings on grain yield of rice crop

Date of transplanting	Yield (t/ha)	Age of seedlings (days)	Yield (t/ha)
24 June	6.36	24	4.94
30 June	7.19	28	5.33
6 July	6.77	32	5.31
12 July	6.69	36	5.65
18 July	4.55	40	5.47
CD at 5%	0.71		0.44

Transplanting 5-6 weeks old seedlings with sufficient water standing in the field, proved beneficial. Salinity under this situation remains low for the young rice seedlings to establish well. Working with long duration rice variety (SR 26 B) under coastal salinity conditions in West Bengal, Bhattacharyya (1977) observed that July planting was the best. Delayed planting in late August or early September with this variety may be adopted only if jute-rice rotation is followed.

Hoeing

The alkali soils are highly dispersible and prone to spontaneous slaking and crust formation after irrigation or rain. In the initial years of reclamation, sealing at the soil surface occurs following irrigation to wheat. The hard layer so formed restricts the entry of water and disrupts the normal exchange of gases between the soil atmosphere around the root zone and free atmosphere. Moreover, upon drying, this surface seal becomes very hard and puts lot of pressure on the young seedling by wrapping their delicate stems. In a field study, hoeing after first and second irrigation significantly increased plant height, shoot and root dry weight, tillers, straw and grain yield of wheat (Table 11.14). One and two hoeing resulted in 17.4 and 28.3% increase in grain yield of wheat, respectively (Sharma *et al.*, 1981). Therefore, in the initial years of reclamation, the hard surface layer, formed after first and second irrigation should be broken by hoeing.

Time of First Irrigation to Wheat

Compared to 21 days in a normal soil, the first irrigation to wheat in alkali soils should be given at 30 days after sowing. The maximum yield of wheat was recorded when 5 irrigations were given at different crop growth stages or when irrigation was given at 1.0 IW/CPE ratio (Ratio of irrigation water to cumulative pan evaporation).

Table 11.14. Effect of hoeing on yield and yield components of wheat

Number of hoeing	Plant height (cm)	Shoot dry weight (g/m row)	Root dry weight (mg/ core)	Number of tillers/m row	Yield (t/ha)	
					Grain	Straw
0	83.3	125.0	385	88	4.44	3.10
1	90.3	150.2	500	102	5.17	3.64
2	93.0	165.5	675	111	5.85	3.98
CD (5%)	2.5	13.7	--	5	0.56	0.42

Seeds/Seedlings Treatment

One of the main causes of the poor germination or crop stand in alkali soils is the disruption in the permeability of roots due to which it fails to maintain balanced ionic absorption. The sodium ions (predominant in such soils) get free access inside the cell due to the deranged membrane permeability and become toxic to plants. The calcium ions on the other hand, are reported to maintain the cell membrane integrity viz-a.- viz permeability and act as antagonist to the toxic sodium ions in plant tissues and growing media, are not absorbed in that proportion (Mengel and Kirkby, 1979; Clarkson and Hanson, 1980). The plants which are able to maintain chemo-osmotic potential in their cells under salt stress are able to withstand salinity and alkali stresses. The chemical, physical and agro-physical treatments of seeds and seedlings have been reported to induce salt tolerance (Henckel, 1954; Strognonov, 1964). Priming, a simple, low-cost, easily performable task, can make a good impression on farmers livelihoods enhancing the crop emergence rate, improving stand establishment, decreasing the total crop duration and getting higher yields. Unfortunately, consistent results on this score have been lacking.

Chaudhuri and Wiebe (1968) reported that the pre treatment with calcium salts increased the salt tolerance of wheat in saline conditions. Shannon and Francois (1977) did not find any evidence of increasing salt tolerance at germination in cotton. The germination was hastened by one day only by the pre-treatment of cotton seeds with salt solution as well as with distilled water. In a field experiment on saline soil (EC_e 12.6 dS/m) irrigated with saline water (EC 10.5 dS/m), the highest yield of wheat was obtained when seeds pre-soaked in 3% Na_2SO_4 solution were used (Puntamkar *et al.*, 1971). Manchanda and Bhandari (1976) observed that grain yield of wheat and barley was not significantly affected when their seeds were pre-soaked in salt solution. However, pre-soaking in the low salinity tube well water resulted in a higher grain yield than seeds pre-soaked in NaCl or Na_2SO_4 solution.

Sodium was found to accumulate progressively more while potassium and calcium decreased with the increase in sodicity in rice. Over-night treatment of 35-day old seedlings prior to their transplanting with calcium chloride and calcium nitrate showed favourable effect in overcoming the observed imbalances in ionic uptake (Qadar *et al.*, 1983). Qadar *et al.* (1980) reported that the rice seedlings, treated with 1% $CaCl_2$ solution for 16 h before transplanting, increased the grain yield of rice by 48% at soil pH_2 of 9.8 and 103% at soil pH_2 of 10.1 (Table 11.15).

Table 11.15. Effect of seedlings treatment on the grain yield of rice

Treatment	**Yield (kg/plot) at pH_2 values**		
	8.5	**9.8**	**10.1**
Control	11.05	6.39	2.70
$CaCl_2$	11.58	9.48	5.49

Seeds treated with 1-120 mM H_2O_2 for 8 h curtailed the mean germination time (MGT) in saline (150 mM NaCl) medium being less than the control (water). Level of H_2O_2 in seedlings arising from H_2O_2-treated seeds was markedly lower than salinized controls, suggested the operation of antioxidant system in them. These seedlings exhibited better photosynthetic capacity, particularly the stomatal conductance, thus improving the leaf gas exchange due to stomatal component of photosynthesis. Moreover, H_2O_2 treatment improved leaf water relations and maintained turgor (Wahid *et al.*, 2007). Iqbal and Ashraf (2006) concluded that seed priming with kinetin (Kin) consistently resulted in beneficial effects on growth and grain yield in two varieties of wheat. The mechanism by which Kin priming induced salt tolerance has not yet been established.

Seed priming with various organics and inorganics such as coconut water, fermented milk water, fermented rice water, Ascorbic acid, KCl, KNO_3, $ZnSO_4$, $CaCl_2$ and water (control) was used to test their impact on seed germination (Galahitigama and Wathugala, 2016). Under saline condition, highest seed germination in rice was observed in KNO_3 primed seeds (95%) and lowest was recorded in ascorbic acid treated seeds (82%). Other chemicals such as $CaCl_2$, $ZnSO_4$ and KCl also showed positive response for seed germination compared to control treatment. The plant growth under saline condition revealed that priming with 2% $CaCl_2$ enhanced the initial growth of rice seedlings and resulted in highest plant height, number of leaves, root volume and shoot dry weights. Priming other chemical solutions also gave positive effect on plant growth compared to control.

Other Practices

Micro-relief even of the order of few cm can result in increasing the salt content on the raised spots and excessive leaching at the low spots. Proper land shaping before cropping therefore, can help correct these elevation differences. Laser land levelling seems to be a better alternative for land levelling in land reclamation programs. On the contrary, land levelling resulting in the formation of shallow profiles or exposure of an impervious layer close to surface may enhance salinization.

Tillage is a mechanical operation, carried out to prepare seed beds and to enhance soil permeability. In alkali and saline soils where ESP or salinity decreases with depth, inversion tillage may be useful as it results in physical reclamation of salt affected soils. On the contrary, if this operation is improperly executed, it might form a plough layer or turn a highly salty soil horizon (inversion) bringing it at the soil surface. Proper monitoring of soil salinity or ESP profile is needed to avoid such eventualities.

Since most saline/alkali soils have sufficient amount of salts in the profile, leaving the land fallow for a long time results in the upward movement of salts. Once the reclamation process begins, the land should be cropped continuously without long gaps between the crops. Otherwise it may result in resodification/resalinization of the reclaimed lands.

The role of micro-organisms in plant growth promotion, nutrient management and disease control is well known and well established. These beneficial micro-organisms colonize the rhizosphere/endorhizosphere of plants and promote growth of the plants through various direct and indirect mechanisms (Nia *et al.*, 2012; Ramadoss *et al.*, 2013). Thus, microbes can serve as abiotic stress alleviation tool in crops grown in saline/alkali soils. Application of phospogypsum @ 25% GR in combinations with vermicompost, mulch and CSR-B-1 strain of *Bacillus* sp. isolated from alkali soils of Rae Bareli district helped cultivation of banana with higher yields in alkali soils of Uttar Pradesh. The bio-stimulant acts as a nutrient mobilizer, soil vitalizer, plant protectant against soil borne diseases and growth enhancer for crops grown in normal and salt affected soils. It prevents environmental pollution caused due to indiscriminate use of fertilizers and pesticides and creates profitable livelihood security to poor farmers owning alkali and marginal lands. The technology now known as CSR-BIO has been commercialized and is being refined over the years.

NUTRIENT MANAGEMENT

Since both salinity/alkalinity and fertility status are known to limit crop yields, it becomes necessary to identify the most limiting factor and take corrective actions accordingly. At a given level of salinity or alkalinity, plant growth and yield can be enhanced by proper plant nutrition, yield loss being more under poor than under adequate nutrition. On the contrary, if the fertility is adequate and salinity is the limiting factor, increasing the fertility status may not result in further increase in the yield (Bernstein *et al.,* 1974). Following are some of the factors that need to be considered while taking any decision on plant nutrition:

- The nutrient or nutrients needed
- Amount of the nutrients to be applied
- Chemical form of the nutrients
- Placement of the nutrients in the soil
- Timing of application

The analysis of barren saline or alkali soils for macro or micro-nutrients may not be helpful in deciding fertilizer requirements. Initially, these soils may contain enough nutrients but deficiency of all or some of the nutrients may occur after the application of amendments and/or leaching. Therefore, soil samples should be drawn and analyzed for the available nutrients prior to cropping.

Alkali Soils

Nitrogen

High pH, high exchangeable sodium, high $CaCO_3$, negligible to low organic matter content and inhospitable soil physical environment of alkali soils affect the transformation and availability of native and applied fertilizer nutrients. Alkali soils, in general, are deficient in Ca, N and Zn, but are rich in available P and K. The grain yield data of experiments conducted for two years on rice and wheat crops in alkali soil indicate that during the initial stages of reclamation, ammonium sulphate is the best source of nitrogen both for rice and wheat crops, next in the order of merit being urea in the rice crop and calcium ammonium nitrate in the wheat crop. Notwithstanding, the relative merits of ammonium based fertilizers in alkali soils, the cost differential makes urea the most used fertilizer by the farmers. The optimum dose was found to be 150 kg N/ha for both crops, which is about 25% more than the recommended dose for normal soils (Dargan and Gaul, 1974a; Table 11.16).

Nitrogen is applied in 3 splits to both the crops (rice-wheat). For rice, 1/2 of the N should be applied as basal while 1/4 each at tillering and panical initiation stages. Besides basal dose of nitrogen should be applied before puddling under pre-submergence conditions as it will reduce the ammonia volatilization losses and enhance nitrogen use efficiency (Swarup and Yaduvanshi, 2004). In wheat 1/2 of the N should be applied as basal while 1/4 each after 21 and 45 days of crop growth (Yaduvanshi and Dey, 2009).

Table 11.16. Effect of doses and sources of nitrogen on crop yield (t/ha)

Treatments	**Rice**		**Wheat**	
	1971	**1972**	**1971-72**	**1972-73**
Dose of N (kg/ha)				
60	3.23	33.2	14.7	19.6
90	36.5	39.3	17.9	30.8
120	44.4	45.8	21.4	34.7
150	50.7	55.2	26.5	40.6
180	49.9	60.0	26.1	41.0
C.D. at 5%	5.5	5.3	4.0	2.6
Source of N				
Ammonium sulphate	51.9	51.2	24.1	36.0
Urea	41.6	44.4	18.3	30.4
CAN	34.7	44.4	21.5	33.6
C.D. at 5%	4.1	3.9	1.7	2.0
Soil pH_2	10.1	10.0	9.4	8.9

Phosphorus

Researches conducted at CSSRI revealed that there is no response to added phosphorus in alkali soils during early years of reclamation. The rice crop responds to phosphorus application only when the Olsen's P in 0-15 cm layer drops below 7.5 – 8.0 kg P/ha nearly five year after cropping. Wheat being a relatively deep rooted crop, it is able to forage P from the lower layers to meet its P requirement. As such, it does not respond to applied P for another 3 to 5 years. Long-term field studies on a gypsum amended alkali soil (pH_2 9.2, ESP 32) with rice, wheat and pearl millet cropping sequence and NPK fertilizer use for 20 years (1974-75 to 1993-94) revealed that phosphorus applied @ 22 kg P/ha to either or both rice and wheat crops in rotation significantly enhanced the grain yield of rice after 1978, when Olsen's extractable P in 0-15 cm soil depth had come down to 12.7 kg/ha, which is very close to widely used critical soil test value of 11.2 kg P/ha. Wheat responded to applied P when available P came down close to 8.7 kg P/ha in 0-15 cm soil depth and nearly close to critical level (11.2 kg P/ha) in the lower depths (15-30 cm).However, few other studies indicate that alkali soils are not always high in available phosphorus and significant increase in yields of some crops is obtained with application of P fertilizer. It is attributed to decrease in Olsen's extractable P of surface soil due to its movement to lower sub-soil layers, uptake by the crop and increased immobilization when these soils are reclaimed by using amendments and growing rice under submerged conditions.

Potassium

Alkali soils of Indo-Gangetic plains generally contain very high amounts of available K. Application of K fertilizer to either or both the crops had no effect on yields of rice and wheat. Lack of crop response to applied K in these soils is attributed to:

- High available K status due to presence of K bearing minerals and
- Large contribution of non-exchangeable K (97%) towards total K uptake by plants.

Crop response to applied K fertilizers is not observed, possibly because of the presence of micaceous minerals and illite, which are capable of releasing sufficient K to meet the crop needs. Studies conducted so far suggest that application of K fertilizer to rice-wheat system can be avoided without any adverse effect on crop productivity and/or K fertility status of the soil. The contribution of the non-exchangeable K towards total potassium removal was about 94.9% in the absence of applied K, which decreased to 69.9% with application of

K. The decrease was about 50.7% when application of K was combined with organic manures.

In an experiment having rice-wheat cropping sequence, eight treatment combinations of N, P and K were tried since the *kharif* of 1974. In some treatments N, N + P and N+P+K were given to both the crops and in some treatments, application of P and P+ K were missed in either rice or wheat or in both crops and were compared with absolute control. In both crops taken in sequence for first three years, effect of nitrogen applied at 100 kg N/ha has been found to be highly significant and with missing of either P or both P+K (50 kg P_2O_5/ha and 50 K_2O/ha) there was no reduction in yield as compared to nitrogen alone. But in subsequent years, effect of P was found to be significant in case of rice crop only. These results suggest that indiscriminate use of P and K fertilizers should be avoided to reduce cost of cultivation on these soils (Table 11.17).

Table 11.17. Effect of missing application of P or both P + K on grain yield (t/ha)*

Fertilizers		Rice 1974	Wheat 74-75	Rice 1975	Wheat 75-76	Rice 1976	Wheat 76-77	Rice 1977	Wheat 77-78	Rice 1978	Wheat 78-79	Rice 1979
Rice	Wheat											
Control	Control	3.81	0.84	3.29	1.20	4.11	1.39	3.70	0.70	3.28	5.00	3.25
N	N	6.64	4.11	5.98	3.80	8.15	4.88	7.81	2.48	7.47	3.20	7.45
NP	NP	6.56	3.71	6.16	3.71	8.34	4.77	8.15	2.28	7.25	3.07	8.25
NP	N	6.63	4.15	5.77	3.78	7.79	4.56	8.06	2.62	7.61	3.11	8.70
N	NP	7.17	3.90	5.79	3.82	8.50	4.88	7.98	2.25	7.64	3.10	7.70
NPK	NPK	7.08	4.05	5.78	3.71	8.05	4.72	8.06	2.35	7.96	3.29	8.30
NPK	N	6.45	4.02	5.78	3.75	8.72	4.36	7.90	2.53	7.41	3.29	8.20
N	NPK	6.85	4.14	5.78	3.66	8.04	4.49	8.19	2.39	7.39	3.07	7.55
CD at 5%		0.96	0.83	0.98	0.33	0.13	0.59	0.68	0.38	0.89	0.37	1.02

C = Control; * Nutrients kg/ha = N = 100; P_2O_5 - 50, K_2O - 50, Soil pH_2, before rice 1974 = 9.2; after wheat 1979-87. Source : Dargan and Chhillar (1979).

Saline Soils

For saline soils, a general recommendation has emerged that 150 kg N, 60 kg P_2O_5 and 50 kg K_2O per ha should be applied for wheat. For other crops it may be reduced to 120 kg N, 40 kg P_2O_5 and 40 kg K_2O (Singh, 2009). Chloride dominated salinity responds to higher applications of phosphorus. However, these are not firm values as the actual values depend upon the crops and resources of the farmers. For example, in barley under coastal saline soils, application of 30 to 60 kg N/ha gave the highest yield (Biswas, 1976, Biswas, 1979, Bandyopadhya and Biswas, 1979). Nitrogen at the rate of 50 kg/ha in case of semi dwarf high yielding rice varieties gave the highest yield. About 66 to 75% of the total N is applied at tillering stage and rest at boot leaf stage in case of high yielding rice varieties. Nitrogen should be applied in the form of sulphur coated urea (a slow release N-fertilizer), that increases nitrogen use efficiency and gives higher grain yield (Biswas and Bandyopadhya, 1979).

Studies have also shown the potential of foliar application of nitrogen to rice and wheat crops. Dargan *et al.* (1973, 1975) found that foliar application of urea at 20 kg N/ha in two sprays, i.e., 10 kg N/ha at a time, was superior to even 60 kg N/ha through soil in case of rice crop while in wheat, foliar application at 20 kg N/ha was better than 40 kg N/ha given through soil. However, foliar application has not become popular for applying major nutrients. The technique has limited application for micro-nutrients mainly because of logistic problems.

Micro-nutrients

The macro-nutrients in alkali soils must be supplemented with micro-nutrients especially Zn and to some extent Fe and Mn (Swarup, 1976). Rice crop invariably suffers due to Zn deficiency, the affected plants show stunted growth, poor tillering, delayed maturity and low grain yield. Efficiency of applied Zn depends upon the degree of amelioration achieved. When the recommended dose of amendments is applied, 10 to 20 kg $ZnSO_4$ /ha is enough to get optimum yields of crops. At low level of gypsum application, the plants suffer more due to excess of Na and deficiency of Ca and are incapable of utilizing Zn resulting in poor yields (Singh *et al.*, 1987).

Since the total requirement of Zn by the plants is small, its continuous application results in build-up of DTPA extractable Zn in reclaimed alkali soils. Based on extensive field studies, it has been observed that once the Zn build-up exceeds 1 mg/kg DTPA extractable Zn, it is enough to apply Zn only to rice crop in rice-wheat cropping sequence to get the optimum yields. Soil application of zinc is better than other methods such as foliar spray, dipping of seedling roots in zinc oxide suspension and seed treatment etc. (Singh, 1985). Basal application of zinc mostly at the transplanting stage of rice and sowing of other crops, is more beneficial. In wheat, zinc may be applied up to 30 days of growth without substantial reduction in the yield but application of zinc after 50 days of sowing results in a significant reduction in the yield. It is possible to prevent the occurrence of Zn deficiency in rice grown on alkali soils by the application of FYM and/or *Sesbania* green manure (Swarup and Yaduvanshi, 2000). Organic amendments like press mud, poultry manure and farmyard manure can effectively supply zinc from the native and applied sources to rice crop in a saline alkali soil.

To compare the efficiency of soluble and sparingly water soluble Zn carriers at 11 and 22 kg Zn/ha, a field experiment was conducted on a Zn deficient alkali soil. Zinc application markedly increased the yield and Zn uptake by rice. The relative efficiency of Zn varied in the order; zinc sulphate > mixture of multi-micronutrient cations > zinc oxide = Zn-frits. The efficiency of zinc oxide and Zn-frits at 22 kg Zn/ha was equal to that of 11 kg Zn/ha from zinc sulphate (Nayyar and Takkar, 1980).

The application of zinc sulphate has been found to be a very important practice in alkali soils as is evident from the data in Table 11.18. Application of zinc sulphate at 45 kg/ha increased the grain yield of rice crop by about 2.0 t/ha, the mean increase of the six treatments being 1.33 t/ha compared to no zinc application. Similarly in the following *berseem* crop, application of 25 kg/ha zinc sulphate increased the *berseem* fodder yield by 11.5 t/ha (mean of 6 treatments) as compared to no zinc application. However, more recent field experiments, have suggested that application of 10 kg $ZnSO_4$ per ha to each crop of rice and wheat or 20 kg $ZnSO_4$ per ha to rice increased the grain yield significantly over no application of Zn (Singh and Chhabra, 1978).

Table 11.18. Effect of zinc, gypsum and FYM on the yield of rice and *berseem* fodder (t/ha)*

Treatments (t/ha)	*Berseem* 1970-71	Rice 1971		*Berseem* 1972-73	
		No Zn	Zn	No Zn	Zn
Control	0.15	5.42	7.21	68.3	85.0
FYM 25t	0.83	6.61	7.84	79.6	93.0
FYM 50 t	1.74	7.68	8.45	86.1	101.7
Gypsum 11 t	9.49	6.69	8.86	94.3	103.6
Gypsum 11 t + FYM 25 t	29.48	7.74	9.22	93.5	100.6
Gypsum 11 t + FYM 50 t	31.89	8.37	8.89	105.5	112.5
Mean		7.08	8.41	87.9	99.4

*Soil pH_2 initial = 10.5; final = 8.4; Source : Dargan (1979)

INTEGRATED NUTRIENT MANAGEMENT

Disproportionate and inappropriate continuous application of chemical fertilizers has adversely impacted the soil fertility and resulted in nutrient hunger in soils. As a result crop yields have stabilized or even declined over the years. Integrated nutrient management comprising of a good blend of organic and inorganic fertilizers can help to sustain crop production in soils and especially in reclaimed lands. Amongst the organic input farmyard manure (FYM) and green manure (GM) are the most commonly used inputs. The nutrient contents present in these two organic inputs are shown in Table 11.19 and Table 11.20.

Table 11.19. Average composition of FYM

Constituents	Minimum	Maximum	Average
Per cent dry matter			
Organic matter	8.54	58.60	31.67
Mineral matter	54.19	91.46	73.40
Nitrogen	0.35	1.88	0.93
Phosphorus (P_2O_5)	0.12	2.40	1.00
Potash (K_2O)	0.17	2.42	1.31
Lime (CaO)	1.71	11.02	5.74
Magnesia (MgO)	0.58	2.19	1.13

Constituents	Minimum	Maximum	Average
Organic carbon (C)	1.49	30.90	11.84
C-N ratio	4.5	18.30	9.50
pH value	7.20	7.90	7.36
Parts per million			
Copper (Cu)	1.89	8.06	4.63
Zinc (Zn)	2.41	10.81	5.26
Manganese (Mn)	47.90	100.70	83.65
Boron (B)	0.03	0.11	0.06

Source : Anonymous (1971)

Table 11.20. Nutrient composition of *dhaincha* crop

Basis	Percentage		
	N	P_2O_5	K_2O
When moist	0.7	0.1	0.6
When dry	2.8	0.4	1.4

The positive impacts of organic inputs are attributed to several changes in the chemical properties, immediate being addition of certain nutrients to the soil. Besides increasing the availability of plant nutrients, repeated and adequate applications of manure increase the humus supply of the soil, thereby augmenting the storage capacity of nitrogen. Evidences exist to show that application of farmyard manure and/or green manure help build-up nitrogen status of the soils and over the years increases crops yield. FYM also helps to meet a part of the micro-nutrient requirement of the crops (Table 11.21). Above all, application of organic inputs helps to promote the desirable physical and biological properties.

Immediate physical effects are less marked, although increases in aeration and the infiltration of water in alkali soils are observed quickly. Increased crop growth, resulting from manure applications, tends to promote aggregation and aids in retaining a good physical condition by protecting the soil against crust formation due to beating action of rain drops, particularly in alkali soils. The addition of manures also causes a rapid increase in the biological activity in the soil.

Table 11.21. Effect of micronutrients and farmyard manure on the grain yield of rice and wheat (t/ha)

Treatment (mg/kg)	FYM (t/ ha)	Rice		Wheat	
		1983	1984	1982-83	1983-84
Control	0	3.92	5.28	2.10	2.01
	10	4.18	5.43	2.25	2.20
Fe 20	0	3.95	5.30	1.95	1.98
	10	4.25	5.50	2.15	2.25
Mn 20	0	3.99	5.35	2.10	2.10
	10	4.27	5.58	2.35	2.28
Zn 5	0	4.15	5.48	2.25	2.22
	10	4.46	5.68	2.40	2.39
Fe + Mn + Zn	0	4.35	5.57	2.18	2.28
	10	4.66	6.07	2.35	2.42
CD (p = 0.05)					
Micro-nutrient		0.21	0.20	0.14	0.15
FYM		0.26	0.23	0.15	0.17

Source: Swarup (1991)

Since FYM is scarce, costly and less stable, green manuring is usually practised on saline or alkali soils. This primarily increases the nitrogen content of the soil and to some extent also the phosphorus availability. Application of green manure enhances organic matter (OM) content of the soil. The sap of *dhaincha* leaves has a pH value of 4.0, and being strongly acidic, it exerts a marked influence in neutralizing the high pH of the alkali soils. The extensive root system improves the soil permeability, liberates carbon dioxide and organic acids which help in dissolving calcium from the impure $CaCO_3$ present in the alkali soils in the form of *kankar*. In this process, the high ESP is also reduced. In addition, the *dhaincha* plant itself contains a very high quantity of calcium (34.2% CaO on ash basis). Apart from its alkali tolerance, this plant can also withstand moderate salinity and waterlogged conditions. It is grown during summer months (May-June), a lean period in the rice- wheat cropping sequence. Normally, 45-50 days old *Sesbania* crop, when it has attained a height of 1.5 to 1.8 m, is incorporated into the soil. Longer growth period renders the plant hardy to make its decomposition difficult. A fifty days old *Sesbania aculeata* is able to supply about 4.2 t/ha dry matter, 90 kg N, accumulated mostly through biological N fixation, 11 kg P and 90 kg K. A major advantage of integrated nutrient management is that it helps to reduce the volatilization losses in urea applied to rice crop in reclaimed alkali soils (Table 11.22).

Table 11.22. Ammonia losses from integrated nutrient management system in rice fields in a semi-reclaimed alkali soil

Treatment combination	**Urea N loss per split**			**Total urea N lost**	**% Urea N loss**	**pH**
	I[st]	**2**[nd]	**3**[rd]			
Control (No fertilizer application)	1.23	-	-	1.23	-	8.56
N_{120}	8.49	8.21	6.76	23.46	19.55	8.49
N_{120} P_{22}	8.28	7.35	6.70	22.33	18.61	8.48
N_{120} P_{22} K_{42}	8.14	7.24	6.65	21.75	18.13	8.45
N_{120} P_{22} K_{42} + GM	5.82	5.20	5.06	16.08	13.40	8.10
N_{120} P_{22} K_{42} + FYM	6.73	5.74	5.28	17.75	14.79	8.15
N_{180} P_{39} K_{63}	12.12	10.60	9.48	32.20	17.89	8.49
Mean	8.26	7.39	6.66			
CD (P = 0.05)						
Treatments	0.51	0.91	1.19			
Stage of urea application	0.32					

The overall beneficial effects of integrated nutrient management in terms of increased yield and/or fertility status of the soil are revealed in Table 11.23. The data show that yield level was higher when organic manure in the form of farmyard manure (FM) or green manure (GM) was applied along with chemical fertilizers at half the rate (N 60 P 13 and K 21 per ha), than the soil applied chemical fertilizers alone @ N120 P26 K42. Maximum yield of the crop was obtained when organic manure was applied with N120 P 26 and K42 through chemical fertilizers (Table 11.23). The yield was equal to the yield obtained with N180 P39 and K36 applied solely through chemical fertilizers (Table 11.23). Similar statement can be made regarding organic carbon and available phosphorus in the soil.

Table 11.23. Effect on yield and soil fertility status with integrated nutrient management in rice –wheat system

Treatment	**Yield (t/ha)**		**Organic carbon (g/kg)**	**Available nutrients (g/kg)**		
	Rice	**Wheat**		**Phosphorus**	**Potash**	**Zinc**
Control	3.25	1.36	3.0	9.8	205	0.75
N_{120}	4.67	3.25	3.1	8.2	198	0.74
N_{120} P_{26}	5.46	4.99	3.1	13.6	210	0.75
N_{120} P_{26} K_{42}	5.54	5.07	3.2	14.0	240	0.76
N_{60} P_{13} K_{21}	4.70	3.37	3.0	11.2	220	0.74
N_{60} P_{13} K_{21} + GM	5.56	3.62	3.2	15.8	250	0.85
N_{120} P_{26} K_{42} + GM	6.52	5.29	3.4	18.5	258	0.86
N_{60} P13 K_{21}+ FM	5.09	3.63	3.5	16.0	246	0.87
N_{120} P_{26} K_{42} + FM	5.94	5.13	3.6	19.0	252	0.89
N_{180} P_{39} P_{63}	6.01	4.96	3.2	19.8	250	0.74
CD	0.55	0.48	0.3	2.4	10	0.11

Source: Swarup (2000)

All studies have conclusively proved that application of organic and inorganic fertilizers in an integrated manner is a proven helpful strategy for sustainable and higher crops yield during reclamation of salt affected soils. Application of gypsum with organic amendments is more likely to be beneficial to improve the soil properties of salt affected soils and sustaining crop yields. Evidences have been generated to show that organic materials such as molasses, press mud and crop residues can also be used for integrated nutrient management.

MOLASSES AND PRESS MUD

Sugar industry waste products such as molasses and press mud are used to reclaim alkali soils and as a source of plant nutrients. Dhar (1935) was the first to use molasses. Later on Dhar and Mukherji (1936) analyzed and reported the composition of molasses (Table 11.24). With addition of molasses, CO_2 and some organic acids are produced because of the oxidation of carbohydrates especially at the time of early stages of decomposition. The acidity so produced neutralizes the high pH in alkali soils and reduces the high ESP upon release calcium through the dissolution of *kankar*. The press mud generated from industries that use sulphitation process to purify juice is most suited for use in alkali environments. Both molasses and press mud are a good source of organic matter and help to improve soil physical and biological properties. These organic inputs with appropriate fertilizer doses have given encouraging results in alkaline water irrigated lands as well.

Table 11.24. Composition of molasses

Constituents	Percentage
Carbohydrate	60-70
Potash (K_2O)	4-5
Lime (CaO)	2
P_2O_5	0.5
N	0.5
H_2SO_4	0.5
$Fe_2O_2 + Al_2O_2$	0.5
Water	20-30

CROP RESIDUES

Large quantities of crop residues from rice-wheat cropping system are available. It is assessed that on an average 1 ton of rice residue contains approximately, 6.1 kg N, 0.8 kg P and 11.4 kg K. For wheat, the values are approximately, 5.1 kg N, 1.2 kg P and 10.5 kg K. All the rice and a part of wheat residue is burnt causing immense loss of nutrients besides causing environmental pollution in the form of toxins and greenhouse gases. On the other hand, if the residues are returned to the soil, it will provide much needed organic matter, energy and nutrients to the soil. The results of a study show that incorporation of wheat residue 50 days prior to rice transplanting either alone or with green manuring or with sulphitation press mud (SPM) with recommended dose of NP improved rice yield as compared to recommended dose of N and P fertilizer alone (Yaduvanshi, 2017; Table 11.25).

Table 11.25. Effect of crop residue management on yield (mean of 3 years) and soil properties of semi-reclaimed alkali soil irrigated with poor quality water

Treatments	**Grain yield (t/ha)**		**pH**	**OC (%)**	**Available nutrients (kg/ha)**		
	Rice	**Wheat**			**N**	**P**	**K**
N_0 P_0	1.05	0.87	9.05	0.28	142	26.8	201
N_{90} $P_{19.5}$ (75 % NP)	2.69	2.31	8.98	0.26	143	25.4	198
N_{120} P_{26} (100 % NP)	3.49	2.94	8.97	0.27	144	27.9	193
100 % NP + Wheat residue burnt	3.72	2.99	8.78	0.32	158	29.3	220
100 % NP + Wheat residue incorporated	4.34	3.19	8.65	0.33	157	29.4	223
100 % NP + Wheat residue incorporated + GM	4.45	3.35	8.70	0.33	157	29.0	220
100 % NP + Wheat residue incorporated + SPM	4.41	3.34	8.79	0.34	156	29.2	219
CD (0.05)	0.58	0.45	0.10	0.01	5.65	1.69	9.40

■■■■

CHAPTER - 12

CEREAL AND OILSEED CROPS

INTRODUCTION

RICE

Rice (*Oryza sativa* L.) is a staple food for more than 50% of the global population. It meets 43% of calorie requirement of more than two third of the Indian population. Rice is grown in almost all the states of India. As per statistics of 2014-15, it occupies an area of 44.11 M ha up from 30.81 M ha in 1950-51 The production is assessed at 105.48 million tons with a productivity level of about 2.39 t/ha, which is rather low compared to 6.71 t/ha of China, Vietnam (5.57 t/ha), Indonesia (5.15 t/ha), and Bangladesh (4.38 t/ha).

Climate

The climatic factors like temperature, day length and humidity largely affect rice production. The higher yields are obtained in warm temperate regions having prolonged sunshine, high humidity and assured water supply. Rice crop requires 20ºC to 35ºC temperature throughout its life period. In the northern region of India where winter temperature is low, rice cultivation is limited mostly to one crop during *kharif* whereas in the eastern region, mean temperature throughout the year is favourable for rice cultivation.

There are three main rice growing seasons in India, namely, summer, *kharif* and winter. The summer crop (*aus*) is usually sown by direct seeding in April and is harvested in July. *Kharif* (*aman*) crop is transplanted from the middle of June to the end of July in northern region and from July to mid-August in eastern region and is harvested from end of September to end of November. The winter (*boro*) crop is transplanted from end of December to end of January and is harvested in April. During summer and winter season, usually short duration varieties are sown whereas in *kharif* season all types of varieties having long, medium and short duration are sown.

Soils

Rice is grown under diverse soil and climatic conditions. The crop is adapted to all types of soils, but is particularly suited to the clay and clay loam soils. Rice is able to tolerate a wide range of soil reaction, but it does have a preference for acidic soils. Rice is an ideal crop to grow in alkali soils having pH_2 up to 9.6. On soils having higher pH_2 (>9.8) alkali tolerant varieties should be grown or some amendment should be applied to lower down the pH (Fig. 12.1). Semi-aquatic nature of the plant requires submerged moisture regime for optimum growth. Since submergence is known to lower soil pH, transplanted rice is known to be highly tolerant to alkali conditions Moreover, submergence is easy to practise in alkali soils due to their poor water transmission characteristics. The rice is known as a reclamative crop as it sets in natural reclamation process in calcareous alkali soils that has been explained in previous chapters. Rice can also be grown in saline soil (EC_e 5 dS/m) without any amendment but salt tolerant varieties perform better even in such soils as have EC_e up to 10.0 dS/m.

Fig. 12.1. Testing of salt tolerant rice varieties in Tamil Nadu (left) and CSR 30, a salt tolerant variety at CSSRI, Karnal (right)

Cropping Patterns

Some of the commonly adopted rice based cropping rotations in alkali soils are:

(i) Rice - Wheat - *dhaincha* green manure

(ii) Rice - Wheat

(iii) Rice - *Berseem/Shaftal*

After growing rice for about 3 years *raya* (*Brassica juncea*) can also be grown successfully in the *rabi* season in place of wheat and *berseem*.

In saline soils, rice can be grown in one of the following three rotations:

(i) Rice - Rice - Rice

(ii) Rice - Barley/Wheat - Mesta

(iii) Rice - Rice – *Dhaincha*

(iv) Rice- Mustard/*Raya*

Varieties

There are thousands of rice varieties available to suit location specific soils and agro-climatic conditions. Some of the varieties for coastal saline soils have been reported by Bhattacharyya (1976, 1979). All high yielding varieties can be grown successfully even during initial stages of reclamation. *Basmati* rice can also be grown under late transplanting conditions, especially from the second year of reclamation. Salt tolerant varieties developed by the CSSRI, Karnal can be successfully grown during initial years of reclamation (Table 12.1).

Seed Rate and Seed Treatment

To safeguard against high mortality and poor tillering in the initial years of reclamation, a seed rate of 55-60 kg/ha is required to raise the seedlings to transplant one hectare of alkali land as compared to 30-35 kg/ha for normal soil. Bold and good seeds of a suitable variety should be selected by soaking the seeds in 10% common salt solution. After removing the floating seeds, the selected seeds should be washed 2 to 3 times with normal water. Selected seeds should then be soaked in solution of Ceresan (0.1%) or in 60 liters of water containing 6 g streptocycline and 60 g of Ceresan wet for 8 to 10 hr before sowing, as prophylactic measure against any seed-borne infection.

Table 12.1. Salt tolerance threshold and yield potential of rice varieties developed by ICAR-CSSRI.

Variety	Salient features
CSR10	Its grain yield is about 3 t/ha in salt affected soils (SAS) (pH_2 up to 10.0 and EC_e 9-10 dS/m).
CSR13	It gives about 3 t/ha yield in saline (EC_e up to 9 dS/m) and alkali soils (pH_2 9.8-9.9).
CSR23	It yields fairly well (~4 t/ha) in SAS (pH_2 up to 9.8 and EC_e 9-10 dS/m).
CSR27	It has high yield potential (~4 t/ha) and relatively early maturity (125 days) under salt stress (pH_2 up to 9.8 and EC_e 9-10 dS/m).
CSR30 (Basmati)	Although it gives relatively low (2-2.5 t/ha) yield under high salinity (EC_e 6-7 dS/m) and alkalinity (pH_2 9.4-9.6) stresses, it is a fine grained *Basmati* variety and fetches premium price.
CSR36	It has high yield potential (~4 t/ha) in SAS (pH_2 up to 10 and EC_e up to 11 dS/m).
CSR 43	It has high yield potential (~3.5 t/ha) in SAS (pH_2 up to 10 and EC_e up to 10 dS/m).

Currently direct seeding of rice (DSR) and system of rice intensification (SRI) practices are also becoming popular. It can be adopted in semi-reclaimed lands after about 3 years of land reclamation. The seed rate in such cases will vary accordingly, being higher in direct and slightly lower in SRI cultivation. For direct seeding it is around 80-100 kg/ha (http://www.knowledgebank.irri.org/step-by-step- production/growth/ planting/direct-seeding). Fine grained and *Basmati* cultivars require much less seed rate. Comparative yields in DSR can be obtained by adopting various cultural practices viz., selection of suitable cultivars, proper sowing time, optimum seed rate, proper weed and water management.

Nursery Raising

The rice seedlings should be raised in a normal soil since rice seedlings are sensitive to alkali at 1-2 leaf stage. Select a fertile, well-drained upland piece in the field near the source of irrigation. For transplanting 1 ha area, about 1000 m^2 area is required for nursery preparation. Before preparing the field, 2.5 tons compost or farmyard manure should be applied. For raising a wet nursery, the field should be puddled in standing water with a puddler or a *desi* plough or a power tiller. In case of dry nursery, about 1 or 1.25 m wide raised beds should be left in between the beds. For raising nursery medium-duration and high yielding varieties, the seeds should be sown 35-50 days in advance of the date of transplanting, the normal dates of transplanting in Punjab and Haryana being 10-20 June. For late transplanting or for *Basmati* rice, it can be delayed accordingly. Before sowing, fertilizers at the rate of 1 kg N and 1 kg P_2O_5 per 100 m^2 should be broadcast and thoroughly mixed in the nursery beds. In case of wet nursery, the seed should be broadcast on the bed saturated with water. Keep the seed bed saturated for the first five days and then as the seedlings grow, gradually increase the level of water up to 5-6 cm. In dry nursery, the treated seeds should be sown in lines and covered with mixture of farmyard manure and soil. Another dose of nitrogen @ 1 kg/100 m^2 should be applied at 15 days after sowing. The nursery should be kept free from weeds and should be regularly irrigated.

Field Preparation

Reclamation of alkali soils starts from the top and slowly proceeds to deeper depth because gypsum application has been limited to the gypsum requirement of top 15 cm of the soil layer. Therefore, shallow ploughing of the land without inverting the soil is recommended. The land should not be tilled in too wet or too dry condition. At proper soil moisture content, 3 to 4 shallow cross ploughing should be done to pulverize the soil. After last ploughing, the field should be levelled with a *sohaga* (leveller). Puddling i.e. ploughing in standing water, an important operation in normal soil for transplanting of rice, is not recommended during the initial years of reclamation (2 to 3 years after gypsum application). The seedlings should be transplanted in the well prepared fields immediately after irrigation. Some time, due to some unavoidable reasons, the surface of field becomes hard before transplanting. Under such conditions, field should be ploughed in standing water to shallow depth with a harrow. This will loosen the soil to facilitate easy transplanting.

Age and Number of Seedlings

The transplanting should be done when the seedlings are 35 to 40 days old as compared to 25-30 days in case of normal soils. For better yield, more than 40 days old seedling should be used under delayed transplanting. About 4-5 seedlings per hill should be transplanted, at 15 cm row to row and plant to plant spacing. After 2-3 years, the spacing can be increased to that in normal soils and number of seedlings may also be reduced to 3-4 per hill.

Application of Fertilizers

During the initial years of reclamation, the alkali soils need higher dose of nitrogen, i.e. 150 kg N/ha for medium duration high yielding varieties and 120 kg N/ha for short duration high yielding varieties. Ammonium sulphate, during the early stage of reclamation, is better source of nitrogen but due to its non-availability and also higher cost, urea may be used profitably in alkali soils.

To reduce nitrogen losses through volatilization and denitrification, nitrogenous fertilizers should be applied in split doses. In rice, nitrogen should be applied in 3 equal splits viz. (1/2+1/4+1/4) or (1/3+1/3+1/3) at transplanting, at 3 weeks and remaining at 6 weeks after transplanting. Nitrogen efficiency can be further increased, if it is applied in 4 equal splits, i.e. one fourth at transplanting, and remaining in three equal splits at 3, 6 and 8 weeks after transplanting.

Application of 10 kg N/ha as a foliar spray (3.0% urea solution) and remaining as soil application has been found beneficial over the application of total dose of nitrogen as soil application. If rice crop is taken after *dhaincha* green manuring, the total dose of nitrogen can be reduced to about 70-75 kg N/ha as *dhaincha* green manuring enables as much rice yield as obtained with an application of 80 kg N/ha. There is no need to apply phosphorus and potassium in initial 4-5 years of reclamation. The P and K should be applied at the time of transplanting, if soil tests show deficiency of these elements.

In fertilizer trials at Canning (W.B.), nitrogen at the rate of 100 kg N/ha in case of semi dwarf high yielding rice varieties gave highest yield. Of the total dose of nitrogen, 66% to 75% is to be applied by tillering stage and rest at boot leaf stage in case of high yielding rice varieties. Application of potash and phosphrous generally gives no response. In case of long duration indica rice varieties, N at the rate of 40 kg/ha and P at the rate of 20 kg/ha gave the best response.

Application of Zinc

Application of zinc is essential in the alkali soils especially in the initial stages of reclamation when the pH is more than 9.5 (Fig. 12.2). Application of 40 kg/ha of zinc sulphate wholly at the time of transplanting is the optimum dose for high yielding rice varieties in the first year of reclamation. Later on dose may be decreased to 20-25 kg/ha of zinc sulphate/ha. If zinc sulphate is applied to both rice and wheat crops then 10 kg $ZnSO_4$/ha is sufficient. After about 3-4 years when the soil is improved, it should be applied on soil test basis. In coastal saline soil (EC_e 4.6 dS/m in the 0-15 cm depth), 10 kg $ZnSO_4$/ha has been found to be the optimum dose.

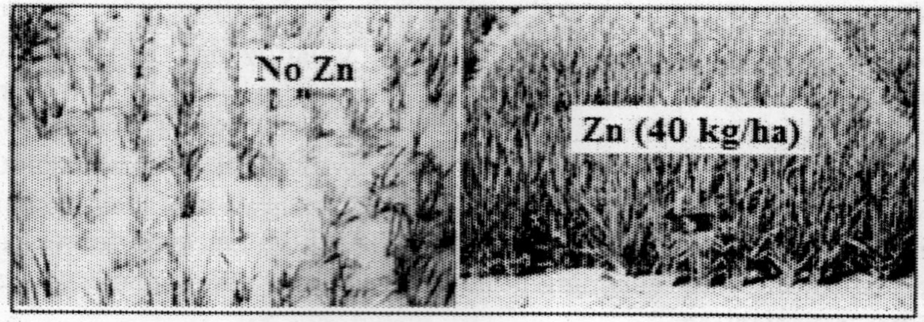

Fig. 12.2. Effect of Zn application on performance of rice crop in alkali soils

Water Management

High yielding dwarf rice varieties require only shallow submergence (5 ± 2 cm) as deeper submergence reduces the yield. The crop generally requires about 15 to 20 irrigations after transplanting. Water should be kept standing except for one or two days after about 8-10 days of transplanting. At this stage if water is present in the field, it should be drained out and irrigation should be withheld for about 2-3 days for better development of root system. Water should also be drained out prior to fertilizer application. Thereafter, about 5 to 7 cm submergence should be maintained. It has also been observed that during period of reduced water availability, ponding of irrigation water one day after disappearance of ponded water results in as much grain yield as in the continuous submergence. There are certain stages namely seedling establishment stage, vegetative or tillering stage, panicle initiation stage and milking and grain development stage, when water requirement is high. The crop should not experience water shortage at these stages. The water should be completely drained from the field 10-12 days before the harvest of crop.

The areas where alkali soils occur extensively, usually receive a rainfall of about 700-800 mm annually. Most of this rainfall occurs in rainy season (July to September). All the rainfall can be stored in the rice fields and the farmers may not worry much about the drainage except when the storms are quite heavy such as rainfall exceeding 15 cm received in one day. In such events excess water should be drained into a surface drainage system provided for this purpose. The storing of rainwater in the fields helps in reducing the irrigation water requirement of the crop (Gupta and Narayana, 1974; Gupta and Pandey, 1979).

In coastal saline soils, a continuous ponding of 2.5 cm water on soil surface during *kharif* season and continuous submergence of soil with a head of 10 cm water during *rabi* season, should be adopted for maximum grain yields of rice.

Inter-culture and Weed Control

Weeds are not a serious problem in alkali soils at least during the first 2-3 years of reclamation. If algal growth is noticed, rice weeder can be used to minimize it or the standing water can be replaced with fresh water. The weed problem after 2-3 years of cropping can be controlled by hand weeding. Grassy weeds can be controlled by applying Machette @ 30 kg granules/ha in 4-5 cm standing water, 3-4 days after transplanting or spray with 2 kg liquid active ingredient in 400 to 600 liter of water per ha. The water from the field should be drained out before the application of herbicide.

Plant Protection

The incidence of pest and diseases in rice grown in alkali soils is no different than normal soils. The methods to control insects, pests and disease are therefore the same. BHC 10% @ 20-25 kg/ha should be dusted against grass hopper, leaf-hoper, rice hispa, rice bug, rice thrips etc. Against stem borer, 0.1% Methylparathion (Metacid 50 EC) @ 750 l/ha should be sprayed 20-30 days after the first spray. Sometimes brown hopper causes considerable damage to the crop when the crop is near maturity. It could be prevented by broadcasting Furadone @ 8 to 10 kg/ha in standing water.

To control diseases like blast, bacterial blight, leaf spot, sheath blight etc. use disease free seed, treated with Agrosan G.N. or Cersean @ 2.5 g per kg of seed. For leaf spot, crop should be sprayed with any copper fungicide containing copper oxychloride using a concentration of 2 to 3 g/l of water. The spray may be repeated after every 7 to 10 days if the rainy and humid weather persists.

Yield

The rice yields are normally quite good in alkali soils. In the very first year of reclamation with improved agronomic and cultural practices, an average grain yield of 4.5-5.0 t/ha can be obtained in case of medium duration, high yielding varieties and 4.0-4.5 t/ha in case of short duration, high yielding varieties. The yields increase appreciably in the subsequent years. From second year onwards yields up to 6.0 t/ha and 5.0 t/ha can be obtained from medium and short duration varieties, respectively. Yields approach to yield of the normal soils within 2-3 years after reclamation both in alkali and saline soils.

WHEAT

Wheat (*Triticum aestivum* L.) is a major food crop and is cultivated extensively throughout the world (Fig. 12.3). Wheat in India is a *rabi* season cereal crop and is placed second to rice. It occupies an area of 31.47 M ha (2014-15) producing 86.52 million tons of grains with an average yield of 2.75 t/ha. Wheat requires a combination of factors including cool climate with moderate rainfall, flat and well drained plain areas, fertile friable loam and heavy inputs in the form of irrigation, HYV seeds, fertilizers and mechanization. Being semi-tolerant to salts, it can be cultivated in alkali and saline soils.

Fig. 12.3. Performance of high yielding wheat variety in a semi-reclaimed alkali land

Wheat is grown mainly in cropping sequences like Rice-Wheat, Sorghum-Wheat, Pearl millet-Wheat, Maize-Wheat, Pulse-Wheat, Cotton-Wheat, Soybean-Wheat etc. in different parts of the country under irrigated condition. Under rain fed condition, Fallow-Wheat is most common, but sometimes short duration pulse crops precede wheat.

Climatic Conditions

Wheat is grown in all types of climatic conditions i.e. tropical, sub-tropical and temperate. In India, major wheat area is under sub-tropical region. The cool and sunny winters are very conducive for growth of wheat crop since the crop needs cool climate during the early part and hot and dry weather at the time of maturity. Warm temperature during the early growth stage reduces tillering and promotes diseases. Too much of rain and cloudy weather results in heavy incidences of rusts. The sowing of wheat crop normally begins in the September-October in Karnataka, Maharashtra, Andhra Pradesh, Madhya Pradesh and West Bengal; October-November in Bihar, Uttar Pradesh, Punjab, Haryana and Rajasthan and Nov.-

December in Himachal Pradesh and Jammu & Kashmir. The crop takes about 180 days in the hills of northern India, 150-60 days in the North West, 120-140 days in Uttar Pradesh, Bihar and Western India and 100-120 days in West Bengal, for its maturity.

Soils

Wheat can be grown in a variety of soils except highly alkali, highly saline and waterlogged soils. Soils with sandy loam to clay loam texture and moderate water holding capacity are ideal for wheat cultivation. In alkali soils, wheat can be grown fairly if ESP is less than 30 or pH_2 is below 9.3. If the ESP is high, it is desirable to apply gypsum and grow rice as the first crop. Subsequently wheat can be grown without any additional application of gypsum. It can be cultivated in saline soils if EC_e is less than 5 dS/m although at this EC_e germination may be affected, crop being sensitive to salts at germination and initial establishment stages. Although, other crops can be grown following rice in the reclaimed land, farmers prefer to grow wheat.

Field Preparation

Wheat crop requires a well-pulverized seedbed having optimum soil moisture for good and uniform germination. After the harvest of rice crop, plough the field and irrigate. Once the field is in workable condition plough the field 2-3 times. Carry out the planking operation after every ploughing. If rice is harvested late, and there is sufficient moisture in the field, then make use of the residual moisture and sow the wheat by ploughing and planking 2 to 3 times. In this case there is no need to apply pre-sowing irrigation as it would delay the sowing operation. If the rice is harvested in the month of November and moisture is insufficient to sow wheat, pre-sowing irrigation should be applied without ploughing the field so that field comes in ploughable condition at an early date. Remember, working in too wet or too dry condition will deteriorate the physical properties of the soils and affect productivity.

Currently, in semi-reclaimed lands no till, minimum till, raised bed and furrow planting are also being adopted on large scale. Under these conditions, field preparation method will vary accordingly.

Varieties

There are hundreds of high-yielding short stature wheat varieties developed by the ICAR and State Agriculture Universities suitable for different zones, regions, states and sowing conditions. One can choose any one of them depending upon the salt status. Some of the varieties are: HD 2009, WH 157, HD 1982, HD 2329 for timely sown and HD 1553, HD 1981, WL 343 for late sown condit-ions. For Tamil Nadu some of the varieties are: HW 971, HUW 318, HW 1085 and HW 2044. In salt affected soils, Kharchia 65 is one of the tested salt tolerant varieties which can be grown in alkali as well as in saline soils. However, its yield potential is very low and susceptibility to rust is very high except in dry areas of Rajasthan. Some varieties developed by CSSRI, Karnal can be used under var-ious agro-climatic conditions (Table 12.2). Besides, for saline soils, WG 357, C-306 and HD 2009 are also suitable than other types.

Table 12.2. Salt tolerance and yield potential of wheat varieties developed by ICAR-CSSRI

Variety	Salient features
KRL 1-4	A selection from Kharchia-65, it gives 2.5-3.5 t/ha grain yield in alkali (pH_2 9.2-9.3) and saline (EC_e 6-7 dS/m) soils.
KRL 19	It produces 2.5-3.5 t/ha in alkali (pH_2 up to 9.3) and saline (EC_e up to 7 dS/m) soils.
KRL 210	Its yield potential is about 5.5 t/ha in normal soils and 3-5 t/ha in salt affected (pH_2 9.2-9.3 and EC_e 6-7 dS/m) soils.
KRL 213	It yields well (3.3 t/ha) under salt stress (pH_2 9.3-9.4 and EC_e 7-8 dS/m) and gives good performance in areas having alkaline and saline (EC_{iw} 15 dS/m; RSC 12-14 me/l) groundwater.

Seed Rate and Seed Treatment

The seed rate varies with variety as well as with time of sowing. High seed rate between 100 and 125 kg/ha should be used to have optimum plant population. In case of late sowing, and in salt affected soils it should be increased by 20-25%. To safeguard against the soil and seed borne diseases, the seed should be treated with Thiram or

Agrosan GN or Captain @ 2.5 g/kg seed. For the control of diseases like Bunts and Smuts, seed treatment is suggested with Vitavex, Bavistin, Thiram or Agrosan GN @ 2.5 gm per kg of seed.

To protect the crop from termites and white ants particularly in rainfed area, treat the seeds with Endosulphan 35EC or Chlorpyriphos 20 EC @ 700 ml per 100 kg of seed, mix the chemical in 5 liters of water and spray over the seed followed by seed drying overnight before sowing.

Time and Method of Sowing

The time of sowing is important to obtain good yield of wheat. It depends upon the variety as well as the local conditions. The optimum time for wheat sowing is the first fortnight of November. The average temperature at the time of sowing should be around 20-25°C. If sowing is delayed beyond 15th November, one can expect a gradual decline in the yield. Wheat sowing after middle of December results in very poor yield. If delay is beyond control, varieties HD 1553, HD 1981 and WL 343 should be preferred.

Sowing should always be done under optimum moisture condition, since even slightly less moisture affects the germination. The seed should be sown in lines 20 cm apart, at a depth of 5 cm with bullock or tractor driven seed drill. Seed-cum fertilizer drill is preferred to ensure uniform placement of seed and fertilizers at the optimum depth. If seed drill is not available, the seed can be sown by '*Kera*' or '*Pora*' method.

Fertilizer application

The fertilizer application should normally be on the basis of soil tests. If soil test reports are not available, fertilizers may be applied at the following rates:

Production condition	Fertilizer requirement (kg/ha)		
	Nitrogen (N)	**Phosphorous (P)**	**Potash (K)**
Irrigated timely sown	120-150	40-60	Based on soil test
Irrigated late sown	80-100	40-60	Based on soil test
Rain fed timely sown	60	20	Nil

The time of fertilizer application depends upon the nature of the fertilizer and the type of soil. The phosphatic and potassic fertilizers should be applied as basal while nitrogenous fertilizers as basal and top-dressing mode in split doses. In medium to heavy soils, application of nitrogen may be made in two splits i.e. half at the time of sowing and half after the first irrigation. In light soils, nitrogen is applied in three splits i.e. one-third at sowing time, one-third after first irrigation and remaining one-third after second irrigation. Sometimes splits are one-half at sowing time, one-fourth after first irrigation and remaining one-fourth after second irrigation. Nitrogen use efficiency in wheat grown in alkali soils can be further improved by applying 150 kg N in four equal splits that is ¼ basal + ¼ at first irrigation + ¼ at second irrigation and ¼ at the third irrigation.

In salt affected soils, 25% more nitrogen (150 kg/ha) is applied as compared to the dose recommended for normal soils (120 kg/ha). Urea and ammonium sulphate are equally efficient as source of nitrogen (Singh and Sharma, 1984). Nitrogen application should not be missed at the time of sowing and at panicle initiation stage.

Apply 10 kg $ZnSO_4$ /ha to the wheat crop at sowing time. Basal dose of nitrogen and full dose of zinc sulphate should be drilled after final preparatory tillage. Phosphorus and potassium do not respond for first 4-5 years in alkali soils. These should be applied only if the soil test shows their deficiency. For saline soils these should be applied either on soil test basis or as per normal soils.

Sulphur

The deficiency of sulphur is being reported from some wheat growing areas, particularly where Rice –Wheat cropping is being practised continuously. Recent studies at PDCSR, Modipuram indicated wide spread deficiency of sulphur in the soils of Rice-Wheat growing belt of western Uttar Pradesh. The sulphur deficiency is managed by application of fertilizers such as super phosphate or ammonium sulphate or even gypsum.

Irrigation

Alkali soils, being structurally unstable, are prone to spontaneous slaking and crust formation following an irrigation event. It not only impedes water intake but also affects the water storage capacity of the soils. Besides, effective water storage is mostly less in alkali soils owing to the restricted root depth, roots being concentrated only in the upper few centimeters of soils. Moreover, because of the low hydraulic conductivity, the movement of stored water to plant roots is very slow resulting in rapid drying of surface soil layer. For these reasons, wheat in alkali soils requires more frequent irrigations of smaller depths than crops grown in normal soils. To obtain good yield of wheat, 5-6 shallow irrigations would suffice. Heavy irrigation proves harmful to the crop because of water stagnation, particularly at the early stages of crop growth. If excess water stands on the land surface, it should be removed within a day to minimize adverse effect on crop yield (Table 12.3).

Table 12.3. Response of various *rabi* crops to short-term water stagnation in semi-reclaimed alkali lands

Duration of water stagnation (Days)	Yield (t/ha)		
	Mustard	Wheat	Barley
Drained	1.43 (100)*	4.01 (100)	3.65 (100)
1	1.31 (92)	3.70 (92)	3.52 (96)
2	1.20 (84)	3.32 (83)	3.39 (93)
4	1.12 (78)	2.93 (73)	3.18 (87)
6	1.02 (71)	2.43 (61)	2.75 (75)
CD (p=0.05)	0.22	0.28	0.28

* Values in parenthesis are the relative yield

The time of first irrigation to wheat grown in alkali soil is also different than normal soil. While first irrigation is normally given 21 days after sowing (DAS) in normal soils, irrigation at this time results in the yellowing of seedlings. It is attributed to the surface concentrated salts moving into the root zone. Therefore, the first irrigation to wheat sown in the first fortnight of November should be applied after 28-30 days of sowing. Under late sowing conditions, the first irrigation should be delayed by additional one week. Second irrigation should be applied after 20-25 days of the first irrigation.

Frequency of irrigation can be increased in the months of February and March. For optimum water availability, following irrigation schedule may be followed.

Four irrigations	At CRI*, tiller completion, boot and milk stages
Five irrigations	At CRI*, tiller completion, late jointing flowering and milk stages
Six irrigations	At CRI*, tiller completion, late jointing, flowering, milk and dough stages

* CRI = crown root initiation

Inter-culture Operations

As already discussed, a hard surface layer (crust) formed in these soils not only adversely affect the young seedling, it also restricts the normal exchange of gases from the root zone to the atmosphere. Therefore, two hoeings should be given one each after first and second irrigation. The hoeing helps in improved soil aeration and better root development.

Weed control

In alkali soils, not much weed problems is anticipated during the first two years of reclamation. But from third year onwards, the weed problem becomes as severe as in normal soils. Dominant weeds which infest wheat crop are: *Phalaris minor (Mandusi), Chenopodium album (Bathua), Melilotus alba* (Wild *Senji*) and *Polypogon monspliensis* (Rabbit foot grass). These weeds can be controlled by applying the weedicides such as: Dosanex, Arelon, Tolean, Graminon and Delron. Any one of these weedicides can be applied @ 1.0 kg a.i/ha after 4 to 5 days of first irrigation.

For an effective weed control following chemicals can be sprayed by making a solution in 400 to 600 liters of water/ha.

Narrow Leaved Weeds: Pendimethalin (Stomp) @ 1.0 kg a.i./ ha, 2 - 3 days after sowing (pre-emergence), followed by Isoproturon (Tolkan/Graminon/Arelon) @ 0.75 kg a.i./ ha, 30-35 days after sowing or Metaxuron (Dosanex/Hexamar/Hilnex) @ 1.5 kg a.i./ha

30-35 days after sowing or Leader @33 g/ha at 35 days after sowing or Topic @400 g/ha at 35 days after sowing.

Broad Leaved Weeds: 2,4-D @0.4kg a.i./ha at 30-35 days after sowing.

Mixed Narrow and Broad Leaved Weeds: Mixture of Isoproturon and 2,4-D at the recommended doses of each or Isoguard-plus @1.2 kg a.i./ha 30-35 days after sowing.

Termite control

Termites damage the crop soon after sowing and later at the time of maturity. For termites control, use Aldrin 5% dust or BHC 10% dust at the rate of 20-25 kg/ha and mix it well in the soil at the time of last ploughing before sowing. Against surface grass hopper and army worms, BHC 10% dust should be applied @ 25 kg/ha. Against wheat aphids and jassids, apply 0.5 liter Folithion or Thiodan or one liter Ekalux 25 EC in 800 liter of water/ha. For the control of brown rust, black rust or yellow rust spray, use 0.2% Zineb or Dithane M-45.

Yield

Good grain yield of wheat can be obtained from alkali soils under reclamation if soil pH_2 ranges from 9.0 to 9.3. About 1.8 to 3.0 t/ha yield can be obtained in timely sown wheat. If pH_2 ranges between 9.0 and 8.5 or below, yields can be 3 to 4 t/ha. Within 2-3 years, the yield may be as high as in normal soils of the region.

BARLEY

Barley (*Hordeum vulgare* L.) is an important *rabi* cereal crop of India (Fig. 12.4) occupying an area of 0.69 M ha with a annual production of 1.60 million tons with per hectare productivity of 2.32 t/ha (2014-15). Its cultivation is mainly confined to Uttar Pradesh, Rajasthan, Punjab, Madhya Pradesh, Haryana, Bihar, Himachal Pradesh, West Bengal and Jammu Kashmir. In cereal-cereal rotation barley is grown as maize—barley, sorghum—barley, pearl millet—barley, rice—barley etc.

Fig. 12.4. A view of barley crop

Climatic Requirement

Barley is grown in sub-tropical climatic conditions. The crop requiring 12-15°C temperature during growing period and warm and dry weather at maturity with temperature around 30°C. It cannot tolerate frost at any stage of growth. Any incidence of frost at flowering is highly detrimental to yield. The crop possesses very high degree of tolerance to drought and saline condition and as such can be cultivated in low rainfall regions and in saline lands.

Soils

Barley grows best on well drained loam soils but can also be raised on poor soils such as sandy, saline and alkali soils. Water logged and acidic soils are not good for barley. In alkali soils with pH_2 in the range of 9.3 to 9.5, barley performs better than wheat. In highly saline soil (EC_e 10-15 dS/m) barley grows better than wheat and can be fitted in rotation with rice crop. Being salt resistant, its cultivation has become possible on saline coastal areas of Sunderban in West Bengal and saline black soils of canal irrigated areas of northern Karnataka.

Varieties

Most farmers use low yielding old barley varieties like Manjula, Azad, Jagriti (U.P.), BH 75 (Haryana), PL 172 (Punjab), Sonu & Dolma (HP). To get better yields, an appropriate variety should be selected as per the requirement of the area, climatic and environmental conditions and post-harvest use of the grains. For industrial uses, proper malt type varieties need to be selected. Varieties like Ratna and DL-3 are suitable for alkali soils and can yield up to 2.5 t/ha in pH_2 range of 9.0 to 9.3. The varieties Amber, DL-88, DL-120 and DL-3 Jyoti have been found suitable for saline soils, and can give a yield of 1.0 - 1.8 t/ha in EC_e range of 10 to 15 dS/m. Under coastal saline conditions in Sunderban area, varieties yielding more than 2.0 t/ha in salinity range of EC_e 6.4 - 19.8 dS/m are DL-48, RD-137 Narendra Barley-1 (UP) and the varieties yielding more than 3.5 t/ha in salinity range of 1.6-5.7 dS/m are Amber, Jyoti and Azad. Salt tolerant varieties RD 2794 and ND B 1173 are suitable for irrigated, timely sown conditions and can yield 3.0-3.5 t/ha.

Dual purpose barley is now becoming popular. Besides grain, it is utilized as a source of green fodder (Singh *et al.*, 2014). The crop can be cut at 50-55 days after sowing for green fodder and the regenerated crop can be utilized for grain purposes. In Northern hills the cut for the green fodder can be taken at 70 DAS. Barley varieties RD2715, RD2035 and RD2552 in plains and BHS 380 in hills can be sown as dual purpose barley with good yield of the green fodder (20 to 25 t/ha) and grain (2.5 to 3.5 t/ha). Ratna variety developed at IARI, New Delhi is highly tolerant to saline, alkaline conditions and yields about 15.0 t/ha of green fodder after 65 days of sowing and regenerated crop yielding 2.0 t/ha of grain. If one does not go for green fodder, it yields about 2.5-3.0 t/ha.

Agronomic Practices

A well prepared and well levelled seed bed is required to raise a good crop of barley. The field should be ploughed 3-4 times followed by planking after each ploughing. The seed rate is 100 kg/ha but is slightly higher up to 120 kg/ha under late sown condition and in salt affected soils to ensure a better crop stand. The seed should be soaked in water overnight at normal temperature for better and quicker germination. To control loose smut, seeds should be treated with Vitavax or Bavistin @ 2 g per kg seed; for the covered smut with 1:1 ratio mixture of Thiram + Bavistin or Vitavax @ 2.5 g per kg seed. To control termites, seed should be treated with 150 ml of Chloropyriphos (20 EC) or 250 ml Formathion (25 EC) in 5 liter of water for 100 kg seed.

Under assured irrigation in salt affected soils, the crop is sown during the first fortnight of November for optimum yields. However, sowing is usually delayed because this crop succeeds rice crop especially in salt affected soils. One can anticipate gradual reduction in yield with increasing delay up to the end of December. The crop is sown in rows by *Kera* in soils with enough moisture. The spacing between rows is kept at 20 cm for the normal sown crop while it is reduced to 18 cm for the late sown crop. A planking operation is required after sowing of the crop.

Fertilizer Application

Fertilizer requirement is less for barley than wheat. The recommended doses are as follows:

Production condition	Fertilizer requirement (kg/ha)	
	Nitrogen	**Phosphorous**
Irrigated timely sown	60	30
Irrigated late sown	60	30
Rain fed plains	30	20
Rain fed hilly region	40	20

Generally alkali soils do not require phosphorus and potash for first 4-5 years of reclamation. Afterwards these may be applied on the soil test basis.

In irrigated crop, half of the nitrogen and full dose of phosphorous should be applied as basal and remaining half of the nitrogen should be top dressed after first irrigation or 30 days after sowing. In light soils, one third of nitrogen and full dose of phosphorous should be applied as basal, one third of nitrogen after first irrigation and rest one third of nitrogen after second irrigation. In rain fed condition, full NPK should be applied as basal at the time of sowing. Apply $ZnSO_4$ @ 20 kg/ha if Zn deficiency is expected or its symptoms appear. Other micronutrients should be applied as per soil tests.

Irrigation

Generally, barley requires 2 to 3 irrigations for higher yield. Depending upon the water availability, suitable stages for irrigation are as follows:

Numbers of Irrigation	Crop stage
Two	1st at crown root initiation (25-30 DAS)
	2nd at panicle emergence (65-70 DAS)
One	Tillering stage (35-40 DAS)

The malt barley may require 3 irrigations to ensure better yield, grain uniformity and better grain quality. Depending upon the water availability, irrigation may be applied at suitable stages. In dry areas of Rajasthan, the numbers of irrigation may go up to 5-6. The crop requires 4-5 irrigations in alkali and 3-4 in saline soils.

Crop Protection

To safeguard the crop against cut worms and termites, the soil should be treated with Aldrin 5% dust or Heptachlore 5% dust at 25 kg/ha at sowing. Aphids sometimes cause significant damage through sucking the sap from the leaves, stems and ear heads. The infested crop should be sprayed with Manazon W.P. @ 400 g/ha or Dimecron at 270 ml/ha or BHC 50% W.P. at the rate of 2.8 kg/ha in 1000 liters of water.

The diseases like covered smut, root rot and foot rot can be controlled by seed treatment with Agrosan GN at 2-2.5 g/kg seed. Stem rust can be controlled by dusting sulphur at 18-25 kg/ha. The disease resistant varieties should be grown to safeguard against loose smut and yellow rust.

Harvesting and Threshing

In Punjab, Haryana and Western UP, the crop is harvested from end of March to early April. The crop should be harvested in the early hours of the day to avoid shattering and breaking of the ears.

PEARL MILLET (*BAJRA*)

Pearl millet (*Pennisetum typhoides* L.) is one of the major coarse grain crops and is generally grown under dry farming conditions (Fig. 12.5). It is grown all over the country except in Assam. It is a main food and fodder crop in Rajasthan and is cultivated on large scale in Maharashtra, Gujarat and Uttar Pradesh. It is mostly grown in tropical and temperate climate. The crop occupies a total area of 7.12 M ha and produces 9.05 million tons of grain (2014-15) with a crop productivity of 1.27 t/ha.

Fig. 12.5. A view of the pearl millet crop in the field (left) and close up of pearl millet cob (right)

Soil Conditions and Varieties

Bajra is generally grown on low fertility soils such as sandy soils. It can also be grown on black cotton soils, alluvial soil and sandy loam but generally heavy soils are not suitable for the crop. Sandy loam well drained soil is the most suitable. It is tolerant to salinity and can be grown in saline soils having EC_e 5 to 10 dS/m and in alkali soils having pH_2 up to 9.2, if transplanted. In direct seeding, it performs well only if $pH_2 < 9.0$.

Varieties

Varieties PHB 10, PHB 14, PHB 50 and BJ 104 are suitable for alkali soils. Variety PHB 14 and BJ 104 can be grown both in saline and alkali soils. ICRISAT composite ICMV 221, TNAU cumbu hybrid CO9 are important hybrids.

Agronomic Practices

To ensure good germination, a fine seed bed should be prepared by giving 2-3 ploughings, each followed by planking. The optimum sowing time is around the middle of July. However, if it is to be sown by transplanting, the seed should be sown in the nursery bed in early July and transplanted after about 25 days. In alkali soils, sowing by transplanting is better than direct seeding. Gypsum @1.0 t/ha per 0.1 unit of pH_2 should be applied in soils if is higher than 9.0. About 5 kg seed/ha is sufficient in direct seeding but in transplanting, half of this seed rate would suffice. For raising seedlings for an area of one ha, 40-50 m^2 well manured area near a water source is required. The area should be free from water stagnation.

Following treatment is made to remove ergot affected seeds and to prevent primary infection of sclerotia.

- Dissolve one kg of common salt in 10 liters of water.
- Drop the seeds into the salt solution
- Remove the ergot and sclerotia affected seeds, which will float.
- Wash seeds in fresh water 2 or 3 times to remove the salt on the seeds.
- Dry the seeds in shade.
- Treat the seeds with three packets (600 g) of the Azospirillum inoculant and 3 packets (600 g) of phosphobacteria or 6 packets (1200 g) of azophos.

In direct seeding, the seed should be sown in rows 45 cm apart at a depth of 2.5 cm. The sowing should be done in optimum moisture to obtain good germination. Within 15 days after germination, thinning should be done leaving about one to two plants at a plant to plant spacing of 15 cm. The uprooted seedlings are used for gap filling, if any. Inter-culture operation should be performed three to five weeks after sowing.

In transplanted crop, following activities are undertaken.

- Pull out the seedlings when they are 15 to 18 days old
- Adopt row to row and plant to plant spacing of 45 x 15 cm for all the varieties
- Plant seedlings on the side of ridge, half way from the bottom. Depth of planting should be 3 to 5 cm
- Root dipping with bio-fertilizers helps. Prepare the slurry with 5 packets (1000 g/ha) of Azospirillum inoculant and 5 packets (1000 g/ha) of phospho-bacteria or 10 packets of azophos (2000 g/ha) in 40 liters of water and dip the roots of the seedlings 15 - 30 min before planting.

Fertilization

Apply NPK fertilizers as per soil test recommendations. In the absence of these recommendations, follow the blanket recommendation of 70:35:35 kg N, P_2O_5, K_2O/ ha for all varieties. For hybrids, apply 80 kg N, 40 kg P_2O_5 and 40 kg K_2O per ha. Apply the recommended N in three splits as 50:25:25 but full dose of phosphorus and potassium should be applied as basal. Combined application of azospirillum and phosphobacteria or azophos along with 75% of the previously recommended level of N and P may be adopted under the rain fed conditions. P and K generally do not give response in alkali soils and can be avoided.

Irrigation

Bajra is a rain fed crop but in saline and alkali soils it needs light and frequent irrigations in the absence of rainfall. Enough moisture should be present at the time of flowering and the seed development. Standing water is harmful to the crop and therefore, excess water should be drained out within few hours.

Plant Protection Measures

Grasshoppers and hairy caterpillars can be controlled by dusting with 10% BHC at the rate of 25 kg/ha. Stem borers and pink borers can be controlled by spraying parathion 50 E.C., at the rate of 500 ml/ha. Sometimes diseases like downy mildew, ergot and smut cause considerable damage. Downy mildew infested plants and ergot infested ears should be removed and burnt. Seed treatment with 20% common salt solution, washing with fresh water and then treating with Agrosan GN or Thiram or Ceresan is helpful in controlling various diseases.

Harvesting and Yield

The crop matures for harvesting after 100 to 120 days of sowing. The mature plants are cut with sickle and stalked in the field for drying. Threshing is done by bullocks or beating the ear heads by sticks or by trampling with stone rollers. The threshed grains are cleaned and dried in the sun for a week before storage. Grain yield of 0.8 to 1.0 t/ha can be obtained under favourable conditions. Effect of gypsum application on the crop performance of four cultivars of pearl millet in alkali soils is shown in Table 12.4 (Singh *et al.*, 1988a).

Table 12.4. Effect of levels of gypsum on dry forage yield (t/ha) of sorghum and pearl millet

Gypsum (t/ha)	pH_2 at sowing (0-15 cm)	Sorghum					Pearl millet				
		S109	S136	S120	JS20	Mean	LC2	LC5	LC6	LC7	Mean
Control	10.1	-	-	-	-	-	-	-	-	-	-
6.25	9.7	10.2	90.2	80.6	10.7	90.7	70.9	70.4	80.2	60.6	70.5
12.5	9.4	12.8	90.6	11.3	12.7	11.3	9.0	9.8	8.0	9.9	9.0
18.75	9.2	9.8	10.9	12.6	13.3	11.6	7.4	9.1	9.4	10.5	9.1
25.0	8.8	11.5	11.9	13.8	12.7	12.5	10.1	12.9	12.3	10.8	11.5
Mean		10.8	10.4	11.5	12.3		8.4	9.8	9.5	9.4	
CD (p = 0.05)		Gypsum=1.30; Crop=1.5					Gypsum=1.20; Crop=1.0				

MUSTARD (*Brassica juncea*)

Mustards are members of the Brassicaceae family, and are among the earliest cultivated plants. Amongst these, *Brassica rapa,*

B. napus and *B. juncea* are grown predominantly for oil and seed meal. More than 85% of area under rapeseed mustard in India is occupied by Indian mustard *B. juncea* (L.) alone. India is the second largest country in rapeseed mustard production. In 2014-15, rapeseed and mustard was cultivated on an area of 5.8 M ha with a total production of 6.28 million tons with a productivity of 1.08 t/ha. The productivity is much below than that of the other *Brassica* growing countries. For example; in UK, France and Germany, the average productivity of rapeseed mustard is two to threefold higher than India. Even the world average is also more than 50% higher than that of India. Rape and mustard crops are semi-tolerant to salinity and alkalinity (Fig. 12.6). Out of these crops; *raya* performs better than *toria, sarson* and *taramira*. The crops are cultivated in Punjab, Haryana, Rajasthan, Uttar Pradesh, Bihar, Gujarat, Assam, Madhya Pradesh and West Bengal during *rabi season*. In south India, *raya* is grown from July to November as a rain fed crop.

Fig. 12.6. Evaluation of salt tolerant *brassica* varieties at Karnal

Soil and Climate

Loam and sandy loam soils are good for *raya* although it may be grown on all types of soils. It can tolerate EC_e of 5 to 10 dS/m. In high pH soil, the crop does not perform well. However, with application of gypsum at 50% GR, the crop can be grown, successfully. Gypsum should be applied at the rate of 1.0 t/ha per 0.1 unit of pH_2 higher than 9.0. Singh *et al*. (1988b) showed that oil seed crops can also be raised successfully even as first crop on highly alkali soils, if reclamation has to be commenced from the winter season (Table 12.5). After the harvest of oilseed crops, pearl millet and sorghum crops can be grown during summer (April to June) as forage crops.

Table 12.5. Effect of gypsum level on the yield of oilseed crops (t/ha)

Gypsum (t/ ha)	pH_2 at sowing (0-15 cm)	*Toria*	Rape seed	*Raya*	*Taramira*	Mean
Control	10.2	-	-	-	-	-
6.25	9.7	0.51	0.82	1.02	0.47	0.71
12.50	9.4	0.86	0.91	1.14	0.58	0.82
18.75	9..2	0.86	0.91	1.24	0.59	0.90
25.0	8.8	0.87	0.98	1.27	0.62	0.93
Mean	-	0.78	0.91	1.17	0.57	
CD ($p = 0.05$)		Gypsum =0.07 and Crop=0.08				

Initial pH_2 (0-15 cm) = 10.4

Mustard is grown in sub-tropical climate. Dry and cool climate with temperature ranging between 10°C and 25°C are the optimum climatic requirement for this crop. It is susceptible to frost.

Varieties

The promising varieties of *raya* are, Punjab bold, RL 18 and RLM 1 in Punjab, Raya Parkash in Haryana and Varuna (T-59) in Uttar Pradesh and Bihar. Other high yielding varieties are: Pusa agrani, Kranti, Pusa Vijay, Pusa mustard 27, Pusa Karishma, Sita. In breeding trials, varieties T-62, T-53 performed well upto pH_2 of 9.1 and in agronomic trial RLM 1 gave higher yield than RL 18. Recently, the Genetic Engineering Appraisal Committee (GEAC), India's regulator for transgenic products, has recommended the genetically modified mustard seeds named Dhara Mustard Hybrid 11 (DMH-11) for commercial release, although it is being opposed by the anti-GM crop groups. Over the years CSSRI, has developed several high yielding salt tolerant varieties (Table 12.6). Any of these varieties can be used to get higher yields under stressed environments of alkalinity/salinity. Being a cross-pollination crop, improved varieties do not remain stable in their yield potential.

Table 12.6. Salt tolerant mustard varieties developed at CSSRI, Karnal

Variety	Salient features
CS 52	Its seed yield is about 1.5 t/ha in saline (EC_e 6-8.5 dS/m) and alkali (pH_2 up to 9.3) soils.
CS 54	It gives about 1.6 t/ha seed yield in SAS (EC_e 6-9 dS/m and pH_2 up to 9.3).
CS 56	Its seed yield is above 1.6 t/ha in saline (EC_e 6-9 dS/m) and alkali (pH_2 up to 9.3) soils.
CS 60	Its seed yield is in the range of 1.9-2.2 t/ha in saline (EC_e < 12 dS/m, irrigation water EC 16 dS/m) and alkali (pH_2 up to 9.5) soils.In normal soil yield ranges from 2.5 – 2.9 t/ha.

Agronomic Practices

The field should be prepared thoroughly by ploughing and planking to secure fine seed bed. Recently, zero and minimum tillage technologies have been perfected for the cultivation of this crop as well. In salt affected soils, *raya* needs a fairly moist bed for good germination. The best sowing time in north India is first fortnight of October. For *raya* crop it is desirable to complete the sowing operation by November end. When rapeseed-mustard is grown as intercrop, time of sowing depends upon cultivation of main crop. In alkali soils since *raya* is generally taken after rice crop, sowing can be done up to early November. About 4-6 kg seed/ha is required for sowing. To protect seed from soil borne pest and disease, seeds are treated with Thiram@3g per kg of seed. For even distribution of seed, it is usually mixed with soil before sowing. The seed is sown by drill or *kera* attached to a plough in rows 30 cm apart and 4-5 cm deep. Thinning is done after about three weeks of sowing to maintain plant spacing of 10 to 15 cm in a row. The crop needs about 2 hoeing, the first one after 20-25 days of sowing.

Fertilization

Raya requires about 80 kg N/ha (Table 12.6). If ammonium sulphate is available it should be preferred over other nitrogenous fertilizers. Phosphorus and potash are generally not required in alkali soils. Half the dose of nitrogen should be applied at sowing and remaining half with first irrigation. Apply 15 to 30 t/ha farm yard manure or well decomposed cow dung to the soil at the time of field preparation.

Table 12.6. Nutrient requirement (kg/ha)

Crop	Nitrogen	Phosphorus	Potash
Toria	60	20	On soil test basis
Raya and *Gobhi Sarson*	100	30	15

Irrigation

Thin and frequent irrigations should be applied in alkali soils. For good growth, crop requires about three irrigations at an interval of three weeks after sowing. In the initial stages of land reclamation, shallow and frequent irrigations (4-5) are required, whereas in semi-reclaimed soil, crop needs about 3-4 irrigations.

Control of Insect Pests

The crop should be sprayed with 2 kg BHC 50 WP in 500 liters water or apply 10% BHC dust at 25 kg/ha against mustard sawfly. To control aphids, Thiamethoxam@200 gm or Quinalphos or Oxydemeton@600 ml or Chlorpyriphos @500 ml in 250-375 liters water ahould be applied per ha. To control painted bug, 750 ml Malathion 50 EC in 750 liters water/ha should be sprayed. If infestation of hairy caterpillar is observed in the field, spray Malathion 5% dust@40 kg/ha or Dichlorvos@500 ml/ha in 250-375 liters of water.

Harvesting and Yield

The crop is harvested when pods turn yellow. Owing to shattering nature of *raya*, care should be exercised for its timely harvest. In North India, *raya* is ready for harvest in March. Since seeds are very small, threshing should be done on cemented threshing floor. The seeds should be cleaned and dried for 7-8 days before storage. In high pH soil, yield up to 1.5 t/ha can be obtained with gypsum application. However, in alkali soil having pH_2 8.5-9.0 yield up to 2.0 t/ha can be obtained.

■■■■

CHAPTER 13

SUGAR AND FIBER CROPS

SUGAR BEET

Sugar beet (*Beta vulgaris* sub sp. maritime) is grown in many parts of world mainly for sugar, fodder and vegetable (Fig. 13.1). Having halophytic ancestors, it is quite salt tolerant and probably the highest biomass producing crop in saline environments. Studies in desert landscape of Egypt revealed that feeding sheep and goats on fodder beet roots can maintain animal productivity, save about 50% of ration concentrate feed mixture and increase milk yield (Mousa, 2011). Sugar beet is also an important sugar crop in temperate countries. In India its cultivation is gaining popularity, because of its short period of growth of six months as compared to sugarcane which occupies the land for the whole year. Thus, instead of one crop of sugarcane, two crops a year, one of sugar beet and one food crop can be harvested from the same piece of land. Besides, it has high sugar content (15-17%), high sugar recovery and high purity and is able to withstand drought, salinity and frost. In sub-tropical semi-arid conditions, 1 kg of sugar can be produced with only 1.4 m^3 of water as compared to 4 m^3 of water for sugar produced with sugarcane. Its by-products provide nutritious cattle feed during the hot months of the year when green fodder is not readily available. In Tamil Nadu it is being propounded as a promising energy crop and alternative raw materials for the production of ethanol. For the first time in Punjab, 4,000 ha of land have been earmarked for the cultivation of sugar beet.

Climate and Soil Requirements

Sugar beet performs well under cool temperate climate. Thus, it has the potential of cultivation during the *rabi* season in north India.

Tropical sugar beet is sown from September to November and harvested during March and May. Sugar beet grows best in well drained loamy and sandy loam soils but can also be raised successfully in saline and alkali soils. It is susceptible to salinity at germination but later on it can tolerate EC_e up to 15 dS/m. With special planting techniques which reduce salinity hazards around seed zone, quite high yields can be obtained in saline and alkali soils (Kanwar and Singh, 1968). To lower down high pH (pH_2 >9.5), gypsum should be applied @ 1.0 t/ha per 0.1 unit of pH_2, higher than 9.5.

Fig. 13.1. Field view of sugar beet (left) and view of the harvested beet (right)

Varieties

Two varieties namely Erotype E from West Germany and Maribo Resistapoly from Denmark have been found to be quite promising for cultivation in alkali soils. The varieties Romanskaya, Kawep Precopoly, Maribo Magnapoly and USH-9 have been found suitable for saline soils. Hybrids Cauvery, Indus and Shubhra are found to suit the tropical conditions of Tamil Nadu.

Agronomic Practices

The fields should be well prepared at optimum soil moisture conditions. Planking after each ploughing is required to achieve the fine tilth. Ridges 12-15 cm high are prepared at 50 cm interval.

Already treated and pelleted seed are marketed in packets of about 120,000 seeds weighing less than 4 kg. Mid-October is the best time to sow this crop. The seeds are sown on one side of the ridge, half way between the top and the bottom of the ridge opposite to the sun direction on the north-western slope. The pre-soaked seed are dibbled at a distance of 12.5 cm to a depth of 2.5 cm. After sowing, thin irrigation is given in furrows in such a way that water does not reach the place where seeds are sown. If that happens, there is a possibility of crust formation, which might hinder germination. The seed zone should remain moist so that salts move to the top of the ridge. The germination takes about 5-6 days. After 25-30 days of sowing, thinning is necessary with optimum population being 1,00,000-1,20,000 /ha. Plant-to-plant distance is maintained at 25 cm. Seedlings uprooted during thinning may be used to fill-up the gaps, if any. Earthing operation is taken up after about two months of sowing. Irrigation is given at 10-12 days intervals. In Punjab irrigation scheduling at IW/CPE ratio of 0.8 was found optimum (Saini and Brar, 2018). Irrigation may be stopped 15-20 days prior to harvest as it helps sugar accumulation.

Fertilization

Nitrogen content of the soil has to be high for getting good yield of sugar beet. Compost or FYM is applied at the rate of 15-20 t/ha. The nitrogen is given at 150 kg/ha. Phosphorus is applied only if soil tests show deficiency. Full dose of P and 2/3rd dose of nitrogen should be given at sowing and remaining nitrogen may be given in two equal splits at thinning and earthing up.

Harvest and Yield

Harvesting begins in April but actual period depends upon the date of sowing. Late sown crop is harvested even in May. The harvesting must be completed before the onset of hot weather. A root brix reading of 15 to 18% indicates the maturity of beet root for harvest. Enough soil moisture at the harvest time facilitates easy harvesting. The average root yield of tropical sugar beet is 80-100 t/ha. Depending upon the method of sowing, root yield in Punjab varied from 55 to 70 t/ha, planting sugar beet in two rows per bed giving the highest yields (Saini and Brar, 2018). After uprooting the plants, the tops are removed from the roots. Store in shaded conditions but transport the roots to the factory within 48 h for processing. Sugar beet tops being rich in nitrogen can be used for green manuring. Sugar beet tops and pulp are also good cattle feed. Pulp contains 10% crude protein and 60% digestible nutrients.

SUGARCANE

Sugarcane (*Saccharum officinarum*) belongs to bamboo family (Fig. 13.2) and is the main source of sugar, *gur* and *khandsari* in India. It is one of the most important food-cum-cash crops. About 2/3 of the total sugarcane produced in India is consumed in making *gur* and *khandsari* while only one third of it reaches the sugar factories. Besides, it also provides raw material for manufacturing alcohol. A part of it is also used as fodder. The important sugarcane growing States are UP, Maharashtra, Gujarat, Bihar, Punjab, Haryana, Assam, Karnataka, Tamil Nadu and Andhra Pradesh. Area under this crop in the country is 5.07 M ha with a cane production of 362.33 million tons having the productivity of 70.52 t/ha (2014-15).

Fig. 13.2. A view of the sugarcane in field (left) and harvested cane (right)

Climatic Conditions

Sugarcane, a tropical plant, requires a warm humid climate for its growth. It is also grown in sub-tropics in India. Bright sunshine, well distributed rainfall, well-drained soil and adequate soil moisture are very conductive for proper crop growth. Moderate rainfall of 800-1000 mm supplemented by timely irrigation ensures good cane yield.

Soil Conditions

Sugarcane grows well on different types of soils ranging from sandy loam to clay loam. However, it thrives best on well drained loamy soil. At a pH_2 of 9.0, yields up to 40 t/ha of fresh cane can be obtained. With special planting techniques and with combined application of gypsum and compost, it can also be cultivated on soils having pH_2 10 or more. In that case yields may be low initially. Sugarcane is considered to be moderately sensitive to salinity (Maas, 1990). Sugarcane growth and yield were not affected by root zone salinities less than about 2 dS/m, but no millable stalks were produced at salinities in excess of 10 dS/m (Wiegand *et al.*, 1996).

Land Preparation

Plough the land 3-4 times length and breadth wise to secure fine tilth. The fields should be properly levelled. Prepare furrows 25cm deep and 75cm apart for short duration and 90 cm apart for the medium duration varieties. In hilly areas, prepare pits in rows along the contour at a spacing of 30 cm in the row and 75 cm between the rows.

Seed Rate and Treatment

At 90 cm row to row spacing and @ 12 setts /m long row, seed requirement is about 8.5-12 t/ha. Following are few guidelines.

Single budded setts (12 setts/m): 134,620 -135,900 setts /ha

Single budded setts (end to end): 78,740-80,000 setts/ha (40-50% seed saving)

Two budded setts (end to end): 67, 3000-68,600 setts/ha

Three budded setts: 44,500-45,700 setts/ha

Select top ends of the mature healthy disease free canes up to top 2/3 of total length and cut into setts of two/three eye buds. Treat the setts for 5 min in 0.1% Carbendazim (100 g in 100 liter of water) before planting. If termites problem is expected, 25 kg BHC 10% powder/ha should be mixed in the soil 3-4 days before planting.

Varieties

Salt tolerant sugarcane clones identified in *Saccharum barberi* are: Katha (Coimbatore), Kewali-14-G, Khatuia-124, Kuswar, Lalri, Nargori and Pathari16. Similarly, salt tolerant clones in *S. sinense* are: Khakai, Panshahi, Reha, Uba – Seedling16 and in *S. robustum*: IJ-76-422, IJ-76-470, 28 Ng 251, 57 Ng-201, 57 Ng 231, Ng 77-34, Ng 77-55, Ng 77-136, Ng 77-34, Ng 77-55, Ng 77-160, Ng 77-167, Ng 77-170, Ng 77-221 and Ng 77-237. Some of the salt and drought tolerant varieties are listed in Table 13.1.

Table 13.1. Salt and drought tolerant varieties developed in India

Varieties	Tolerance/resistance characters	Developed at
Co 95003, Co 93005, Co 97008, Co 85019, Co 99004, Co 2001-13	Salt tolerant	ICAR-SBI, Coimbatore
Co 8145, Co 88019, Co 94008, Co 99004, Co 2001-13, Co 2001–15, Co 0212, Co 06027	Salt stress and drought/rain fed	ICAR-SBI, Coimbatore
Co 0238, Co 05011	Salt stress and drought	ICAR-SBIRC, Karnal
Co 395, Co 453, Co 87263, Co 975, Co 453, Co 1148	Water logging and salt stress	ICAR-SBI, Coimbatore
Bo 99, Bo 106, B 37172	Salt stress and drought/ rain fed	RAU, Pusa, Bihar
Copant 97222, Copant 90223	Water logging/rain fed	GBPUAT, Pantnagar
Co 290, Co 86249, Co 94008 and D 109	Multiple- abiotic stress	ICAR-SBI, Coimbatore
Cose 96234	Multiple stress	UPCSR, Shajahanpur

Time and Method of Planting

In north India, best time for planting is from middle of February to middle of March for spring crop and end of September to first week of October for winter crop. In Tamil Nadu, Andhra Pradesh, Maharashtra and part of Karnataka, cane is planted between December and February. Some of the methods of planting are as follows (Ram *et al.*, 2011).

Conventional flat planting

In flat planting, deep furrows are opened by a plough 60 to 75 cm apart. The cut setts are planted end to end with the eye buds facing sideways and covered with 4 to 5 cm of soil. Thereafter the field is levelled up with a beam.

Dry planting in ridges and furrows

Ridges and furrows are made at 90 cm distance using a tractor drawn-ridger. The trenches are 20-25 cm deep. Sugarcane setts are planted manually. Setts are covered with soil using a spade followed by light irrigation in furrows. Subsequent irrigations are given as and when required. Spray Atrazine @ 5 kg/ha in 350-400 liter water after 2-3 days of planting. This method is better than the conventional flat system in terms of germination percentage, water use efficiency, aeration, inter-culture operations and yield.

Paired row planting in trenches

Trenches are opened using a tractor mounted trench-opener at 150 cm distance. Sugarcane is planted in paired-rows at 30:30- 90-30:30 cm spacing. Interspaces between trenches are used for taking intercrops. Trench planting gives higher yield and return than ridges and furrow system

Ring pit method

About 2,700 circular pits of 60 cm diameter are prepared to a depth of 30 cm using a tractor mounted digger. A gap of 60 cm is left between adjacent pits as well as on other sides. Before planting, pits are filled with loose soil and FYM to a depth of 15 cm (Yadav *et al.*, 2012). About 20-22 two-budded setts are placed per pit and covered with soil to a height of 2-4 cm (Fig. 13.3). Some of the advantages of pit method are: possible to take 2-3 ratoons; 25-50% higher yield than ridges and furrow method; Possible to use drip irrigation to enhance water use efficiency.

Fig. 13.3. Ring-pit method of planting sugarcane

Sustainable Sugarcane Initiative (SSI)

It is a new initiative based on the principle of ´more with less´ that improves cane productivity while reducing the cost of production. The basic principles of this practice are (Mani *et al.*, 2017):

- Raising nursery using single budded chips
- Transplanting 25-35 days young seedlings
- Maintaining wide spacing of 1.5m*0.6 m
- Using drip method of irrigation and fertigation
- Practicing intercropping

Planting in Alkali Soil

Ridges are made at the spacing of about 75 cm. Gypsum at the rate of 0.1 t/ha per 0.1 unit of pH_2 higher than 9.0 and compost at the rate of 25 t/ha are applied and mixed with soil in furrows in between ridges about a fortnight before planting. Thereafter, irrigation is applied in furrows. When the soil comes in workable condition, it is loosened before planting. The cut setts are planted end to end in furrows 4 to 5 cm deep and covered with soil. Thereafter, irrigation is applied in furrows. Since the place of planting remains wet, the salts move to the untreated and drier soil on the top of adjoining ridges and thus furrows having seed setts remain comparatively free from salts resulting in a good crop stand.

Fertilizer Application

Sugarcane is a heavy feeder of nutrients and therefore, adequate manuring is essential. For mild alkali soils, apply press mud 5 t/ha and for acid soils apply dolomite 500 kg/ha or calcium carbonate 750 kg/ha. Phosphorus should be applied on the basis of soil test especially during the initial years of reclamation of alkali soils. The crop needs about 200 to 225 kg N/ha. Ammonium sulphate, if available should be preferred over other sources of nitrogen. Half the dose of nitrogen and the full dose of phosphorus, if soil test shows its deficiency, should be applied as basal application. The remaining half dose should be given in two equal splits after two months of planting and before the onset of monsoon in end-June. Light/partial earthing up should be done at 90 days after planting. The final earthing up may be done during June end or before the onset of monsoon.

Inter-culture and Irrigation

Blind hoeing should be performed at about 10-12 days after planting and then 2-3 hoeing may be given later after crop germination. Generally, weeds are not a problem in alkali soils for first few years.

Irrigate the crop 8-10 times depending upon the distribution of rains. In early growth periods, irrigate the crop at frequent intervals. However, too much moisture or water stagnation should be avoided especially during germination and early growth phases. First irrigation is given in furrows just after planting and thereafter light and frequent irrigations are given at about one week interval during the pre-monsoon period. During monsoon and in the post monsoon period irrigations are given as and when required. Irrigation is withheld about 15 days before the crop is ready for harvest.

Plant Protection Measures

Insect pests cause damage to the crop beginning from April to November. Against black bug and pyrilla, spray 5-7.5 kg BHC 50 WP or 1 liter malathion 50 EC in 1000 liters water/ha. Malathion also controls mites which generally appear during April-June. To control over attack the infested shoots should be collected and destroyed weekly.

Harvesting and Yield

In north India, spring crop generally matures in December depending upon the variety. The sugar content continues to increase until about end of March. Crop should therefore, be harvested during this period. The yield depends upon stage of reclamation. In high pH soil with improved technology, yields up to 40 t/ha can be obtained in furrow and ridge method of sowing. In reclaimed alkali soil when pH_2 comes down to about 8.5, yield up to 100 t/ha of stripped cane can be obtained.

Climate change may impact the sugarcane cultivation scenario in India and abroad. Studies in India, Mauritius, South Africa and Trinidad have shown that for every two degree centigrade rise in temperature, there is a every likelihood of 30% or more drop in sugarcane productivity. On the other hand, efforts to develop transgenic sugarcane can help to increase yields, reduce production costs, improve sugar quality and reduce the environmental impact of sugarcane cultivation.

COTTON

Cotton belongs to the family Malvaceae, Tribe: Gossypieae, Genus: Gossypium. Cotton (*Gossypium* Spp.) is an important fiber crop of India (Fig. 13.4). It is generally a rain fed crop on the moisture retentive and heavy black soils but needs irrigation when grown on alluvial soil. Cotton in India is cultivated in Punjab, Haryana, Rajasthan, Gujarat, Madhya Pradesh, Maharashtra, Andhra Pradesh, Telangana, Karnataka and Tamil Nadu, besides, in small areas in non- traditional states like Uttar Pradesh, Odisha, West Bengal, Assam and Tripura (MoA&FW, 2017).In India, cotton was cultivated on 12.82 M ha area in 2014-15, with a production of 34.80

million bales (170 kg each) @ 462 kg/ha of lint. India is the only country which grows all the four species of cultivated cotton i.e. *Gossypium arboreum* and *G. herbaceum* (Asiatic cotton), *G. barbadense* (Egyptian cotton) and *G. hirsutum* (American upland cotton) besides hybrid cotton. Cotton is also the first crop in India, where genetically engineered technology (GE), has been commercialized in the form of insect-resistant Bt cotton. Bt cotton area, which was hardly 0.029 M ha (0.38%) out of 7.67 M ha in 2002-03, increased to 11.94 M ha out of 12.82 M ha in 2014-15 showing more than 93.14% adoption within a span of thirteen years. Cotton yield in Gujarat, Andhra Pradesh, Punjab and Tamil Nadu are equivalent to world average of 650-750 kg lint/ha.

Fig. 13.4. A view of the cotton crop

Cotton plant not only provides the fiber for making cloth but its seeds contain about 18 % edible oil. The deoiled cake containing 40 to 42% protein can be excellent cattle, poultry and fish feed. Even the cotton stalks, which are currently used as fuel, can be converted into pulp. Good quality craft paper and corrugated boxes etc. can be made out of this pulp.

Climatic and Soil Requirements

Cotton is grown in sub-tropical climate. It requires mean annual temperature over 16ºC, well distributed rainfall, abundant sunshine during the period of boll maturation and harvesting and frostless season of 6 to 8 months. Adequate soil moisture is necessary during all stages of crop growth. Bright sunny days ensure good quality of the produce. Cotton can be grown successfully on all types of soils with pH values ranging from 5.5 to 8.5. However, too sandy, saline and waterlogged soils are not good for cotton cultivation. It grows well in the Indo Gangetic alluvium, black and light black soils of central India and Deccan plateau and red lateritic soils of eastern and southern India. Cotton can tolerate EC_e up to 10 dS/m and pH_2 up to 9.5 but the yield level is very low at pH_2 above 9.5. Cotton is sensitive at germination but later on its tolerance to salts increases. Gypsum should be applied in high pH soils at the rate of 1 t/ha per 0.1 unit of $pH_2 > 9.0$. Good drainage to drain excess water during rains is essential for waterlogged saline soils.

Varieties

Varieties like Acala 1517, Stoneville, Coker 100-6, Kh 33/1146, G 27, J 34 and Krishna are considered to be salt tolerant types. Since now more than 93% area is under Bt cotton, readers may refer MoA&FW (2017) to know popular Bt cotton hybrids for various coton growing zones of India. For Punjab recommended hybrids of Bt Cotton are RCH 134, RCH 317, MRC 6301, MRC 6304.

Agronomic Practices

The field should be prepared by giving 4-5 ploughing, each followed by planking. A levelled seed bed is necessary for efficient irrigation and drainage. The sowing season varies from tract to tract. In major part of the country, it is sown as *kharif* crop from March to May under irrigated conditions and June-July under rainfed conditions with the commencement of monsoon. If sowing is completed during April under Punjab conditions, it helps to escape the attack of American bollworm to a great extent. In southern India, cotton is sown from September to November.

Seed Rate and Treatment

Seed rate depends upon size of the seed, spacing, and method of sowing. The usual seed rate is 12-15 kg in arboreum (Desi types) and hirsutum cotton (American types). Acid treatment is required to dlint the seeds. For this purpose, mix 400 g commercial grade concentrated sulphuric acid with 4 kg cotton seed for American cotton and mix 300 g commercial grade of sulphuric acid with 3 kg seed of *desi* cotton in glass container. Stir vigorously for two to three minutes with a thick wooden or glass rod. To prevent seed borne pathogens, seed should be soaked in 0.1% Ceresan Wet along with 50 ppm of streptocycline for 6-8 h before sowing. The treated seed should be rubbed with fine earth, cow dung or ash to ensure uniform distribution.

Method of Sowing

Line sowing using a cotton seed drill is a common practice. American cotton is sown at a spacing of 60 to 75 cm from row to row and 45 cm from plant to plant. In *desi* cotton, row spacing is kept at 60 cm while plant spacing is 30 cm.

In some irrigated areas and areas of heavy rainfall, ridge and furrow method is followed to economize the use of water and to ensure perfect drainage of excess rain water. This method is also useful in salt affected soils. Ridges are made about 20 cm high at distance of about 60 cm. Seeds are dibbled on the sides of the ridges at a height of 10 cm from the base of ridge. Light irrigation is applied in furrows which keeps the seed zone wet so that salts move to the top of the ridge. The seed zone becomes comparatively free from salts resulting in good germination.

In another method seedlings are raised in polythene bags and transplanted in the field. Since cotton seed is sensitive to salts at germination, this method is also useful for saline and alkali soils.

Thinning, Hoeing and Weeding

Extra and weak plants are thinned out within a fortnight after germination. Weed problem is generally less in salt affected soils as only few weeds emerge. First hoeing should be done before first irrigation and subsequent hoeings are done after each irrigation or rain event. The spray of basalin @ 2000 ml/ ha in 200 liter water before sowing or Treflan @ 2000 ml/ha before germination or Stamp @ 2.5-3.5 l/ha dissolved in 200-250 liter water after sowing or

Duron @ 500 g/ha or gramaxon @1500 ml/ha after germination are used for weed control in Haryana.

Manures and Fertilizers

Cotton responds very well to the application of nitrogen but phosphorus and potash generally do not give any response on alkali lands. The nitrogen dose varies with the species and varieties. Hirsutum types (American cotton) respond to 60 to 120 kg N/ha whereas arboreum types (*desi* cotton) need less nitrogen from 30 to 60 kg/ha. In alkali lands, N should be applied in three equal splits at sowing, thinning and at commencement of flowering.

Irrigation

Cotton requires 800-1000 mm water during its growing season. A part of it is met through rainfall. About 4 irrigations supplying 240 mm of water are needed under the Haryana conditions. The crop should not be stressed during flowering and boll formation period as it may result in shedding of flowers. Irrigation should not be given after the opening of about 1/3rd bolls. In salt affected soils, cotton needs light and frequent irrigations so that root zone may remain comparatively free from salts. Since cotton does not tolerate water logging, excess rainwater from the fields should be drained out as soon as possible.

Plant Protection

Avoid growing lady's finger, *moong* bean, pigeon pea, castor, *dhaincha* in and around the cotton fields to avoid simultaneous build-up and spread of pests and diseases to cotton. Jassids and bollworms are the most serious pests of cotton. Jassids damage the crop during June-August while bollworms become serious with the start of flowering. Leaf roller may become serious during August-September if weather remains cloudy. Seed should be acid delinted to kill the pink bollworm larvae and bacterial pathogen of angular leaf spot. To control jassids, crop should be sprayed with 1 liter of malathion 50 EC or 600 ml of rogor 30 EC or 750 ml of metasystox 25 EC in 500 liters of water per hectare. Three such sprays may be necessary. For the control of bollworms, spray 650 ml of folithion or

1200 ml of thiodon 35 EC in 750 liters water/ha at fortnightly intervals two weeks after start of flowering. To safeguard against diseases, copper oxychloride 50 WP at 1875 g/ha can be mixed with insecticides.

Sometimes, yields are reduced because of poor boll opening. Defoliants help to hasten the boll opening. Application of Dropp @ 200 g/ha at the time of 60% boll opening stage contributes significantly in increasing the effective number of open bolls resulting in higher cotton yield.

Harvesting and Yield

In north India, picking starts from the beginning of October. Second picking is done in early November, third in end-November and last by 10th of December. The cotton –picking period from mid-September to November must have bright sunny days to ensure good quality produce.

Problems

The agronomic requirements of different Bt cotton hybrids have not been studied thoroughly which hinder the realization of maximum yield potential of these hybrids by the farmers. The area under spurious/underscript hybrids of Bt cotton and even F2 seed of Bt cotton hybrids is quite high which creates favorable environment for the development of insect-pests and diseases. There is a need to spread the word that one should not grow unapproved hybrids of Bt cotton to realize the maximum benefits.

■■■■

CHAPTER 14

FORAGES AND GREEN MANURE CROPS

FORAGE CROPS

Fodder crops, the plant species that are cultivated and harvested for feeding the animals, are cultivated on about 8.3 M ha on individual crop basis. These are fed in the form of forage (cut green and fed fresh), silage (preserved under anaerobic condition) and hay (dehydrated green fodder). The total area under sorghum amongst the *kharif* crops (2.6 M ha) and *berseem* (Egyptian clover) amongst the *rabi* crops (1.9 M ha) occupy about 54% of the total cultivated fodder area. Other crops grown for fodder are: *Senji* (Sweet clover), *Shaftal* (Persian clover), *Metha* (Fenugreek), *Lobia* (cowpea), *Guar* (Clusterbean), *Jai* (Oat), *Jau* (Barley), *Jowar/Chari* (Sorghum), *Bajra* (Pearl millet), *Makka* (Maize), *Makchari* (Teosinte), and *Chara sarson* (Chinese cabbage). Besides these cereal and legume fodder crops, several grasses such as Karnal grass, hybrid Napier grass, para grass, guinea grass and blue-panic grass (see Chapter 15) and trees or their leaves such as Subabul, *Sesbania* and Gliricidia are used as fodder (ICAR, 2011). Some of the alkali tolerant forage crops are reported in Table 14.1. This and the next chapter discuss some of the fodder crops (*Rabi* season) and grasses that can be adapted to salt affected soils during and after the reclamation.

BERSEEM

The *berseem* (*Trifolium alexandrinum* L.) or Egyptian clover is one of the most valuable productive and nutritious *rabi* fodder crops in sub-tropical regions such as irrigated areas of the Punjab, Haryana, Delhi, UP, Uttrakhand, Rajasthan and other parts of western and northern India (Fig. 14.1). It can be sown alone or in combination with other species. The crop provides green fodder from end of November to mid-May in 5 to 6 cuttings. It contains 17% crude protein and 25.9% crude fiber; the total digestible nutrients content being 60-65%. It is very useful for the milch cattle being very rich in protein. Since *berseem* contains saponins, it leads to bloating in the ruminants, if fed in excessive quantities. It is mixed with grass (rye grass) or with a winter cereal such as oats to make high quality silage.

Table 14.1. Promising forage crops for alkali soils

pH*	Grass/forage crop (Botanical name)
Rainy season, *kharif*	
9.6	Sorghum (*Sorghum bicolor*), Sudan grass (*Sorghum sudanence*), *Dhaincha* (*Sesbania aculeate*)
9.2	Teosinte (*Euchlaena americanum*)
8.8	Cowpea (*Vigna unguiculata*), Guar (*Cyamopsis tetragonalob*)
Winter season, *rabi*	
9.6	Oat (*Avena sativa*), Barley (*Hordeum vulgare*), *Shaftal*/Persian Clover (*Trifolium resupinatum*)
9.2	*Berseem* (*Trifolium alexandrinum*)

*The pH values at which yield reductions is 50% or less

Fig. 14.1. A field view of the Egyptian clover crop

Climatic Requirements and Soil Conditions

The *berseem* needs mild temperature (15°C to 25°C) for germination and establishment and thereafter dry and cool climate for normal growth. Growth is restricted during very cold and frosty periods. For seed production, it requires dry weather and high temperature at maturity time. Although suited to plains, it can be cultivated at altitudes of 1,215 m above mean sea level.

Berseem thrives well on well drained, medium to heavy soil. Light, sandy and acidic soils are not suitable for this crop. It improves physical, chemical and biological properties of soil and thus improves its fertility status. It can withstand moderate alkalinity and salinity and is thus a very useful crop in the reclamation of salt affected soils. The *berseem* can be grown as first crop on alkali soils after treatment with gypsum but it should preferably succeed rice crop for better yield. If cultivated as the first crop, gypsum @ 1.0 t/ha per 0.1 unit of pH_2 higher than 9.0 should be applied to the soil. In saline soils rhizobial symbiosis and nodulation may be affected at or above an EC_e of 5 dS/m.

Varieties

During eighties only two varieties, namely, *Fahli* and *Mescavi* were being cultivated until a new variety BL-1 was evolved at Punjab Agricultural University, Ludhiana, which produces more tillers and yields higher than *Mescavi*. Its first cutting is ready about a week earlier and finishes about two weeks later as compared to *Mescavi*. Now many varieties such as BL 10, BL 42, Wardan, BL 22, HFB 600, Varda, JB-1, 2 and 3, UPB 103, Pusa giant, Khadarvi, Chindwara and BL 180 have been developed. There are many more varieties that need to be selected in a particular region.

Field Preparation

The field should be prepared thoroughly by giving 3-4 ploughing followed by planking to secure a fine and levelled seed bed. The proper levelling is very important in salt affected soils. The field should be laid out in small sized plots (check basins) for uniform spread of irrigation water.

Seed Rate and Inoculation

Salt affected soils require higher seed rate at 30 to 35 kg/ha as compared to normal soils which need 20 to 25 kg/ha. Sometimes seed gets mixed with the seed of *Kasani* (*Dichorium intybus*) which can be removed by treating the *berseem* seed with one per cent salt solution. Since *Kasani* seeds are lighter, these will float on the surface and can be removed.

In soils, where *berseem* is grown for the first time, inoculation of the seed is essential because it requires special species of bacteria (*Rhizobium trifolli*) for its proper growth. For this purpose, a solution of 1.5 liter of 10% brown sugar (*gur*) is prepared and mixed with three packets of *berseem* culture. This paste is mixed with 30 to 35 kg seed with the help of fingers. The seeds are rolled thoroughly to give a fine coating on every seed. The treated seed is dried in shade before sowing.

Time and Method of Sowing

The optimum sowing time is from last week of September to first week of October. However, in salt affected soils where *berseem* follows rice, sowing is delayed until the end of October. The sowing should generally be done in the evening. The seeds are broadcast in the standing water when the weather is calm. It is followed by raking so that a fine layer of earth settles down on the seeds. If high wind prevails at the time of sowing, it is better to broadcast the seeds evenly on dry land, immediately followed by raking and irrigation.

Manure and Fertilizers

Application of farmyard manure or compost at the rate of 25 t/ha at the time of field preparation is quite useful practice especially in salt affected soils both from reclamation point of view and better crop yield. At the time of sowing, nitrogen at 25 kg/ha, phosphorus at 75 kg/ha and zinc sulphate at 10 kg/ha should be applied. Another dose of nitrogen at 25 kg/ha can be given after first cutting. Potash application is not necessary in alkali soils.

Pest and Disease Control

Avoid raising the crop nearby tomato, gram and late sown wheat fields to control gram caterpillar attack. If infestation is observed, spray Chlorantraniliprole 18.5 SL @ 125 ml/ha or Spinosad 48 SC @ 150 ml/ha in 200-250 liters water. To control stem rot, a seed borne disease, remove and destroy the affected plants. Spray Carbendazim @ 1000 g in 500 liters of water/ha.

Irrigation and Forage Yield

The crop requires frequent irrigations. Normally, the water should not be allowed to stand for a long time but as far as possible, keep the soil moist to suppress salts. The first irrigation may be given within 3-4 days of sowing. Afterwards thin irrigations of about 4 cm depth at 7 to 10 days interval are given depending upon weather. Each cutting must be followed by irrigation.

The first cutting should be taken after 60 days of sowing even if the growth is less. Generally, yield from first cutting is poor in salt affected soils but it picks up later as pH or salinity are reduced to some extent. Subsequent cuttings are taken at about 40 days interval during winter and 30 days interval in spring. In all 4-6 cuttings can be taken. The total yield in the first year of reclamation of alkali land is about 30 to 40 t/ha after first crop of rice. However, in the second year of reclamation, when pH comes down to near about 8.5 to 9.0, yields are as high as 100 t/ha or more.

Berseem for Soil Reclamation and Fertility Enhancement

The growing of *berseem* after rice crop is a very useful practice in the reclamation of salt affected soils. It adds to the soil fertility by increasing the organic carbon, total nitrogen and available phosphorus. It improves the physical condition of the soil due to root residues left in the soil. The compact soil is loosened, the soil aggregation is better and the bacterial activity of the soil is accelerated. The grain yield of rice crop succeeding *berseem* is higher than followed by wheat.

SHAFTAL

Shaftal (Trifolium resupinatum L.) or Persian clover, also known as *bhukal*, is a nutritious leguminous fodder, liked by animals. High soluble carbohydrate, high protein content and low non-digestible fiber content makes it a good feed for animals. *Shaftal* is a multi-cut fodder like *berseem* and is considered a better soil-builder than *berseem* and sweet clover (*senji*). It can also be grown in open rangelands for grazing. It is a late season crop. It is used as a mix with oats and ryegrass to increase the overall productivity. It retains excellent feeding value as dry standing hay during dry weather. It should be fed mixed with some dry fodder as unmixed green fodder may cause tympanitis in cattle.

Fig. 14.2. A close view of *shaftal* plants

Agronomic Practices

The method of cultivation, its soil and climatic requirements are almost the same as those of *berseem*. It is tolerant to seasonal water logging and mild salinity. Compared to *berseem*, the crop performs better in alkali soils. However, it does not require bacterial inoculation of seed or soil and can be grown successfully on new land. Its seed rate is half that of *berseem* crop (about 15-17 kg/ha).

The seed rate can be reduced proportionately when sown as a mix with mustard and/or oat.

Commonly grown varieties of shaftal for northern India are Shaftal 48, SH 69, SH 48. Irrigation requirement is similar to that of *berseem* but the nutrient requirement is quite less. At the time of sowing, 15 kg of nitrogen and 50 kg phosphorus per ha is applied as basal dose. In alkali soils Zn @ 10 kg $ZnSO_4$ may be applied

Insect Pest Control

Spray Flubendiamide@50 ml or Quinalphos@1000 ml/ha in 375 liters of water to control hairy caterpillar that destroys the crop leaves and stem. Spray 125 ml Chlorantraniliprole 18.5 SL or 150 ml Tracer to control gram caterpillar. Repeat these sprays after 10 days, if required. For the treatment of downy mildew, spray Zineb 75WP @1000g/ha. Apply two to three sprays at interval of 10 days.

Harvesting and Yield

Crop is ready for first cutting in 55-60 days. Other cuttings are taken at interval of 30 days. It gives about 4-6 cuttings from December to May. As a first crop in highly alkali soil, it yield about 40 - 50 t/ha after adding gypsum @ 50% of GR. In the second year of reclamation and subsequently, one may shift to *berseem* as its yield potential is much higher than that of *shaftal* when the soil pH_2 is 9.0 or below. An N-fixing legume, Persian clover improves soil fertility and is considered to be a better reclamative crop for alkali soils than *berseem* (Kumar and Abrol, 1986).

GREEN MANURING CROPS

Green manuring is the practice of growing a suitable crop, generally a legume that is turned into the soil while green for improving soil productivity. The practice of green manuring was widely prevalent in India. However, the emphasis on it decreased after the sixties due to introduction of intensive farming system with easy availability of chemical fertilizers. With the energy crisis of the mid-seventies, green manuring has again become popular, mainly due to the limited supply of organic matter from other sources and the high cost of chemical fertilizers. Green manure crop supplies readily decomposable organic matter and nitrogen. In alkali soil, green manuring is extremely important not only for supplying the nutrients to crop but also for buffering action of soil, and for improving the physical condition. In saline soils, it provides a good crop cover during intervening periods between the two crops.

The crops most commonly used for green manuring are: *dhaincha* (*Sesbania aculeata*), cluster bean (*Cyamopsis terragonoloba*), cowpea (*Vigna sinesis*), sannhemp (*Crotalaria juncea*) and *guar* (*Cyamopsis psoraloides*) (Table 14.2; http://agritech.tnau.ac.in/org_farm/orgfarm_green%20manure.html). *Moong* bean is also used for its beans as well as for green manuring. Among these crops, *dhaincha* has been found to be the most beneficial. Uppal (1955) reported that out of three green manure crops tried, viz. *dhaincha*, sannhemp and *guar*, *dhaincha* was found to be the most useful. Dhawan (1964) reported that *dhaincha* green manure crop, in addition contained the highest amount of calcium, having 34.2% of CaO on ash basis. Although, it contains lesser nitrogen than *guar* but because of its high yield potential, it is preferred on the basis of its overall total nitrogen content. In a barren alkali soil (pH_2 10.6), amended with gypsum @ 7.5 t/ha, *dhaincha* gave a fresh biomass of 16 t/ha (Rao, 1986). The plant has in general, a vigorous growth habit and in partially reclaimed alkali soil, 20-24 t/ha fresh biomass is obtained in 45 days with the plant reaching 1.2 m in height. The plants are capable to nodulate vigorously in almost all types of soils including highly alkali soils. *Dhainha* contains 0.7% N, 0.1% P_2O_5 and 0.6% K_2O on moist basis (Dargan *et al.*, 1975).

Table 14.2. Green manuring crops, dry matter yields and nitrogen accumulation in soils

Crop	Age (Days)	Dry matter (t/ha)	N accumulated
Dhaincha (Sesbania aculeate)	60	23.2	133
Sannhemp *(Crotalaria juncea)*	60	30.6	134
Cow pea *(Vigna sinesis)*	60	23.2	74
Pillipesara (Vigna trilobata)	60	25.0	102
Cluster bean *(Cyamopsis terragonoloba)*	50	3.2	91
Sesbania (Sesbania rostrata)	50	5.0	96

DHAINCHA (*Sesbania aculeata*)

Dhaincha, also known as *Jantar*, is mainly grown for green manuring (Fig. 14.3). It is a quick growing leguminous plant and can be grown on all type of soils. It gives best results when grown in sandy loam to loamy soils. It can withstand adverse conditions of water logging, salinity and alkali. In waterlogged fields, it grows to a height of 1.5 to 1.8 m in a short time. It is an ideal green manure crop for rice growing soils.

Climate and Sowing Season

Land is prepared by giving 1-2 ploughings and plankings. *Dhaincha* is grown during the summer and the rainy seasons in north India. The summer crop is usually sown from end of April to early May under irrigated conditions after the harvest of *rabi* crop and is green manured after 7 to 9 weeks of growth from the middle of June to early July and the rice crop is transplanted soon after. In the *kharif* season cultivators usually prefer growing of some *kharif* crop like rice, maize and millets etc. However, if sowing season of *kharif* crop is over, *dhaincha* is sown either for seed production or green manuring. Green manuring is done before middle October to give 3-4 weeks´ time for its decomposition. A *rabi* crop, usually wheat is sown in November.

Fig. 14.3. *Dhaincha* for green manuring

Seed Rate and Method of Sowing

In highly alkali soils having pH_2 higher than 9.5, *dhaincha* cultivation is rather difficult. When assured irrigation facility is available, its sowing can be taken up as first crop during the process of reclamation in the month of April. In that event after levelling, gypsum @ 50% GR is broadcast uniformly. It is then mixed with the soil by shallow cultivation followed by planking. The field is then irrigated so that gypsum may react with the soil and soluble salts if present may leach down to some extent. After a few days, when this water is absorbed in the soil, thin irrigation is given and the *dhaincha* seeds at the rate of 50 to 70 kg/ha are uniformly distributed in the shallow standing water. Germplasm CSD 137 and CSD 123 screened at CSSRI, Karnal are tolerant to alkali, saline and waterlogged alkali soils (Table 14.3).

Table 14.3. Salt tolerant *dhaincha* germplasm

Dhaincha (Sesbania)	CSD 137	Its green biomass at 55 days is 43.2 and 33.0 t/ha in normal and in saline (EC_e 7 dS/m) and alkali (pH_2 up to 9.8) soils respectively.
	CSD-123	Its green biomass at 55 days is 35.0 and 28.0 t/ha in normal and in saline (EC_e 7.5 dS/m) and alkali (pH_2 up to 9.6) soils respectively.

For green manuring crop, use row spacing of 20-22.5 cm or broadcasting method but for seed purpose a spacing of 45 × 20 cm is preferred. Before broadcasting, the seeds are soaked in water for 8 to 12 hr. After sowing, thin irrigations are given at frequent intervals to keep the land wet. In spite of all these precautions, growth of the crop remains poor in high pH soil but after about one month, when the pH is somewhat reduced, the crop picks up growth.

Keeping in view these difficulties, the growing of *dhaincha* as a first crop is generally not recommended in high pH soils but it may be sown as a third crop after taking rice in *kharif* and wheat in *rabi*. After harvesting wheat, land is prepared by giving 2-3 ploughing and planking and then seed is sown in the standing water as mentioned previously. If the time is short, the crop can be sown without ploughing the land. In such a case, a thin irrigation is applied and the seeds are broadcast in the shallow standing water.

Fertilizer Application and Irrigation

A starter dose of nitrogen at 20 kg N/ha can be given as basal application. In normal soils, the green manure crop should be supplied P at the rate of 40 to 50 kg P_2O_5/ha. Application of phosphate fertilizers is not necessary in initial years of alkali land reclamation.

The *dhaincha* crop in salt affected soils need thin but frequent irrigations. As first crop, it requires about 20 irrigations. However, as third crop after taking rice and wheat, the crop can be grown successfully with about 10 irrigations. The rainy season crop usually requires no irrigation, if cultivated for green manure.

The *dhaincha* crop makes maximum growth in about 8 weeks after sowing. The standing crop is harvested and is then buried in the soil with a disc plough or with a tractor drawn disc harrow. In order to make the field available for the timely transplanting of rice crop, the crop can be buried even earlier at 6 to 7 weeks after sowing and the rice crop can be transplanted immediately thereafter. Being succulent and having a low C:N ratio, it decomposes easily and quickly. The standing water in the rice crop takes care of decomposition without any harm to the crop (Bhardwaj *et al.*, 1977). A growth period of about 45 days and no decomposition period between incorporation and transplanting resulted in better yield as well as saving in the time period of cultivation. Longer growth period renders the plant hardy and difficult to decompose. Besides, allowing time for decomposition of a succulent 45-days old *dhaincha* green matter results in greater losses of nitrogen through volatilization and less nitrogen recovery by the rice crop (Table 14.4).

Table 14.4. Effect of age and period of decomposition of *dhaincha* on rice yield

Age of *dhaincha* (days)	**Period of decomposition (days)**	**% N in the rice plant 25 days after transplanting**	**Rice grain yield (t/ha)**
45	0	2.73	5.93
45	10	2.09	5.69
45	20	1.99	4.39

Yield

Dhaincha grown during summer season adds 10 to 20 t/ha of green matter/ha depending upon the stage of reclamation whereas the yield of rainy season crop may be 20 to 25 t/ha. While CSD 137 gives a green biomass of 43 t/ha in normal and 33 t/ha in salt affected soils, CSD 123 gives a green biomass of 35 t/ha and 28 t/ha at 55 days in respective soils. *Dhaincha* contributes about 80 kg nitrogen to the rice crop (Fig. 14.4). About 50% of the added nitrogen is converted into readily available ammoniacal form in 4-6 days and the rest in 10-20 days of application under the normal temperature conditions that prevail during the main rice crop.

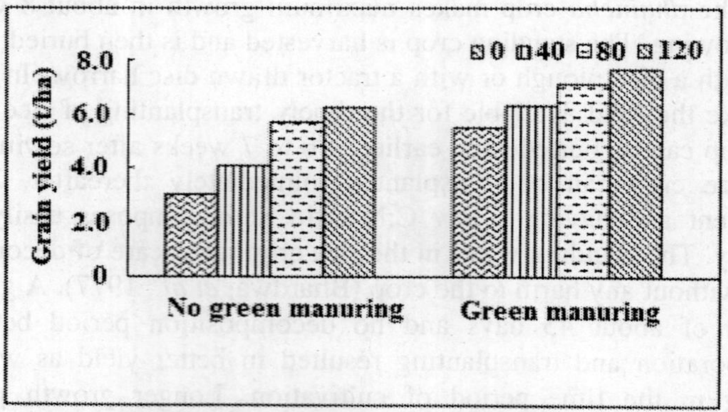

Fig. 14.4. Nitrogen contribution through green manuring (kg/ha) with *dhaincha* to rice crop

Depending upon the soil and rice variety, green manuring with *dhaincha* has been found to increase yields by 30 to 60% over no N application (Rao, 1986). These results show green manuring with *dhaincha* is beneficial in alkali soils since it not only increases rice yield but also saves almost half the dose of nitrogen. The cost of growing the green manure crop (*dhaincha*), availability of irrigation water during summer months and other economic constraints generally influence the decision of a farmer to include a green manure crop in the crop rotation.

■■■■

CHAPTER 15

GRASSES AND FOREST PLANTATIONS

GRASSES

The country, by 2020, is likely to face an estimated shortage of 728 million tons of green fodder and 157 million tons of dry fodder. Besides, the pasture lands which have already reduced from about 70 M ha in 1947 to just about 38 M ha are being diverted to cultivated crops and getting degraded. Cultivation is expanding into rangelands that receive precipitation at 30-50 cm/annum. The others have either already degraded or are in the process of degradation with average carrying capacity of less than 1 ACU (adult cattle unit). Village common lands, fallows and wastelands especially the salt affected lands can form an important source of forage to the village cattle, if used for cultivating forage grasses. Recent studies have proved that several forage grasses can be grown successfully on these deteriorated lands with minimum inputs (Table 15.1). One cultivated, the para grass (*Bracharia mutica*) and one wild, the *Diplachne fusca* locally called Karnal grass grow well in alkali soils even without application of chemical amendment. The latter can also tolerate water logging and soil salinity. These grasses have reclamative properties such that after few years of cultivation, less tolerant crops can be successfully grown on these lands.

Karnal Grass

Karnal grass (*Diplachne fusca*), also known as Australian grass and *kallar* grass, grows quite satisfactorily in alkali soils having pH_2 as high as 10.5 (Fig. 15.1). It is most tolerant to alkali conditions and it can come on its own at relatively lower elevations in alkali fields

Table 15.1. Promising forage grasses for alkali soils

pH*	Grasses (Botanical name)
10.4	Karnal grass (*Leptochloa fusca*), Rhodes grass (*Chloris gayana*), Gatton panic (*Panicum maximum*), Bermuda grass (*Cynodon dactylon*)
10.0	Para grass (*Brachiaria mutica*)
9.6	Blue panic (*Panicum antidotale*)
9.2	Hybrid Napier (*Pennisetum purpureum*), *Setaria grass* (*Setaria anceps*)
8.8	Anjan grass (*Cenchrus ciliaris*), Guinea grass (*Panicum maximum*), Deenanath grass (*Pennisetum pedicellatum*)

*The pH values at which yield reductions is 50% or less

where monsoon rainwater stands for comparatively longer duration. One can cultivate this grass without or with limited application of amendment. The grass is perennial, erect type and grows taller than other grasses that come up naturally on alkali soils, going as high as 75 to 100 cm. It is quite succulent. It can be propagated through root slips and stem cuttings or by direct seeding. Once the grass is established at a place, replanting is not required since the fallen seeds germinate at proper time in the next season. Planting of the grass is usually done in February-March or in July during the rainy season. The rooted slips/stem cuttings are planted in rows 25 cm apart, with 20 cm spacing between the plants. In the first year, growth of rooted slips is somewhat better than stem cuttings but later on growth becomes almost similar. Irrigation is necessary after planting and afterwards except during monsoon (semi-arid climate). Drainage requirement are minimal and rainwater can be stored in dyked fields for as much as 8 days without sacrificing the yield.

Crop planted in July can give 3 cuts yielding about 18 t/ha but in subsequent years, 6-7 cuts can be taken annually with forage yield of 40 to 45 t/ha at ESP as high as 60. The grass is relished by cattle at all its growth stages. Its dry matter contains 62.1, 21.7 and 8.6% non-digestible fiber, cellulose and protein respectively. The minor variations in these values may occur due to growth stage at the time of cutting and level of soil salinity or alkali. The grass remains dormant during winter months. With cultivation of the grass, the soil pH reduces considerably within 2-3 years. At this time, the land can be used to cultivate salt tolerant rice and wheat varieties.

Fig. 15.1. Performance of Karnal grass in alkali soil

Para Grass

Introduced in India in 1894, it is now extensively grown in various states of India. Para grass (*Brachiaria mutica*), commonly known as buffalo grass, California grass, giant couch grass, water grass and *Pani Wali ghas,* is an exotic, hairy and perennial grass. It spreads rapidly by surface runners and profuse roots at nodes. The leafy material is relished by the livestock, and on dry matter basis it contains 8.1% digestible crude proteins.

Climate and Soil

The grass grows well in warm humid climate. In protected areas, it can persist even with low rainfall. It can be cultivated on all types of soils. However, it thrives well on low lying lands subject to water logging or flooding. As such, it is ideally suited for cultivation in humid regions, water-inundated lands and sewage farms. It is quite profitable to grow on alkali or saline soils where nothing else survives. Like Karnal grass, it also helps to reclaim alkali lands. After levelling, the land should be well prepared to obtain a fine tilth.

Fig. 15.2. Para grass in a normal soil

Agronomic Practices

If irrigation facilities are available, the grass is planted in February-March. Under rain fed conditions, it is planted with the onset of rains in June-July. The grass is propagated by root slips or stem cuttings. The roots are planted 50 cm apart in either direction and irrigation is applied just after planting. About 400 kg of root setts with 2-3 nodes or 40,000 rooted slips per hectare are required. For getting high yields, about 25 t/ha of FYM or compost should be applied as basal application. The fertilizer requirement is 40:50:50 kg NPK/ha. Zinc sulphate at 10 kg/ha should be applied in alkali soils at the time of final preparation. Additional nitrogen should be applied @30 kg/ha after each cut.

Harvest and Yield

The first cut is ready in about 3 months after planting and subsequent cuts can be taken at one month interval. In normal soils, the green forage yield ranges from 130 to 140 t/ha. The growth is quite limited at low temperatures experienced during December to February. In the first year, crop planted in rainy season on highly alkali soil (pH_2 10.0) without any gypsum gives only two cuts with a total yield of about 20 t/ha of green fodder. At 50 ESP, the grass can yield up to 50 t/ha. The crop planted in March can give about 60 t/ha in 5-6 cuttings. In reclaimed soil (pH_2 8.5) quite high yield upto 100 t/ha can be obtained in 6-7 cuts. It can withstand moderate grazing. The first grazing should be deferred till the grass attains the height of

30-60 cm. A slight disadvantage is that it tends to become more fibrous and hairy under stressed conditions.

PLANTATION

Reclamation of alkali soils for arable purposes is technically feasible, economically viable and eco-friendly proposition (Singh, 1995, Singh, 1996). However, lands owned by the village *Panchayats* or by the Government are not amenable to these technologies and have failed to make a dent. It is because under the community ownership, no one owns the responsibility. Resource poor farmers may also fail to take advantage of the technology for arable cropping. This section looks at the possibilities of forest plantations as an alternative land use. Besides meeting the wood requirement such an approach offers an opportunity to increase the forest cover of the country, which presently is far below the minimum requirement of one third of the total land surface. Afforestation of these barren lands will also contribute to the maintenance of healthy agro-ecological system and averting the environment hazards.

Table 15.2. Salinity tolerance of some fuel wood trees

Tolerant (EC_e 25-35 dS/m)	*Tamarix troupii, T. articulata, Prosopis juliflora, Pithecellobium dulce, Parkinsonia aculeata, Acacia farnesiana*
Moderately tolerant (EC_e 15-25 dS/m)	*Callistemon lanceolatuy, Acacia nilotic, A. pennatula, A. tortilis, Casuarina glauca 13144, C. glauca 13987, C. obesa 27, C. glauca (FRI), C. equisetifolia (FRI), Eucalyptus camaldulensis, Leucaena leucocephala, Erescentia alata*
Moderately sensitive (EC_e 10-15 dS/m)	*Casuarina cunninghamiana (FRI), C. Cunninghamiana (Aust), Eucalyptus tereticornis, Acacia auriculiformis, Terminalia arjuna, Pongamia pinnata, Guazuma ulmifolia, Leucaena shannoni, Samanea saman, Albizzia caribea, Senna atomaria*
Sensitive (EC_e 7-10 dS/m)	*Syzygium cumini, S. fruticosum, Tamarindus indicia, Salix* spp., *Acacia deami, Albizzia guachepela, Alelia herbertsmithi, Caesalpinia eriostachyas, C. velutina, Haematoxylon brasiletto*

The relative salt tolerance of forest plants to soil salinity and soil alkali is given in Table 15.2 and 15.3. The data can be used to make a knowledgable decision on the selection of forest plants to green these lands.

Table 15.3. Relative tolerance of forest tree species to soil pH/sodicity

Highly tolerant $pH_2 > 10.0$	**Tolerant pH_2 9.5 -10.0**	**Moderately tolerant pH_2 9.0 – 9.5**	**Sensitive $pH_2 < 9.0$**
Phari Kiker (Prosopis juliflora), Kikar (Acacia nilotica), Pharans (Tamarix articulata), Jor-tor (Casuarina equisetifolia)	*Jangal jalebi (Pithecello – bium dulce), Safeda (Eucalyptus tereticornis), Arjuna* tree *(Terminalia arjuna), Imly (Tamarindus indica), Papri (Pongamia pinnata), Khair (Acacia catechu), Dhak (Butea monosperma), Lasura* (Cordia rothii), *Siris (Albizzia lebbeck), Parkinsonia aculeata*	Kigelia *(Kigelia pinnata), Shisham (Dalbergia sissoo),* Sieamea *(Cassia siamea), Neem (Azadirachta indica),* Siliver oak *(Grevillea robustra),* Dhaincha *(Sesbania sesban), Bakain (Melia azedarach), Susbabul (Leucaena leucocephala), Saunjana (Moringa olifera),* Bamboo *(Morus alba), Siris (Albizia procera), Acacia auriculiformis, Acacia senegal*	*Poplar (Populus deltoides),* Teak *(Tectona grandis), Ficus virena, Bombax ceiba, Hardwickia binnata, Thespesia populnea, Bauhinia variegata, Lagerstro emia speciosa, Cadrella toona*

Tree Plantation Techniques in Alkali Soils

Depending upon the alkali conditions, topography and resources, any of the techniques namely pit method (Fig. 15.3 left), auger hole method (Fig. 15.3 middle), combination of pit and auger hole method (Fig. 15.3 right), ridge plantation and trench plantation can be used.

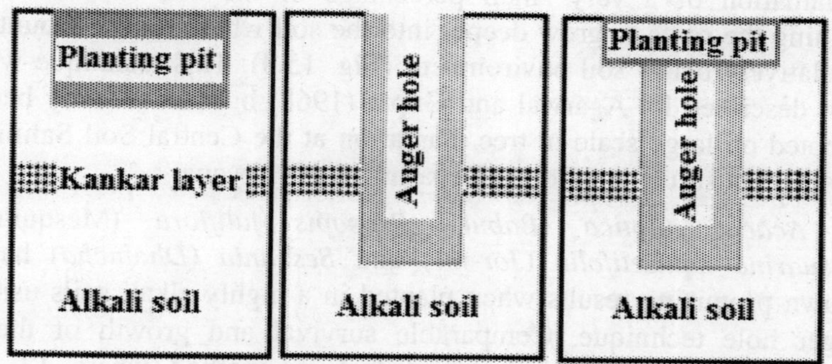

Fig. 15.3. Planting methods in alkali soils

Pit Method

Depending upon the type of soil, and the depth of the *kankar* layer, pits of different sizes are used to plant trees. In alkali soils, the pits of size 90×90×90 cm are dug. The depth of the pits may be increased to 1 m or so if *kankar* layer is expected at about this depth. It will help to break this layer, which may otherwise hinder the proliferation of roots. The soil of the pits is either replaced by a good soil from nearby place or amended suitably by adding gypsum or any other amendment. FYM and basal dose of nutrients is also added at this stage. The results revealed that the species like *Eucalyptus* hybrid, *Prosopis juliflora* and *Acacia nilotica* can be grown successfully by treating the soil of the planting pit with gypsum + FYM. Performance is as good as with the soil replacement.

Auger Hole Technique

This technique of tree plantation has been devised keeping in view that a hard *kankar* layer is always encountered at about 1 to 1.5 m depth. Thus, to pierce this layer, an auger hole of about 15 cm and 1.2 m to 1.8 m deep is made. A mechanically operated auger is normally used since manually preparing the auger holes is quite tedious (Fig. 15.4). Auger holes are refilled either with the normal soil from adjoining areas or properly amending the original alkali soil. For amending the soil, mix 2-3 kg gypsum and 8 kg FYM with the original alkali soil per post hole of 15 cm diameter of 150 cm depth. The technique is cheaper than the pit method as it involves reclamation of a very small percentage of the soil and aims at training the roots to grow deeper into the soil where roots encounter a relatively better soil environment (Fig. 15.3). This technique was first described by Agarwal and Gupta (1968) but has recently been adopted on large scale in tree plantation at the Central Soil Salinity Research Institute and Forest Department of Haryana.

Acacia nilotica (*Babul*), *Prosopis juliflora* (Mesquite), *Casuarina equisetifolia* (*Jor-tor*) and *Sesbania* (*Dhaincha*) have shown promising results when planted in a highly alkali soils using auger hole technique. Comparable survival and growth of these species was observed in conventional pit method where 12 kg gypsum and 24 kg FYM was applied.

Fig. 15.4. Tree plantation in alkali lands: auger holes being made (left) and field with planted trees (right)

Combination of Pit and Auger Hole Methods

While auger hole technique has several plus points, it has few limitations as well. It has been observed that the initial advantage gained in growth of plants in this method during first 2-3 years is evened out in later years. This was particularly observed in the case of *Eucalyptus* plantations. It is attributed to root proliferation mainly in the vertical direction. It is because the pH value remains high in the area surrounding the auger hole, that restricts the horizontal root proliferation. Besides, low fertility status of lower soil layers impacts the nutrient uptake. Therefore, a combination of the pit and auger hole technique is used (Fig. 15.3 right). In recent studies the usefulness of this technique has been established for horticultural plants.

Ridge Plantation

Run-off potential of alkali lands is quite high. The run-off resulting from barren lands causes flooding in low-lying areas. This technique of plantation helps to conserve rainwater within the planted area. In this technique, ridges of different sizes and shape and appropriate spacing are made depending upon the plant species, soil type and climate of the area. While saplings are planted on the ridges, trenches are used to store the rainwater. The biomass of the plants was much higher in the ridge-trench than the flat method of planting. The cost of ridge making and the overall economics might determine the application of this technology by the forest department.

Trench Plantation

In this technique, trenches of different sizes and shapes are opened either manually or mechanically. Trenches are refilled by adding suitable amendments. The plants are planted in the trenches. Mesquite was successfully established with this technique (Singh and Dagar, 1998). Although the technique is not as good as the pit or the auger hole method yet it is cheaper and easier to implement to cover large areas.

Cultural Practices

The success of tree plantations depends to a great extent on the kind of planting technique, soil preparation, use of amendments, time of plantation and selection of suitable species. Besides, well designed cultural practices can help to improve the performance of tree species under the adverse environment. Some practices are as follows.

Planting Time

The two planting seasons for woody species are: February to March and July to September, the latter being preferred under rain fed situations. If assured irrigation facilities are available and the trees are tolerant to cold during the establishment stage, the February-March season is preferred. Deciduous trees like *shisham* and poplar should always be planted in this season.

Tree Spacing

If the primary purpose of plantation is firewood production or energy plantation then 2 m x 1 m or 2 m x 2 m spacing may be adopted. In alkali soils, closer planting covers the risk of high mortality of seedlings. In agro-forestry, the minimum spacing between rows and plant should be kept at 4 m and 2 m, respectively. If the tree is spreading type like *Prosopis, Acacia, Dalbergia* etc. the spacing should not be less than 5 m x 3 m. Further, if the purpose is firewood and timber production, high density plantation may be preferred. After about 3 or 4 years, 50% of the plants may be harvested to meet the firewood needs while the remaining 50% may be kept for timber production.

Irrigation

Irrigation especially during the establishment stage of the woody species is a critical input. Fortnightly irrigation for the first three months, and monthly during the later growth period up to two years, results in significant increase in survival, growth and bio-mass production of some forest tree species (Fig. 15.5). After two years, irrigation may be withdrawn under good monsoon rainfall conditions.

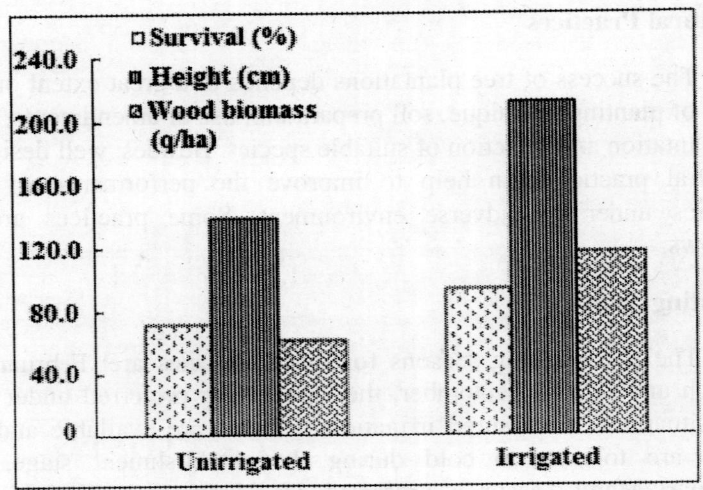

Fig. 15.5. Effect of irrigation on mesquite plants

Nutrient Management

Alkali soils being deficient in N, Ca and Zn, application of N to the non-nitrogen fixing trees improves their growth significantly. A basal dose of N (50 g urea/plant) is mixed in the soil mixture used to fill the auger-holes. Zinc sulphate at the rate of 10 g per auger-hole may also be applied at the time of planting. Later, 50 g urea is applied in two equal splits every year, one during the monsoon season and the second in February-March. Application of phosphorus and potash are not needed as these are present in sufficient quantities. In saline soils, P and K may be applied along with basal dose of N. Application of farm-yard manure does help in meeting the iron needs of the woody species. There is no need to apply N to N-fixing species.

Pruning/Lopping

Pruning is the removal of unwanted side branches of trees which are photo synthetically less active. It is also practised to give a proper shape to the trees and to enhance the growth of vegetation underneath or of inter crops, if cultivated. At the first lopping, all the side branches should be removed up to one meter height, and at the second lopping 2 m stem should be kept free of unwanted side branches. Most of the trees of the tropical regions should be lopped between November and January when they are in dormant phase. The biomass so harvested can be used as firewood. Lopping of *Prosopis juliflora* at 16 and 40 months after planting (at 2 m x 2 m spacing) yielded about 13 t of woody bio-mass/ ha and increased the green forage yield of inter-planted Karnal grass.

Inter-planting of Grasses

The promising grasses for planting along with trees are: *Aeluropus lagopoides, Brachiaria setigera, Chloris barbata, Echinocloa colonum, Dicanthium annulatum, Leptochloa fusca, Sporobolus helvolus, Cynodon dactylon, Phragmites* spp. and *Sida* spp.

Soil Amelioration by Trees

In addition to bio-mass, trees ameliorate and improve the physical, chemical and biological properties of the soil on which these are grown (Singh *et al.*, 2008). Incorporation of organic matter through leaf litter and foliage results in favourable changes in the physico-chemical properties of the soil. Growing leguminous trees like *Prosopis, Acacia* and *Casuarina* ameliorates these soils at a much faster rate than non-leguminous trees, because of the build-up of organic matter and the recycling of important nutrients. It has also been proved that highly alkali soils reclaimed through tree plantations can be successfully used for growing all kinds of agricultural crops after harvesting the trees.

Agro-forestry for Moderately Alkali or Reclaimed Lands

Most promising trees for agro-forestry on reclaimed alkali lands are: *Populus deltoides* (*Poplar*), *Eucalyptus tereticornis* (*Safeda*), *Acacia nilotica* (*Kikar,* Fig. 15.6) and *Tectona grandis* (*Teak*). The suitable inter-crops during establishment stage are rice-*berseem* and rice-wheat. From third year onwards, as tree canopies develop, shade loving species such as *berseem*, guinea grass, turmeric and colcasia perform well. Poplar based agro-forestry is more remunerative because of its faster growth and high wood price in the market compared to *safeda* and *kikar* based systems. The intercrops have been found to aid the growth of the trees.

Fig. 15.6. Performance of *Acacia* in a semi-reclaimed alkali land

Planting Technique for Saline Soils

In general, saline lands are more suited to plant fuel wood than timber species. Even the production of fuel wood depends on the ability of the tree species to withstand salinity, drought and various other adverse environmental conditions. Besides salinity/tolerance of the plants (Table 15.2), site conditions also determine as to which plant species to grow. For example *Eucalyptus camaldulensis* can tolerate saline groundwater and periodic flooding, but cannot be planted close to the sea coast because of its intolerance to windblown salt. However, *Casuarina equisetifolia* is able to tolerate the salt spray and provides valuable fuel but is rarely planted inland due to its relatively slow growth rate and poor coppicing ability.

Unlike alkali soils, furrow planting is most suitable method to plant trees in saline soils. Tomar and Gupta (1987) compared the relative performance of three planting methods i.e. subsurface, ridge-trench and furrow planting. The survival and growth of 5 tree species, 27 months after transplantation revealed that furrow planted trees in general performed better than trees planted with other methods. Other studies have also shown that irrigation in furrows or in circles around the stems (basin) improved the performance of the trees, furrow irrigation being the best.

Cultural Practices

Alike arable cropping, high mortality of saplings and limited growth of trees requires that trees are planted at closer spacing than in a normal land (Singh *et al.*, 1993). The exact spacing of course depends upon the kind of the tree species. It may be noted that a denser canopy resulting from closer row-to-row and plant-to-plant spacing helps to check soil salinization at early growth stage.

Irrigation is necessary for the survival and growth of plants as adverse effects of salinity get blunted with irrigation. Water application methods that prevent excessive accumulation of salts in the root zone are preferred. Good quality water for irrigation leaches down the salts from the soil profile. Frequent irrigations with good quality water therefore, are highly beneficial to raise trees in saline soils as irrigation at short intervals provides a low salinity environment to the young roots (Singh *et al.*, 1993). Wherever possible, irrigations should be applied during dry and hot periods.

Since nutrient status of waterlogged saline soils is generally poor, many times even the most tolerant tree species do not grow well. Application of fertilizers is one way to enhance tree growth. The exact amounts and kinds of nutrients would depend on the kind of salts and salt concentration of the soil. Tomar and Yadav (1980, 1985) reported that application of phosphatic fertilizer significantly benefited the growth of *Eucalyptus hybrid* by counteracting the harmful effect of saline water irrigation. On the other hand, nitrogen application caused a set back to its growth. More studies are needed to critically evaluate the response of fertilizers to growth of tree species in saline soils.

Since, Weeds growth in these soils is limited; weeds should not pose a threat to the planted saplings. However, weeding operation helps to eliminate the competition for limited moisture, necessitating that this operation should be carried out as and when necessary.

The promising grasses for inter-planting with trees are the same as that of alkali lands namely: *Aeluropus lagopoides, Brachiaria setigera, Chloris barbata, Echinocloa colonum, Dicanthium annulatum, Leptochloa fusca, Sporobolus helvolus, Cynodon dactylon, Phragmites* spp. and *Sida* spp.

Horticultural Plants

The most promising planting technique for horticultural plants in alkali soils is the combination of pit and auger hole method. For this purpose, pits of 60 cm x 60 cm x 60 cm dimensions are made manually. In the center of each pit, an auger hole of 20-25 cm diameter and 140-160 cm deep (ensure that it pierces through the *Kankar* pan) is made using tractor mounted augers. These pits cum auger holes are filled back with a mixture of original soil + 8 to 10 kg gypsum + 10 kg FYM + 20 kg river sand. Selection of appropriate plants and varieties is made depending upon the pH/ESP of the soil (Table 15.4, Singh *et al.*, 1994).

To improve upon survival and quality, salt tolerant rootstock is planted, followed by *in-situ* grafting for the desired characters. Some of the fruit species are sensitive to water logging and frost such as pomegranate and sapota. So, proper drainage in the rainy season and protection of the young tender saplings during winter is a pre-requisite for the overall success of the horticulture plants. Apply

Table 15.4. Fruit species and their relative tolerance

Fruit species	**Tolerance**		**Suitable varieties**
	pH	**ESP**	
Aonla - Emblica officinalis	9.5	40	Chakaiya, NA-7, NA-6
Guava - *Psidium guajava*	9.2	30	Sardar (L-49)
Ber - Ziziphus mauritiana	9.5	45	Banarasi, Karaka, Ponda
Bael - Aegle marmelos	9.0	30	Narendra-5, Narendra-9
Mahua - Madhuca latifolia	9.2	-	Narendra selection
Grape - *Vitis vinifera*	9.0	30	Beauty seedless, Perlette
Karaunda - Carrisa carandus	9.2	35	Narendra selection

regular irrigation to keep the orchard soil moist to off-set frost damage. For better growth of the plants, apply 25% higher dose of nitrogen and FYM every year than recommended for normal soils. Grow *Sesbania* during rainy season as a green manure crop to improve soil fertility.

■■■■

CHAPTER - 16

ECONOMIC AND SOCIAL ASPECTS OF RECLAMATION

INTRODUCTION

In order to ensure a well-fed society in India in 2020, a long-term projection was made on the basis of an extreme scenario (high population growth). It was food grain requirement of pegged at 273 million tons. It was considered a challenge, as it required an additional production of 93 million tons over and above the production level of 180 million tons achieved in the year 2000. It is satisfying that according to 4th advance estimates of production of major crops for 2016-17, total food grains production is estimated at a record 275.68 million tons. Clearly, not only the target is exceeded but the same has been achieved 3 years ahead of the stipulated date. On the other hand, we are still keeping our fingers crossed as the self-sufficiency achieved is quite precarious. One/two poor monsoons may pale this achievement. Moreover, CSSRI projections reveal that reclaimed salt affected soils during this period also contributed more than 10 million tons to this total. Therefore, to meet the future requirements, efforts in land reclamation have to be pursued with the same vigour as in the past. Besides, it is foreseen that future challenges to ensure food security in 2050 have to be met with shrinking land and water resources. For example, best quality land and water are being allocated to meet the requirements of the domestic and industrial sectors. Therefore, agriculture has to fend with limited or so called poor quality resources. It is satisfying that recent advances in the field of reclamation have shown that the marginal productivity of currently undeveloped salt affected land resource is much higher than the already developed normal cultivated areas. It calls for increasing investments in land reclamation. Primary investments for this purpose may come from the private sector (farmer), while secondary investments may be shared by public and private (industry) institutions. Critical evaluation of economics and the social environment may only be able to bring together these stakeholders. This chapter describes the methodologies adopted and some case studies that highlight the view that reclamation is an economically viable and socially acceptable proposition.

COST COMPONENTS OF RECLAMATION

The reclamation cost of salt affected soils can be considered under two broad parts.

Capital Cost

The capital cost of reclamation is the sum of all the expenditures required to make the barren soil fit for cultivation. For salt affected soils, items contributing towards capital cost are of a non-recurrent nature. These include the cost of land levelling, *bunding*, provision of irrigation channel system and surface drainage network, cost of groundwater development (installation of tube wells), purchase of chemical amendments and their application, installation of subsurface drainage, leaching etc. Cost of alkali land reclamation in Uttar Pradesh has been assessed at about 50,000 excluding the cost of tube well (Table 16.1, Thimmappa *et al.*, 2013). Subsurface drainage costs are substantially higher and may exceed 110,000/ha in heavy textured soils (Table 16.2).

Table 16.1. Investment on alkali land reclamation

Particulars	Cost*(Rs./ha)	Percentage of total cost
Gypsum @12 t/ha	28800	57.97
Farm development activities	13350	26.87
Gypsum application, irrigation, flushing and leaching of salts	7530	15.16
Total cost (without subsidy)	49680	100.00
Total cost (with subsidy**)	13108	-

Note: *based on 2011 prices, **the subsidy amount given to 'B' class land and the percentage of subsidy varied amongst different components.

Table 16.2. Cost of SSD for a heavy soil with gravity outlet and medium textured soil with pumped outlet

SSD Component	Cost (Rs./ha)	
	Heavy texture soils (30 m spacing) with gravity outlet	**Medium texture soils (67 m spacing with pumped outlet)**
DPR	2,500 (2.2)	2,500 (3.7)
Pipes & fittings	41,989 (37.6)	21,750 (29.7)
Filters	14,776 (13.2)	8,334 (11.4)
Structures	3,785 (3.4)	7,142 (9.8)
Installation cost	36,766 (33.0)	22,953 (31.4)
Dewatering	0 (0.0)	500 (0.7)
Supervision charges	4,000 (3.5)	4,000 (5.4)
Contingency	5,352 (4.8)	3,431 (4.7)
Farmer's training	750 (0.7)	750 (1.0)
M&E	2,000 (1.8)	2,000 (2.7)
Total:	111,500 (100)	73,000 (100)

Note: Figure in parenthesis show the share (%) of total cost

Annual Cost

The components of the annual cost include operation/ maintenance of irrigation and drainage system, repair of dykes and bunds and the cost of raising the crops over such lands (Table 16.3). In alkali soils, besides the cost of raising crops, other cost components are of very little magnitude. The recurrent costs in saline soils reclamation include the annual costs of operation/maintenance of the subsurface drainage system, cost of pumping subsurface water from the drains (in the absence of gravity outlet) and periodic leaching of salts. The amount of water and recurrent expenditure on leaching depends upon the quantity of salts added to the soil, soil texture; quality of water used for leaching and of course the climate.

Table 16.3. Recurring costs in land reclamation

Sr. No.	Item	Major intervention (Alkali)	Major intervention (Saline)
1.	Operation and maintenance	Surface drainage	Subsurface drainage
2.	Crop production (*Kharif*)	Rice	Pearl millet/ sorghum/ cotton/Rice
3.	Crop production (*Rabi*)	Wheat/*Shaftal*	Wheat/barley/mustard
4.	Crop production (*Zaid*/green manuring)	*Dhaincha*	*Dhaincha*

BENEFITS

Like costs, benefits are also assessed to conduct a worthwhile economic analysis. Benefits having direct relevance to economic analysis are:

- Increase in crop yield
- Improved product quality and higher profits
- Reduced cost of farm operations and inputs
- Higher cropping intensity
- Advancement of farming operations for sowing of post monsoon crops resulting in higher yield
- A change in cropping pattern with introduction of more paying crops
- A source of irrigation (Subsurface drainage)

To arrive at the most reasonable numbers for economic analysis, neither costs nor the benefits should be assigned higher or lower values.

ECONOMIC CRITERIA

A project is worthy of investment, only if it is economically viable and financially sound. The judgment on economic and financial viability is made on the basis of one of the following three criteria:

- Net benefits or financial feasibility
- Benefit-cost ratio
- Rate of internal return
- Payback period

For the project to be financially sound, net benefits should exceed the costs, benefit-cost ratio should be more than one and internal rate of return must be higher than the rate of return on investment made in other avenues say fixed deposits in banks. Payback period should be quick with a small gestation period. For a farmer, social benefits become secondary although from state, regional or national point of view, these issues may have great significance.

Net Benefit Analysis: Equated Seasonal Installment (ESI)

Farmers usually borrow the capital for land reclamation. Cost of the borrowed capital can be readily estimated from the known interest rate and the period within which the loan has to be repaid. The equated seasonal/annual installment, (ESI) is calculated by the following formula.

$$ESI = (L * I) * [(1+I)^M / \{(1+I)^M - 1\}] \quad (16.1)$$

Here, M is the loan period in seasons (Years*2), I is the interest rate, (Interest rate per annum /2) /100, L is the borrowed capital in Rs.

For a financially viable project, the seasonal/annual benefits must exceed the seasonal/annual installment.

Example 16.1

A farmer obtains a soft loan of Rs. 100,000 at an interest rate of 7% per annum. He has to pay ESI for 10 years. What should be the minimum additional earnings per season so that he is able to repay the loan installments?

Interest rate =0.07/2 = 0.035 per season

Seasons =10*2 = 20

Incorporating the values in eq. (16.1), we have

$$ESI = (100000 * .035) * [(1+0.035)^{20} / \{(1+0.035)^{20} - 1\}]$$

$$ESI = \text{Rs. } 7036/-$$

It means the farmer must earn an additional income of Rs. 7036/- per season.

Benefit-Cost Analysis

In agriculture, seasonal, annual or long-term benefit-cost ratio (BC ratio) is calculated by simple accounting procedures wherein seasonal costs and benefits are accumulated for the desired period of time. Since there are no initial costs involved, procedure works well. But in the reclamation projects, where there are initial investments, one needs to work out the cost-benefit considering the investment on land reclamation. Usually two methods are employed:

- Fixed and variable cost approach
- Net present worth approach

Fixed and Variable Cost Approach

Fixed costs generally include depreciation, interest and insurance while variable costs include repair and maintenance, fuel and lubrication and cost of crop production. Because of the nature of land reclamation activity, one can neglect depreciation since land value increases after reclamation. The annual interest and insurance (particularly if investments are made through bank loan) is added to the variable cost. The expenditure on fuel and lubrication for the tube well is added to the costs of crop production. The returns from the crops and its byproducts constitute the benefits side. Although the costs and returns change from year to year, depending upon the internal and external factors of production, for a long-term assessment, it can be assumed that the benefits stabilize after the first 2-3 years of reclamation. In reality, they may increase with time.

Net Present Worth (NPW) Approach

Net present worth is also called Net Present Value (NPV). It is the difference between the present values of benefits over the present values of costs. To estimate the NPW, cost involved and benefits realized on any future dates are assigned equivalent present values using discount factors. Both net present worth of benefits and net present worth of costs are calculated. The difference gives the NPW of the project.

$$\text{NPW of benefits} = \sum_{t=1}^{n} B_t/(1+i)^t = \sum_{t=1}^{n} B_t * DF_t \qquad (16.2)$$

Here B_t is the benefits received in t^{th} year, and t is the year say 1, 2, 3 …..n, i is the Dr/100 such that Dr is the discount rate (%), DF_t is the discount factor in t^{th} year (Eq. 16.2). Discount factors for a discount rate of 8% over a period of 30 years are given in Table 16.4. Normally for alkali and saline soils (with deep water table) time t is taken as 10 years while for waterlogged saline soil it is taken as 30 years.

$$\text{NPW of costs} = \sum_{t=1}^{n} C_t/(1+i)^t = \sum_{t=1}^{n} C_t * DF_t \qquad (16.3)$$

Here C_t is the costs incurred in t^{th} year. Other factors are as described for eq. (16.2).

Subtracting NPW of benefits from NPW of costs, we get NPW as:

$$\text{NPW} = \sum_{t=1}^{n} (B_t - C_t)/(1+i)^t \qquad (16.4)$$

If the NPW is positive, it indicates the economic viability of the project.

Table 16.4. Discount factors at a discount rate of 8%

Year	Discount factor	Year	Discount factor	Year	Discount factor	Year	Discount factor
1	0.926	9	0.501	17	0.270	25	0.146
2	0.857	10	0.464	18	0.251	26	0.135
3	0.784	11	0.429	19	0.232	27	0.125
4	0.735	12	0.397	20	0.215	28	0.116
5	0.607	13	0.368	21	0.199	29	0.108
6	0.681	14	0.341	22	0.184	30	0.099
7	0.584	15	0.316	23	0.171		
8	0.541	16	0.292	24	0.158		

Benefit-cost Ratio (B:C ratio)

It is the ratio of the present value of benefits (eq. 16.2) divided by the present value of costs (eq. 16.3). For an economically viable project, the ratio should be more than one.

Internal Rate of Return

The IRR is used to find out the rate of return, which a project is likely to generate over its useful life. This measure is practically used for all economic and financial analysis of reclamation projects by the international financing agencies. The IRR is the discount rate at which NPW is equal to zero. At this rate, BC ratio will be equal to one.

Payback Period (PBP)

Payback period is the length of time (years) that it takes for a project to recover its initial cost out of the cash receipts that it generates. Payback period is calculated by summing up the cash flows successively from the beginning such that the value considering various signs turns out to be nearly zero. The year in which it happens is the year at which investments are fully returned and the year is designated as PBP. The shorter the PBP, more desirable would be the investments in the project.

When the net annual cash flow is same every year, the payback period is given as:

$$PBR = Investments/Annual\ cash\ inflow \qquad (16.5)$$

CASE STUDIES OF AKALI SOIL RECLAMATION

Mishra (1977) reported that reclamation work taken up with complete technology and package of practices is remunerative. In an economic study of a demonstration plot in Hisar district, the yields of rice and wheat in 1975-76 were 4.45 and 3.46 t/ha. The non-recurring expenditure on land development and reclamation came to Rs. 2454.68 /ha. The production cost of rice and wheat came to Rs. 2269.16 /ha and Rs. 2237.30/ha respectively. The returns in terms of produce values of rice and wheat were Rs. 3239.52 and Rs. 4026.10/ha respectively. John *et al.* (1977) in an economic study in

Haryana concluded that (i) a fair return on the investment including repayment is possible even in the first year; (ii) it is possible that the returns will further go up when the reclamation process is complete in a span of few years and the entire reclamation cost is recovered in much shorter period, and (iii) the reclamation of alkali soils will enhance the land value, which itself will be sufficient to compensate the investments made. Recent experiences in Uttar Pradesh also attest to this observation (Table 16.5).

Table 16.5. Land value before and after reclamation in Uttar Pradesh

Category of land	Land value (Rs./ha)	
	Before reclamation	After reclamation
B+	115,000	170,000
B	55,000	155,000
C	30,000	125,000

Singh (1978) evaluated the economics of alkali land reclamation in a cluster of villages namely Kachwa, Sagga, Birnaraina, and Sambhli, in district Karnal. The average values of data on various aspects of land reclamation collected from 68 farmers for the first year of reclamation (1975-76) reveal that the per hectare cost of land preparation and amendment application was Rs. 3,257.90. Of this, the cost of gypsum alone accounted for nearly two thirds. The cost of cultivation of the rice-wheat crops was Rs. 3,334.67 per ha. The net profit was Rs. 790/ha. It means, it will take four to five years to recover the entire reclamation cost. Since one expects higher yield in subsequent years, the gestation period may reduce and the farmer may be able to repay the entire loan in 3-4 years. The following 2 issues were highlighted to show that the reclamation operation is brighter than emerging from the analysis.

- The land leasing cost has been included in the analysis. If the land is owned by the farmer, the benefits will increase accordingly.
- The farmers operate under resource constraints and are unable to fully adopt the recommended technology. It often results in low yields as in the case of the example just discussed. To illustrate, the yield of rice and wheat crops

obtained under farmers´ conditions was 3.1 t/ha and 1.38 t/ha respectively. These yields were much lower than the average yield of 4.8 t/ha of rice and 2.8 t/ha of wheat obtained with the adoption of full technology package on some of the Institute´s demonstration plots in the same area. This suggests that if the farmers adopt the recommended package of practices, they can obtain higher yields of rice and wheat crops with little additional investments. It will increase the profit margin.

The cost-benefit of land reclamation on lands adopted by the Central Soil Salinity Research Institute, Karnal and the land reclaimed by the farmers reveal that the benefit- cost ratio even at the farmer's level of adoption is 3 or more than 3 depending upon the component of the subsidy. With a subsidy of 70%, the B-C ratio is 4.6 (Table 16.7). The B-C ratio can be still higher if the complete rather than partial package is implemented. The cost of reclamation is generally recovered within 3 years as revealed by the payback period. Economics studies carried out by Haryana Land Reclamation and Development Corporation brings out that a fair return on investments including some repayment is possible even in the first year of reclamation.

Table 16.7. Economic parameters of the alkali land reclamation technology

Indicators	**Subsidy level**		
	Without subsidy	**50% subsidy**	**70% subsidy**
As per CSSRI recommendation			
Benefit-cost ratio	4.6	6.0	7.0
Net present worth (10^3 Rs/ha)	66.0	70.4	72.5
Net benefit investment ratio	7.0	11.4	16.7
Payback period (years)	2.0	1.8	1.6
As per farmer's level of adoption			
Benefit-cost ratio	3.0	3.9	4.6
Net present worth (10^3 Rs/ha)	36.7	41.0	43.0
Net benefit investment ratio	3.0	4.6	5.6
Payback period (years)	3.4	2.0	2.6

Chhiba and Singh (1978) worked out the economics of alkali land reclamation in the Operational Research Project, Kaputhala. The soil was highly deteriorated, initial pH_2 of the soil being 10.5. Gypsum @ 10 t/ha was added. A statement of expenditure and returns show that cost of reclamation was recovered within 2 years (Table 16.6). A marginal loss of Rs. 170/- was incurred in the first year when the land reclamation cost is also included towards the total cost. In the second year, not only the total investment is recovered but a net gain of Rs. 3530/- is also made. Apparently, land reclamation in Punjab is a financially viable proposition.

Table 16.6. Expenditure and returns (Rs./ha) statement for ORP Kapurthala

Particulars	**Period**	
	1975-76	**1976-77**
Cost of operations		
Bunding and levelling	540	--
Gypsum application and leaching	1230	--
Green manuring (Growing and burying)	1070	560
Production cost of rice	1990	1340
Production cost of wheat	1970	1260
Gross cost	6700	3160
Sale of rice and its by-products	3730	3950
Sale of wheat and its by-products	2800	2910
Total	6530	6860
Net returns	(-)170	3700
Net gain in 2 years	3530	

Figures are nearest to ten rupees

In a recent study, Thimmapa *et al*. (2013) reported that the land reclamation had a remarkable positive impact on cropping intensity, productivity, income and employment in Uttar Pradesh. Productivity of rice increased from 1.68 t/ha to 3.30 t/ha while that of wheat from

0.94 t/ha to 2.75 t/ha. The net income from rice in post-reclamation period was Rs. 17628/ha as against the loss of Rs. 193/ha in the pre-reclamation period. The B-C ratio of rice, which was 0.99 in pre-reclamation period, increased to 1.89 in post-reclamation period. A similar change was observed in wheat which depicted an appreciable increase in net income (Rs. 15265/ha) after reclamation as against the loss of Rs. 4281/ha in pre-reclamation period. The cropping intensity and labour employment also increased by 14% and 11%, respectively.

Singh (1978) studied the comparative economics of land reclamation by use of pyrites and gypsum at *usar* reclamation project, Dalipnagar, Kanpur. Considering the rates of pyrites and gypsum, it was reported that additional cost of reclamation due to use of gypsum worked out to be Rs. 962/ha for 40% gypsum requirement and Rs. 1,432 for 60% GR. Differences in the rice yield as a result of higher dose of the amendment were insignificant but the yield of wheat increased by 0.3-0.4 t/ha in gypsum treated than pyrites treated plots. Pyrites was observed to be economically viable proposition.

ECONOMICS OF RECLAMATION OF WATERLOGGED SALINE SOILS

The main approach for the reclamation of waterlogged saline soil is to establish a drainage system to facilitate the leaching process and to control the water table at such a level that resalinization of the soil root zone is prevented. The impact of waterlogged saline reclamation on soil salinity at Appikatla in Andhra Pradesh show that drainage reduced EC_e from an initial mean value of 16.2 dS/m to 3.0 dS/m, showing a reduction of about 82% (Fig. 16.1). It resulted in higher rice yields in reclaimed than un-reclaimed areas. Data on crops yield of 4 crops shown in Table 16.8 for the Gohana project site in Haryana also indicate higher crop yields in drained plots (Sharma and Gupta, 2006).

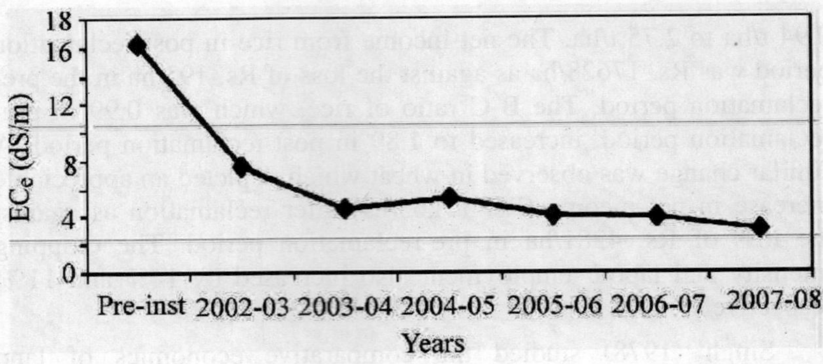

Fig. 16.1. Reduction in soil salinity in subsurface drained plots in Apitkala (Andhra Pradesh)

Table 16.8. Impact of subsurface drainage on crop yields (t/ha)

Crop	Area	Before drainage 1995/1996	After drainage 1999/2000	Percent yield increase
Rice (unhusked rice)	Drained	1.41 ± 0.30	1.74 ± 0.24	21.27± 07.25
	Undrained	1.38 ± 0.26	1.28 ± 0.29	
Pearl millet	Drained	1.05 ± 0.30	1.34 ± 0.22	27.62 ± 07.61
	Undrained	0.92 ± 0.25	0.85 ± 0.28	
Sorghum (fodder)	Drained	23.63 ± 5.44	28.90 ± 4.62	22.30 ± 07.60
	Undrained	22.38 ± 4.30	20.68 ± 3.22	
Wheat	Drained	03.04 ± 0.61	03.61 ± 0.41	18.75 ± 13.99
	Undrained	02.93 ± 0.54	02.52 ± 0.62	

Datta and Joshi (1991) assessed the financial feasibility of investments on subsurface drainage. The B-C ratio and the NPW were estimated at 1.26 and Rs. 4,659 per ha respectively. The IRR of 13.3% was sufficiently higher than the market rate of interest. It may be noted that values for each parameter were higher under scientific experiments than under farmer's fields mainly because the farmers were unable to implement the full package of land reclamation. Similar results have been reported from a number of sites with payback period ranging from 3-4 years (Table 16.9, IDNP, 2002).

Table 16.9. Economic parameters of four drainage projects under varying agro-climatic conditions

Particulars	Konaki	Uppugunduru	Segwa	Lakhuwali
B:C ratio	2.8	2.5	1.7	3.2
NPW (Rs./ha)	36, 655	50,872	74.957	51,500
IRR (%)	32	36	58	39
Payback period (Years)	3	3	3	4

SOCIAL BENEFITS

In many projects, social benefits outweigh the financial benefits. Therefore, such projects are sometimes approved even if these do not stringently satisfy the economic and financial criteria. National and international funding organizations finance schemes with low BC ratio provided such schemes are able to generate employment, rehabilitate the poor and bridge gender disparity. Some projects are approved if these help to improve the hygienic or the environmental conditions of an area.

Employment

A social benefit that must be included in the evaluation of any project is its capacity to generate employment opportunities (Singh, 1979). It has been documented that both alkali and waterlogged saline land reclamation programs generate huge employment. While trained engineers are employed in drainage construction, skilled and unskilled labour is required to undertake various activities in these projects. At the field level, it generates work for the farm family/ landless people. Capacity to generate employment opportunity at the field level due to reclamation of alkali lands is given in Table 16.10. In the first year, the employment generation is due to land reclamation as well as due to agricultural operations on the reclaimed lands. In subsequent years, it is mainly due to agricultural operations alone. The employment generation potential is higher in reclamation of waterlogged saline soils with drainage intervention. Lesser the lateral drain spacing higher is the employment generation (Table 16.11). The employment generation in agricultural operations is of the same magnitude as for an alkali land reclamation program.

Table 16.10. Employment generation at field level due to reclamation of 1 ha of barren alkali land

Year	Particulars	Number of		
		Person days	Bullock days	Tractor hours
I	Reclamation	31	10.0	13.3
	Rice	70	1.0	-
	Wheat	47	7.5	3.4
II	Rice	87	6.9	3.0
	Wheat	35	5.6	3.8

Source: Singh (1980)

Table 16.11. Employment generation in saline land reclamation projects involving pipe drainage

Spacing (m)	Installation of laterals and collectors	Construction of		
		Junction boxes	Sump	Total
25	290	5	14	309
50	170	3	14	187
75	112	2	14	128
100	103	2	14	119

Source: Joshi *et al.* (1987)

Human Health

Canal irrigation has often been traced to be the cause of increased human health problems. It accentuates water borne diseases, either as a result of running water or water stagnation in low lying depressions. The latter is often more serious than the former. Increased incidences of malaria have been reported from waterlogged areas as Anolpheles mosquitoes breed in the standing water. Documented evidence of such a situation is a report on migration of people living in and around Amritsar in the first decade of the 20th Century. The vertical drainage scheme designed to halt the migration of people from Amritsar (India) in 1917 is often quoted as an illustration (Gupta, 2002). The main aim of the scheme was to improve the general hygienic conditions of the area by lowering the water table. As per IWMI reports, government expenditure on health services can be curtailed in the same order of magnitude as the investments made in the land reclamation programs.

Inter-sectoral Terms of Trade

Resource mobilization from one sector to another is an indication of likely strengthening of inter-sectoral terms of trade and economic development. While land reclamation programs provide opportunities to use low quality gypsum, mining of additional gypsum provides employment opportunities. As the current thinking is to use phosphogypsum, distillery spent wash and press mud, all byproducts of the industrial sector, it should help the industries to dispose of their byproduct in a fruitful manner. Likewise, ash from thermal power plants may or may not reclaim the lands, but huge quantities can be disposed on agricultural lands without adversely impacting the soil resource or product quality. Besides, its use helps to minimize the pollution of soil and water resources.

Total cost of a subsurface drainage system is associated with three different sectors (Joshi *et al.,* 1987). The share of the industrial sector dominates the total cost of drainage being as high as 59% (Table 16.12). The share of the rural based non-farm resources occupies the second place with 20.8%. The data suggest that land drainage projects should strengthen both the industrial and the agriculture sector.

Table 16.12. Sectors wise share of total cost of land drainage

Sector	Share in total cost (%)
Industrial	59.1
Rural based non-farm resources	20.8
Rural labour	20.1
Total	100.0

Increase in the Land Value

One of the major secondary benefits of land reclamation is that there is instant enhancement in the land value. The data in Table 16.5 show that all kinds of alkali lands were quoted at higher rates after reclamation although the maximum increase in the value was for the C grade soils, which were lying barren before reclamation. Similar increase in land value was observed for waterlogged saline lands (Fig. 16.2). If this increase in land value is taken into account, it alone takes care of the investments made in land reclamation.

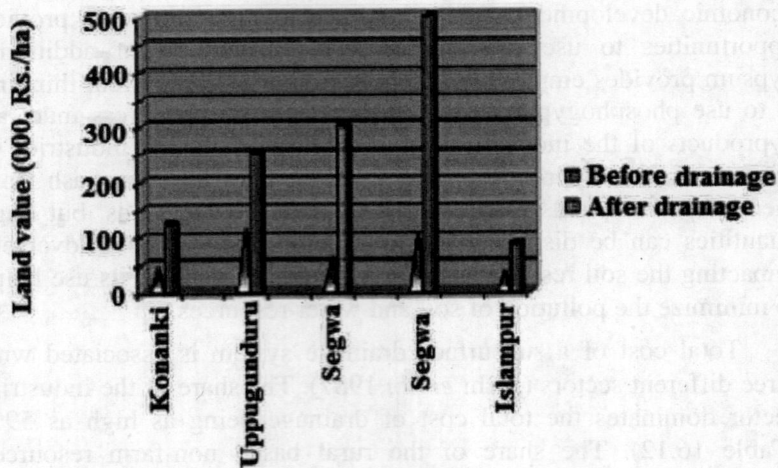

Fig. 16.2. Increase in land value after reclamation at various sites

Source of Irrigation Water

Subsurface drainage studies in India have shown that drainage besides reclaiming saline soils provides a source of water, which helps to replenish the meager water supplies through the canal system. It has been observed that water quality at some sites is better than the tube well water being used by the farmers. At other sites, the drainage water quality improves with time. Within 2-3 years, drainage effluent can be reused within the reclaimed land to provide supplemental irrigations (Fig. 16.3).

Fig. 16.3. Reuse of drainage water at Kalayat (Haryana)

Groundwater Recharge

As the land reclamation process is initiated, there is perceptible improvement in the infiltration rate and the hydraulic conductivity of alkali lands. As such, it not only reduces the total runoff but also the peak rate of runoff. The water retention capacity of the watershed is also higher because of *bunding* and other on-farm development works. All these together reduce flooding and increase groundwater recharge. Thus, a part of the groundwater exploited for rice-wheat cultivation in reclaimed alkali lands is compensated by this process. Current technology of storing rainfall in rice fields could further enhance the groundwater recharge. It may not be out of place to mention that more than 90% of the rainfall occurring in reclaimed catchments can be stored within the area including the rice fields. Similarly, provision of drainage vastly improves the surface and subsurface components in waterlogged saline lands. Thus, the conservation of rainfall in saline lands vastly alters the salinization-desalinization cycle resulting in reduced salinization and increased desalinization. As such, the reuse of drainage effluent is made possible during winter cropping. Besides solving the disposal related problems, the reuse helps to enhance productivity resulting from additional numbers of irrigations in irrigation commands.

Enhanced Status of the Poor in Rural Setting

In a rural setting, farmers having barren salt affected lands are ill-treated mainly because of their poor economic status. They are forced to work on someone else's fields where they are forced to do all kinds of works. With reclamation, these farmers become the proud owners of lush green fields. They are able to engage their family members at their own farm and in some cases even provide employment to few landless labourers. With this transformation, they are now able to play an independent role in socio-politico-economic functions of the village. Farmers, who used to migrate from the village, could stay back to take care of their families. It should halt ill treatment of women folk, which might happen in the absence of male members upon their migration.

PROGRESS OF RECLAMATION

On the basis of reconciled data on area reclaimed by different organizations, gypsum distributed for land reclamation and subsurface drainage installed in different states, it has emerged that more than 2.0 M ha of barren salt affected soils of India have been reclaimed so far. Major reclamation activities have been undertaken in the states of Punjab, Haryana and Uttar Pradesh. Rajasthan leads in reclamation of saline lands through subsurface drainage followed by Haryana and Maharashtra. The reclaimed lands have contributed 15 -16 million tons of food grains to national food basket. The technologies both for alkali and saline lands are now well tested and validated under different agro-climatic conditions of the country. It is believed that reclamation of entire alkali affected area alone can add around 12 million tons of additional food grains to the national pool.

Currently on an average 50,000 ha of salt affected soils are being reclaimed per annum. The massive efforts that have gone in land reclamation have ensured household food security and changed the socio-economic profile of more than 9 million people directly. This has also benefitted a large number of landless labourers indirectly. In spite of the achievements made so far, there is a need to enhance the pace of land reclamation as total salt affected soils currently are at about the same level as reported by Abrol and Bhumbla in 1971. Indirectly, it means that the pace of reclamation and degradation is at about the same level during the last 40 years. Nonetheless, data presented in table reveal that reclaimed lands besides adding to food grains valued at 15, 000 million Rs. has generated 8.5 million person months of employment (Modified from

Tripathi and Sharma, 2011). These results suggest that if barren alkali lands are brought under cultivation, it will create enormous employment opportunities for rural labour, draft animals and equipment and machineries (Table 16.13). It will produce several million tons of food grains to meet the demand of the increasing population of the country. Besides, it is justified from the social audit point of view as its benefits will go ultimately to the rural populace.

Table 16.13. Benefits of reclamation of alkali/saline lands

Parameters	Benefits
Land reclaimed	2.0 million ha
Food grains contribution	15-16 million tons
Agricultural income	Rs. 15,000 million
Employment generation	8.5 million person months/annum

Social Constraints

The reclamation technology has been widely accepted by the farmers. No social inhibitions have been encountered in any region wherever these technologies have spread their wings. Since traditional technologies adopted by few stake holders such as to scrap the top layers, mixing of sand and FYM application have not paid due dividends, clearly there are no competitive technologies. It seems that lack of extension is a major hindrance in the spread of the reclamation technologies. Marginal and small farmers have low level of knowledge regarding salt affected soils and their reclamation. Probably an extensive mobile training network with an agricultural institute in the region as the nodal point might help in spreading the message quickly and effectively.

■■■■

ANNEXURE I

EQUIVALENT WEIGHT OF VARIOUS IONS, SALTS AND AMENDMENTS USED IN RECLAMATION

Ions/Salts	Equivalent weight	Salts	Equivalent weight	Salts/ amendments	Equivalent weight
Calcium	20.04	Magnesium chloride	47.62	Sodium carbonate	53.00
Magnesium	12.16	Magnesium sulphate	60.19	Sodium bicarbonate	84.01
Sodium	23.00	$MgCl_2$ $6H_2O$	101.7	Potassium chloride	74.56
Potassium	39.10	$MgSO_4$ $7H_2O$	123.2	Potassium sulphate	87.13
Sulphate	48.03	Magnesium carbonate	42.16	Potassium carbonate	69.10
Chloride	35.46	Sodium chloride	58.45	Potassium bicarbonate	100.11
Carbonate	30.00	Sodium sulphate	71.03	Sulphur	16.03
Bicarbonate	61.01	Sodium carbonate	53.00	Sulphuric acid	49.04
Calcium chloride	55.50	Sodium bicarbonate	84.01	Aluminium sulphate	111.07
$CaCl_2\,2H_2O$	73.5	Sodium chloride	58.45	Iron sulphate (ferrous)	139.01
Calcium sulphate	68.07	Sodium sulphate	71.03	Calcium carbonate	50.04
$CaSO_4\,2H_2O$	86.09				

BIBLIOGRAPHY

Abrol, I.P. and Bhumbla, D.R. 1973. Field studies on salt leaching in a highly saline-sodic soil. Soil Sci. 15:429- 433.

Abrol, I.P. and Bhumbla, D.R. 1979. Crop response to differential gypsum application in a highly sodic soil and tolerance of several crops to exchangeable sodium under field conditions. Soil Science. 127: 79-85.

Abrol, I.P., Chhabra, R. and Gupta, R.K. 1980. A fresh look at the diagnostic criteria for sodic soils. In: Int. Symp. on Salt Affected Soils. Central Soil Salinity Research Institute, Karnal. February 18-21, 1980. 142-147.

Abrol, I.P., Dahiya, I.S. and Bhumbla, D.R. 1975. On the method of determining gypsum requirement of soils. Soil Science. 120: 30-36.

Abrol, I.P., Dargan, K.S. and Bhumbla, D.R. 1973. Reclaiming Alkali Soils. Bull. 2. Central Soil Salinity Research Institute, Karnal.

Abrol, I.P., Yadav, J.S.P and Massoud, F.I. 1988. Salt Affected Soils and Their Management. FAO Soils Bulletin 39. Food and Agriculture Organization of the United Nations, Rome.

Acharya, C.L. and Abrol, I.P. 1975. Comparative study of soil water behaviour in a sodic and adjacement normal soil. J. Indian Soc. Soil Sci., 23: 391-400.

Acharya, C.L. and Abrol, I.P. 1978. Exchangeable sodium and soil water behaviour under field conditions. Soil Sci., 125: 310-319.

Acharya, C.L. and Abrol. I.P. 1991. Soil Water Relations and Root Water Extraction in Alkali Soils. Tech. Bull. 16. Central Soil salinity Research Institute, Karnal. 96 p.

Acharya, C.L., Sandhu, S.S. and Abrol, I.P. 1979. Effect of exchangeable sodium on the field drying pattern of soil at two evaporative demands. Soil Sci., 127: 56-62.

Agarwal, M.C. and Khanna, S.S. 1983. Efficient Soil and Water Management in Haryana. Bulletin HAU, Hisar. 118 p.

Agarwal, R.R. and Gupta, R.N. 1968. Saline-alkali Soils in India. Tech. Bull. (Agric. Series), ICAR, New Delhi, 15, 221 p.

Agarwal, R.R., Yadav, J.S.P. and Gupta, R.N. 1979. Saline and Alkali Soils of India. ICAR, New Delhi, 286 p.

Agboma, P., Jones, M.G.K., Peltonen-Sainio, P., Hinkkanen, R. and Pehu, E. 1997. Exogenous glycine betaine enhances grain yield of maize, sorghum and wheat grown under two supplementary watering regimes. Journal of Agronomy and Crop Science, 178: 29–37.

Ambast, S.K., Gupta, S.K. and Sharma, D.K. 2015. Laser Land Levelling. Tech. Bull. 01/2015, Central Soil Salinity Research Institute, Karnal. 60 p.

Anonymous. 1971. Handbook of Manures and Fertilizers. ICAR, New Delhi, 80-81.

Anonymous. 1983. Coordinated Project on Management of Salt Affected Soils and Use of Saline Water in Agriculture. Ann. Rep. RBS College, Agra.

ANZECC (Australian and New Zealand Environment and Conservation Council). 2000. Australian and New Zealand Guidelines for Fresh and Marine Water Quality: The Guidelines. Paper No. 4. Agriculture and Resource Management Council of Australia and New Zealand, Canberra. 314 p.

Arora, S., Singh, A.K., Mishra, V.K., Singh, Y.P. and Sharma, D.K. 2015. Kinetics of gypsum and native $CaCO_3$ dissolution and nutrient transformations mediated through organic amendments and microbial inoculants for crop production in sodic soils. Annual Report 2014-15. Central Soil Salinity Research Institute, Karnal, 95-100.

Arora, S., Singh, Y.P., Vanza, M. and Sahni, D. 2016. Bio-remediation of saline and sodic soils through halophilic bacteria to enhance agricultural production. Journal of Soil and Water Conservation. 15:302-305.

Ashraf, M. and Foolad, M.R. 2007. Roles of glycine betaine and proline in improving plant abiotic stress resistance. Environmental and Experimental Botany, 59: 206–216.

Ayers, R.S. and Westcot, D.W. 1976. Water Quality for Agriculture. FAO Irrig. Drain Paper, 97 p.

Ayers, R. S. and Westcot, D.W. 1985. Water Quality for Agriculture. Irrigation and Drainage paper 29, Rev. 1. FAO, Rome. 174 p.

Bajwa, M.S., Hira, G.S. and Singh, N. T. 1983. Effect of sodium and bicarbonate irrigation waters on sodium accumulation and maize and wheat yields in northern India. Irrigation Science. 4:91-99.

Bale, R. and Jennifer, J. 2014. India's Food Security 2050 and Beyond: A Strategic Analysis https://www.mygov.in/sites/default/files/mygov_1460157311424912l.pdf.

Bandyopadhya, A.K. 1971. Effect of rainfall on calcareous saline sodic soils in India. Agrokem. Talaji, 20: 511-514.

Bandyopadhya, A.K. and Biswas, C.R. 1979. Nutrient requirement of high yielding rice in coastal saline soil condition. Ann. Rep. CSSRI, Karnal, p.184.

Bandyopadhyay, S., Dasgupta, S., Khan, Z. H. and Wheeler, D. 2018. Cyclonic Storm Landfalls in Bangladesh, West Bengal and Odisha, 1877–2016: A Spatio-temporal Analysis. Policy Research Working Paper 8316. Environment and Energy Team, Development Research Group and the Environment and Natural Resources Global Practice Group. International Bank for Reconstruction and Development/World Bank, Washington, DC.

Basu, J.K. and Tagare, V.D. 1943. Soils of the Deccan canals, IV. The alkali soils, their nature and management. Indian J. agric. Sci., 13: 157-181.

Batra, L., Kumar, A. and Manna, M. C. 1997. Microbiological and chemical amelioration of alkaline soils by growing Karnal grass and gypsum application. Experimental Agriculture. 33: 389-94.

Bernstein, L. 1964. Salt tolerance of plants. USDA. Inf. Bull., 283, 23 p.

Bernstein, L. and Francois, L.E. 1973. Comparisons of drip furrow and sprinkler irrigation. Soil Sci. 115: 73-86.

Bernstein, L.E., Francois, L.E. and Clark, R.A. 1974. Interactive effects of salinity and fertility on yields of grains and vegetable. Agron J. 66: 412-421.

Beyce, O. 1972. Drainage of salty soils. Irrigation and Drainage Paper 16: FAO Rome: 63-67.

Bhardwaj, K.K.R., Venkat Krishnan, S. and Batra, L. 1977. Studies on manurial value of *dhaincha* as a green manure crop. Ann. Rep. CSSRI, Karnal, p. 36.

Bhargava, G.P., Singhla, S.K. and Gupta, I.C. 1974. Distribution of boron in a salt affected and contiguous normal soils. Ann. Arid Zone, 13: 32-38.

Bhattacharyya, R.K. 1976. Promising rice selections suited to coastal saline soils. J. Soc. exp. Agric. 1: 21-24.

Bhattacharyya, R.K. 1977. Influence of time of planting and spacing on rice yield under coastal soil of sunderbans. Indian Agriculturist. 21: 83-86.

Bhattacharyya, R.K. 1979. Stability of some rice varieties grown in salt affected soils. Curr. Agric. 3: 37-42.

Bhumbla, D.R. 1973. India, Salt Affected Soils (Map), CSSRI, Karnal.

Bhumbla, D.R. and Abrol, I.P. 1975. Effect of application of different levels of gypsum on the yields of rice, wheat and barley grown in a saline sodic soil. Proc. Int. Symp. New Develop. Field Salt Affected Soils, Cairo, 1972, 537-544.

Bhumbla, D.R. and Dhingra, D.R. 1964. Micronutrient status of saline and alkali soils of the Punjab. J. India Soc. Soil. Sci., 12: 255-260.

Biswas, C.R. 1976. Nutrient requirement of different crops in coastal saline soils. Ann. Rep. CSSRI, Karnal, p. 113.

Biswas, C.R. 1979. Nutrient requirement of barley under coastal saline conditions. Ann. Rep., CSSRI, Karnal, p. 179.

Biswas, C.R. and Bandyopadhya, A.K. 1979. Nutrient requirement of different crops in coastal saline soils. Ann. Rep. CSSRI, Karnal, p. 180.

Borojevic, S., Cupina, T. and Krsmanovic, M. 1980. Green area parameters in relation to grain yield of different wheat genotypes. Zeitschrift fu¨r Pflanzenzu¨chtung, 84: 265–283.

Boumans, J.H. and van Der Molen, W.H. 1964. Out watering- behoefts can bevolede gronden in verband met hun zouthuishouding. Landis Tydrschn ft., 76: 880-887.

Bower, C.A., Spencer, J.R. and Weeks, L.O. 1969. Salt and water balance, Coachella Valley, California, Proc. ASCE, 95: 55-64.

Bouwer, H. 1969. Salt balance, irrigation efficiency and drainage design. Proc. ASCE, 95: 153-170.

Branson, R.L. 1960. Proc. 2nd Regional Training Course in Soil Salinity, UNESCO, Lahore.

Celia, M.M., de Souza, S. and Elisabeth F.F. 2012. Effect of salinity on soil micro-organisms Embrapa Environment, Brazil https://www.alice.cnptia.embrapa.br/bitstream/doc/944144/1/2012CL03.pdf

Chandra, S. 1979. Genetics and plant breeding. In: A Decade of Research. CSSRI, Karnal. 80-98.

Chandra, S. 1980. Evaluation and breeding of barley varieties suited to salt affected soils. Ann. Rep., CSSRI, Karnal.

Chandraju, S., Siddappa, C.S., Chidan Kumar and Shree Harsha Kumar, S.S. 2013. Studies on the germination and growth of neem (*Azadirachta indica*) and honge (*Pongamia pinnata)* seeds irrigated by distillery spent wash. International Journal of Pharmaceutical, Chemical and Biological Sciences. 3: 102-106.

Chatterji, R.K. and Sachde, C.B. 1979. Mineralogical characteristics of salt affected and adjacent normal soils. Curr. Agric., 3: 43-52.

Chaudhuri, I. I. and Wiebe, H.H. 1968. Influence of calcium pretreatment on wheat germination in saline media. Pl Soil, 28: 208-216.

Chawla, K.L. 1985. Salt and Water Transport under High Water Table Conditions. Unpublished Ph. D. thesis. Kurukshetra University, Kurukshetra.

Chhabra, R. and Lal, K. 2001-02. Dynamics of silica "and phosphorus during reclamation of alkali soil: Effect of fly ash on the yield of rice-wheat on a reclaimed alkali soil. Annual Report- 2001-02, CSSRI, Karnal. 16-17

Chhiba, I. M., Singh, N. T. 1978. An ex-army man reclaims his alkali land. Indian Farming. 28: 15-17.

Chinchmalatpure, A.K., Ali, S., Kulshrestha, N., Singh, R. K., Bundela, D.S., Kumar, P. and Sharma, D.K. 2015. Intellectual Property Management and Commercialization of ICAR-CSSRI Technologies for Management of Salt Affected and Waterlogged Soils in India. Tech. Bull. 03/2015, Central Soil Salinity Research Institute, Karnal. 72 p.

Chow, V.T. (Ed.) 1964. Coastal ecosystem management. A technical manual for the conservation of coastal zone resources. John Wiley and Sons Inc., New York. 928 p.

Clarkson, D.T. and Hanson, J.B. 1980. The mineral nutrition of higher plants. Ann. Rev. Pl. Physiol., 3: 239-298.

Colding, L.A. 1872. Om Lovene for vardets bevaegelse i jordan, Danske Vidensk selsk skr 5 Rackke Natur vog math Afd. 9B: 563-621.

CSSRI. 1997. *Vision 2020*. CSSRI Perspective Plan, Central Soil Salinity Research Institute, Karnal. 32 p.

CSSRI. 2007. *CSSRI- Perspective Plan : Vision 2025*. Central Soil Salinity Research Institute, Karnal, 127 p.

CSSRI. 2011. *CSSRI Vision 2030*. ICAR- Central Soil Salinity Research Institute, Karnal, 26 p.

CSSRI. 2015. *CSSRI Vision 2050*. ICAR- Central Soil Salinity Research Institute, Karnal, 31 p.

CWC. 2013. Water Statistics. Central Water Commission (CWC), Ministry of Water Resources, GoI, New Delhi. 201 p.

Dagar, J.C. 2011. Bio-drainage for amelioration of water logging. In: Salinity Management for Sustainable Agriculture in Canal command (Dey and Gupta Eds.). CADA, Haryana and CSSRI, Karnal, 54-58.

Dahiya, I.S., Malik, R.S. and Singh, M. 1981. Field studies on leaching behaviour of highly saline-sodic soil under two modes of water application in presence of crops. J. Agric. Sci. Camb. 97:383-389.

Damodaran, T., Sharma, D.K., Mishra, V.K. and Jha, S.K. 2014. Rhizosphere engineering: an innovative approach for sustainable crop production in sodic soils. In: 4th National Seminar on Innovative Saline Agriculture in Changing Environment. 12-14 December at RVSKV, Gwalior.

Dargan, K.S. 1979. Agronomic and cultural practices for crop production in salt affected soils. Curr. Agric., 3: 1-20.

Dargan, K.S., Abrol, I.P. and Bhumbla, D.R. 1974. Performance of rice varieties in a highly saline sodic soil as influenced by plant population. Agron, J., 66: 279-280.

Dargan, K.S. and Chhillar R.K. 1973. Technique for sugarbeet and vegetables in saline alkali soils. Indian Fmg., 22: 13-16.

Dargan, K.S. and Chhillar, R.K. 1975. Sugar beet yields high in saline sodic soils. Indian Fmrs., Digest, 8: 11-13.

Dargan, K.S. and Chhillar, R.K. 1979. Effect of missing application of P or K or both P + K in rice or wheat or in both crops on their yield in semi-reclaimed sodic soil. Ann. Rep., CSSRI, Karnal.

Dargan, K.S., Chhillar, R.K. and Bhardwaj, K.K.R. 1975. In alkali soils green manuring for more paddy. Indian Farming, 25: 13-15.

Dargan, K.S. and Gaul, B.L. 1974a. Optimum nitrogen levels of dwarf varieties of paddy in semi reclaimed soils. Fertil. News, 19: 24-25.

Dargan, K.S. and Gaul, B.L. 1974b. For paddy in alkali soil seedling age and planting date vital. Indian Farming, 24: 11-12.

Dargan, K.S. and Gaul, B.L. and Abrol, L.P. 1973. Growing sugarcane in saline alkali soils. Indian Farming, 23: 9-10, 19.

Dargan, K.S., Gaul, B.L. and Abrol, I.P. 1976a. Effect of gypsum, farm yard manure and method of sowing on the yield of sugarcane in a sodic soil. Indian J. Agron., 21: 349-353.

Dargan, K.S., Gaul, B.L., Abrol, I.P. and Bhumbla D.R. 1976b. Effect of gypsum farm yard manure and zinc on the yield of berseem. paddy and maize in a highly sodic soil. Indian J. agric. Sci., 46: 535-541.

Datta, A., Basak, N., Chinchmalatpure, A.R., Banyal, R. and Chaudhari, S. K. 2017. Land-use influences soil properties of sodic land in northwest India. Journal of Soil Salinity and Water Quality, 9: 178-186.

Datta, K.K. and Joshi, P.K. 1991. Cost and Benefits of Subsurface Drainage for the Control of Soil Salinity and Water Logging. Better Farming in Salt Affected Soils No. 20. CSSRI, Karnal.

Dhar, N.R. 1935. A new method of nitrogen fixation and conservation and reclamation of alkali lands. Annals Indian Acad. Sci., New Delhi

Dhar, N.R. and Mukherji, S.K. 1936. Alkali soils and their reclamation II. Proc. Natn. Acad. Sci. India, 6A: 297-303.

Dhawan, C.L. 1964. Land Reclamation of Saline and Alkali Soils, Vol. III. Publication No. 48, Central Board of Irrigation and Power, New Delhi.

Dhir, R.P. 1977. Saline waters - their potentiality as a source of irrigation. In: Desertification and its Control. ICAR, New Delhi, 130-148.

Doering, E.J. and Reeve, R.C. 1965. Engineering aspects of the reclamation of sodic soils with high salt waters. J. Irrig. Drain. Div. ASCE 91: 59-72.

Doneen, L.D. 1963. Water Quality for Agriculture. Department of Irrigation, University of California, Davis. 48p.

Elmahl, Y.E. and Mustafa M.A. 1980. The effect of electrolyte concentration and sodium adsorption ratio on phosphate retention by soils, Soil Sci., 130: 321-325.

Eaton, F. M. 1950. Significance of carbonate in irrigation water. Soil Sci. 69: 123-133.

Elgabaly, M.M. 1971. Reclamation and management of salt affected soils. Salinity Seminar, Baghdad. FAO Irrigation and Drainage Paper 7: 50-79.

FCI-FAI. 1977. Proc. FCI-FAI Seminar on Use of Gypsum in Reclamation of Alkali Soils. FCI-FAI, New Delhi, 190 p.

Galahitigama, G. A. H. and Wathugala, D.L. 2016. Effects of pre-sowing seed treatments on seed germination and salinity tolerance of rice (*Oryza sativa* L.) seedlings. International Journal of Agronomy and Agricultural Research, 9: 112-117.

Gardner, W.R. and Fireman, M. 1958. Laboratory studies of evaporation from soil columns in presence of a water table. J. Soil Sci. 85: 244-249.

Gaul, B.L., Abrol, I.P. and Dargan, K.S. 1973. Frequent and light irrigations best for saline sodic soils. Indian Farming. 23: 14-15.

Gaul, B.L. and Dargan, K.S. 1978. Effect of doses and frequency of gypsum application and crop production in sodic soil. Indian J. Agron., 23: 1-4.

Garrels R.M. and Christ, C. L. 1965. Solutions, Minerals and Equilibria. Freeman, Cooper & Company, San Francisco, 450 p.

Giannico, G. and Souder, J.A. 2005. Tide Gates in the Pacific Northwest: Operation, Types, and Environmental Effects, Oregon State University, Corvallis, Oregon, USA. 28 p.

Gupta, I.C. 1972. Note on relationship between pH of saturated soil paste and 1 : 2 soil water suspension. Indian J. agric. Sci., 42: 962-63.

Gupta, I.C. 1979a. A new classification and evaluation of quality of irrigation waters for arid and semi arid zones of India. Trans. Isdt and Ucds, 4: 6-12.

Gupta, I.C. 1979b. Note on the effect of leaching and gypsum on the detoxification of boron in a saline sodic soil. Curr. Agric., 3: 231-234.

Gupta, I.C. 1984. Reassessment of irrigation water quality criteria and standards. Current Agriculture. 7:150-155.

Gupta, I.C. and Abichandani, C.T. 1970. Seasonal variations in salt composition of some saline water irrigated soils of western Rajasthan I. Effect of rainfall, J. Indian Soc. Soil Sci., 18: 428-435.

Gupta, I.C. and Bandyopadhya, A.K. 1975. Effect of heating in the reclamation of a salt affected soil. Proc. Int. Symp. New Develop. Salt Affected Soils, Cairo, (1972), 525-528.

Gupta, I.C., Bohra, R.K. and Abichandani, C.T. 1969. Relationships between certain indices of saline water irrigated soils of western Rajasthan, Ann. Arid Zone, 8: 27-32.

Gupta, I. C. and Chandra, H. 1972. Effect of gypsum in reducing boron hazard of saline waters and soils. Annals of Arid Zone, 11: 228-230.

Gupta, I.C., Singhla, S.K. and Bhargava, G.P. 1974. Distribution of lithium in some salt affected soils profiles. J. Indian Soc. Soil Sci., 22: 88-89.

Gupta, I.C., Yaduvanshi, N.P.S. and Gupta, S.K. 2012. Standard Methods for Analysis of Soil Plant and Water. Scientific Publishers (India), Jodhpur. 224 p.

Gupta, R.N. and Narain, B. 1971. Investigations on some physical properties of alluvial soils of Uttar Pradesh related to conservation and management. J. Indian Soc. Soil, Sci., 19: 1-9.

Gupta, R.N. and Singh, N. 1971. Some physical and physico-chemical investigations on clay pan formation in the saline alkali soils of Gangetic basin of U.P. Symp. Soil Salinity, Kanpur, 73-77.

Gupta, S.K. 1985. Salt Movement and Dispersion - convection Equation: Material Balance Concept. Tech. Bull. No. 9. Central Soil Salinity Research Institute, Karnal. 48 p.

Gupta, S.K. 2002. A Century of Subsurface Drainage Research in India. Irrigation and Drainage Systems. 16: 69-84.

Gupta, S.K. 2005. Water logging, soil salinity and land drainage. In: Souvenir National Symposium on Efficient Management for Eco-friendly Sustainable Agriculture. ISWAM, New Delhi, 36-46.

Gupta, S.K. 2011. Case studies on subsurface drainage systems and guidelines for design and evaluation. In: Salinity Management for Sustainable Agriculture in Canal Commands (Dey, P. and Gupta, S.K. Eds.). Central Soil Salinity Research Institute, Karnal. 11-22.

Gupta, S.K. 2018. Drainage Engineering: Principles and Practices. Scientific Publishers (India), Jodhpur. (In Press).

Gupta, S.K. and Gupta, I.C. 2014. Salt Affected Soils: Reclamation and Management. Scientific Publishers (India), Jodhpur. 316 p.

Gupta, S.K. and Khosla, B.K. 1996. Salinity control in the root zone of irrigated soils. Proc. Workshop Waterlogging and Soil Salinity in Irrigated Agriculture. CBIP, New Delhi and CSSRI, Karnal. 163-180.

Gupta, S.K. and Narayana, V.V. Dhruva. 1974. Design considerations for dual purpose ponds. Soil Conserv. Digest, 2:1-7.

Gupta, S.K. and Pandey, R.N. 1979. Crop and water yields as affected by rain water storage in rice fields- A field evaluation. Field Crops Research, 2: 365-371.

Gupta, S.K. and Pandey, R.N. 1980. The leaching efficiency criterion and its evaluation during reclamation of saline soils. Int. Symp. Salt Affected Soils, Karnal, 300-306.

Gupta, S.K. and Pandey, R.S. 1983. A mathematical model for evaluating leaching of saline soils when saline and fresh water are used in conjunction. Indian J. Agric. Sci. 53: 701-706.

Gupta, S. K., Sharma, D. P., Tyagi, N. K. and Dubey, S.K. 2002. *Lavaniya-Kshariya Mirdaon ka Sudhar aur Parbandh*. CSSRI, Karnal. 153 p.

Gupta, V.K. and Dabas, D.S. 1980. Distribution of molybdenum in some saline sodic soils from Haryana. J. Indian Soc. Soil Sci., 28: 28-30.

Haroon, M.A.R. and Bose, S.C.M. 2004. Use of distillery spent wash for alkali soil reclamation, treated distillery effluent for ferti-irrigation of crops. Indian Farming, 53: 48-51.

Hasanuzzaman, M., Nahar, K., Alam, M.M., Bhowmik, P.C., Hossain, M.A., Rahman, M.M., Vara Prasad, M.N, Ozturk, M. and Fujita, M. 2014. Potential use of halophytes to remediate saline soils. BioMed Research International. 2014: 1-12.

Henckel, P.A. 1954. Salt resistance of plants and ways to increase yield. Akad. Nauk Timir Cht. 12 : (fide Bernstein, L. 1962. Salt affected soils and plants. Arid Zone Res. 18: 138-174.

Hermsmeier, I.E. 1979. Drainage practice in Imperial Valley. Trans. ASAE, 21: 105-109.

Hira, G.S. and Singh, N.T. 1980. Irrigation water requirement for dissolution of gypsum in sodic soil Soil Sci. Soc. Am. J., 44: 930-933.

Hooghoudt, S.B. 1937. In Drainage of Agricultural Lands (Eds. Luthin, J.N.) Am. Soc. Agron. Madison, Wisconsin, p. 81.

Horton, R.E. 1940. An approach to the physical interpretation of infiltration capacity. Proc. Soil Sci. Soc. Am., 5: 399-417.

http://agritech.tnau.ac.in/org_farm/orgfarm_green%20manure.html).Opened on 20.03.2018.

http://www.knowledgebank.irri.org/step-by-step- production/growth/planti-ng/direct-seeding). Opened on 22.10.2017.

ICAR (Indian Council of Agricultural Research). 2011. Forage Crops and Grasses In: Handbook of Agriculture. Indian Council of Agricultural Research, New Delhi. 1353-1417.

ICAR-NAAS. 2010. Degraded and Wastelands of India: Status and Spatial Distribution. National Academy of Agricultural Sciences, New Delhi. 158 p.

IDNP. 2002. Recommendations on Water Logging and Salinity Control based on Pilot Area Drainage Research. ICAR-CSSRI/Alterra ILRI Indo-Dutch Network Publication. 100 p.

Informativo CAPEBE 06/12/2010 - Um patrimônio chamado solo. Disponível em. >http://www.capebe.org.br/informativo.php?id=355> Acessed on 13.02.2018

Iqbal, M. and Ashraf, M. 2006. Wheat seed priming in relation to salt tolerance: growth, yield and levels of free salicylic acid and polyamines. Annales Botanici Fennici, 43: 250.259.

Irrigation Commission. 1972. Report of the Irrigation Commission. Ministry of Irrigation, Govt. of India, New Delhi.

Isaac, R.K., Sharma, D.P. and Swaroop, N. 2010. New approaches in reclamation of degraded soils with special reference to sodic soil: an Indian experience. In: Degradation and Desertification: Assessment, Mitigation and Remediation. Springer, UK. 253-266

Jaggi, T.N. 1979. Use of Amjore pyrites as alkali soil amendment. Proc. Sem. Use of sedimentary pyrites in reclamation of alkali soils. Lucknow, 37: 43.

John, P.V., Srivastava, A.P. and Kachwaha, K.S. 1977. Economic benefits of using gypsum for reclamation of alkali soils - Case study of selected farmers in Haryana. Proc. Sem. Use of gypsum in reclamation of alkali soils. Lucknow, 107-122.

Joshi, N. and Sharma, S. 2010. Physico-chemical characterization of sulphitation press mud, composted press mud and vermi-composted press mud. Report and Opinion. 2: 79-82.

Joshi, P.K., Singh, O.P., Rao, K.V.G.K. and Singh, K.N. 1987. Subsurface drainage for salinity control - an economic analysis. Ind. J. Agric. Econ., 42: 198-206.

Jury, W.A., Jarrell, W.M. and Devitt, D. 1979. Reclamation of saline-sodic soil by leaching. Soil Sci. Soc. Am. J. 43:1100- 1106.

Kanwar, J.S. and Singh N.T. 1968. Suitability of sugar beet for saline and saline alkali soils of northern India I. Adaptability. Indian J. agric. Sci., 38: 115-121.

Kelly, J. and Rengasamy, P. 2006. Diagnosis and management of soil constraints: transient salinity, sodicity and alkalinity. The University of Adelaide and Grain Research and Development Corporation, Australia. Available at http://www.arris.com.au/ index.php?id=39

Keren, R. 2000. Salinity. In: Handbook of Soil Science, edited by M. E. Summer, CRC Press, Boca Raton, Fla. G1 – G26,

Khosla, B.K., Dargan, K.S., Abrol, I.P. and Bhumbla, D.R. 1973. Effect of depth of mixing gypsum on soil properties and yield of barley, rice and wheat grown on saline sodic soil. Indian J. agric. Sci., 43: 1024-1031.

Kovda, V.A., van den Berg. C. and Hagan, R.M. (Ed.) (1973). Irrigation Drainage/Salinity, FAO, Rome.

Kumar, A. and Abrol, I.P. 1983. Studies on the reclaiming effect of Karnal-grass and para-grass grown in a highly sodic soil. Indian J. Agric. Sci. 54:189-193.

Kumar, A. and Abrol, I.P. 1986. Grasses in Alkali Soils. Bull No. 11. CSSRI, Karnal, 95 p

Kumar, A., Abrol, I.P. and Dargan, K.S. 1980. Karnal grass grows well in sodic soils. Indian Farming, 30: 14-15.

Kumar, D. and Singh, B. 2003. The use of coal fly ash in sodic soil reclamation. Land Degrad. Develop. 14: 285-299.

Lado, M., Paz, A. and Ben-Hur, M. 2004. Organic matter and aggregate-size interactions in saturated hydraulic conductivity. Soil Sci. Society of America J. 68: 234-242.

Lavado, R.S. Gonzalez Quintana, J.A. and Hevla, G.G. 1978. Content and distribution of lithium in Lapampa soils. Commun. Soil Sci. Pl. Analy., 9: 299-310.

Leather, J.W. 1897. Reclamation of *reh* and *usar* land. Agric. Ledger, 4: 195-232.

Leffelaar, P.A. and Sharma, R.P. 1977. Leaching of a highly saline sodic soil. J. Hydrol., 32: 203-218.

Linsley, R.K. Jr., Kohler, M.A. and Paulhus, J.L.H. 1949. Applied Hydrology, McGraw-Hill Book Co., New York. 689 p.

Lutts, S. 2000. Exogenous glycine betaine reduces sodium accumulation in salt stressed rice plants. International Rice Research Notes, 25: 39–40.

Maas E.V. 1980. Crop tolerance to saline soil and water. In: Prospectus for Bio-saline Research. Proc. US –Pak Bio-saline Res. Workshop, Karachi, Pakistan: 205- 211.

Maas, E.V. 1990. Crop salt tolerance. In: Agricultural Assessment and Management Manual. K.K. Tanji (ed.). ASCE, New York. 262-304.

Maas E.V. and Hoffman G.J. 1977. Crop salt tolerance: current assessment. J. Irrig. Drainage Div. 103:115- 134.

Maharudrappa, K., Baligar, V.C., Prabhakar, A.S. and Patil, S.V. 1972. Effect of gypsum application with and without FYM to Karl. Soil on the yield of Jayadhar cotton. Proc. Sem. Drought, Bangalore, 1968.

Manchanda, H.R. and Bhandari, D.K. 1976. Effect of pre soaking of seeds in salt solutions on the yield of wheat and barley irrigated with highly saline waters. J. Indian Soc. Soil Sci., 24: 432-435.

Manchanda, H.R., Garg, R.N., Sharma, S.K. and Singh, J.P. 1985. Effect of continuous use of sodium and bicarbonate-rich irrigation water with gypsum and farm-yard manure on soil properties and the yield of wheat in a fine loamy soil. J. Indian Soc. Soil Sci. 33: 876-883.

Mandiratta, R.S., Darra, B.L., Singh, H. and Singh, Y.P. 1972. Effect of some cultural, chemical and manurial treatments on the chemical characteristics of saline sodic soils under different crop rotations. Indian J. agric. Res., 6: 61-68.

Mani, G., Selvi, D.T. and Utharasu, S. 2017. Sustainable sugarcane initiative (SSI)- the choicest method of sugarcane cultivation in India. Everyman's Science, LII: 176-185.

Martinez-Beltran J. and Manzur CL. 2005. Overview of salinity problems in the world and FAO strategies to address the problem. Proceedings of the International Salinity Forum, Riverside, California, April 2005, 311–313.

Mays, D.A. and Mortvedt, J.J. 1986. Crop response to soil applications of phospho gypsum. Journal of Environmental Quality, 15: 78–81.

Mehta, K.K. and Yadav, J.S.P. 1977. Phosphogypsum for reclamation of alkali soils. Indian Farming, 27: 6-7.

Mengel, K. and Kirkby, E.A. 1979. Principles of Plant Nutrition. International Patash Institute, Berne, Switzerland.

Minhas, P.S. and Gupta, R.K. 1992. Quality of Irrigation Water - Assessment and Management. ICAR Pub., New Delhi. 123 p.

Mishra, U.N. 1977. Reclamation of alkali soils - role of fertilizer industry Proc. Sem. Use of gypsum in reclamation of alkali soil, Lucknow, 155-169.

MoA&FW (Ministry of Agriculture & Farmers Welfare). 2017. Status Paper of Indian Cotton. Directorate of Cotton Development, Ministry of Agriculture & Farmers Welfare, Nagpur. 182 p.

Mondal, A.K., Obi Reddy, G.P. and Ravisankar, T. 2011. Digital database of salt affected soils in India using Geographic Information System. Journal of Soil Salinity and Water Quality. 3: 16-29.

Mousa, M.R.M. 2011. Effect of partial replacement of dietary concentrate feed mixture by fodder beet roots on productive performance of ewes and doe goats under the conditions of North Sinai. Asian Journal of Animal Sciences, 5: 228-242.

MoWR (Ministry of Water Resources). 2017. A report on Problems of Salination of Land in Coastal Areas of India and Suitable Protection Measures. Central Water Commission, New Delhi. 347 p.

Murthy, R.S., Hirekerur, I.R. and Bhattacharjee, J.C. 1980. The taxonomy of salt affected soils of the Indian Sub continent. Int. Symp Salt Affected Soils, Karnal, 67-76.

Nalamwar, R.V. and Dastane, N.G. 1968. Cultural practices for crop production under soil salinity cum high water table conditions. Proc. Symp Wat. Management, Udaipur, 1966, 251-255.

Narayana, V.V.D. 1977. Drainage of salt affected soils in India. Proc. Indo-Hungarian Sem Management of salt Affected Soils, Karnal, 160-176.

Narayana, V.V.D. 1979. Rainwater Management for low land rice cultivation in India. J. Irrig. Drain. Div. ASCE, 105: 87-98.

Narayana, V.V. Dhruva and Gupta, S.K. 1974. Analysis of canal seepage. Paper presented at ISAE convention held at Bhubaneswar, Odisha.

Narayana, V.V.D., Gupta, S.K. and Tiwari, A.K. 1978. Rainfall and runoff analysis for rain water management in agriculture. Proc. Symp Hydrology Rivers with Small Catchments, CBIP, New Delhi, 2: 43-58.

Narayana, V.V.D., Kalra, V.D. and Singh, O.P. 1980. Design of farm layout for irrigation with limited discharges. Agril. Water Management. 3: 143-151.

Narwal, R.P., Singh, M. and Singh, J.P. 1979. Selenium distribution in some salt affected and normal soils of Haryana. Curr. Agric. 3: 59-62.

Nasir, N. M. 2006. Better Management Practices for Cotton and Sugarcane. WWF - Pakistan, Ferozepur Road, Lahore - 54600, Pakistan.

Nayak, A.K., Sharma, D.K., Mishra, V.K., Jha, S.K., Singh, C.S., Shahabuddin, Md., Srivastava, M., Kale, S.P. and Singh, G.B. 2011.

Phosphogypsum: An Alternate Chemical Amendment for Sodic Lands. CSSRI, RRS, Lucknow. Tech. Bull.1/2011, 42 p.

Nayyar V.K. 1972. Studies on molybdenum and copper in calcareous flood plain soils of Punjab. Ph. D. Dissertation, P.A.U., Ludhiana.

Nayyar, V.K. and Takkar, P.N. 1980. Evaluation of various zinc sources for rice grown on alkali soil. J. of Plant nutrition and Soil Science. 143: 489-493.

Neelakantan, S. and Pahwa, M.R. 1977. Note on the distribution of micro organisms in normal, saline and alkali soils. Curr. Agric., 1: 77-78.

Nia S.H., Zarea M.J., Rejali F. and Varma A. 2012. Yield and yield components of wheat as affected by salinity and inoculation with *Azospirillum* strains from saline or non-saline soil. J. Saudi Soc. Agric. Sci., 11: 113–121.

Nielson, D.R. and Biggar, J.W. 1961. Miscible displacement in soils I. Soil Sci. Soc. Am Proc., 25: 1-5.

Nielsen, D.R., Biggar, J.W. and Luthin, J.N. 1966. Desalinization of soils under controlled unsaturated flow conditions. 6th Int. Cong. Irrigation and Drainage. New Delhi II: 15-24.

Omar, S.A., Abdel-Sater M.A.; Khallil, A.M. and Abdalla, M.H. 1994. Growth and enzyme activities of fungi and bacteria in soil salinized with sodium chloride. Folia Microbiologica, 39: 23-28.

Oster, J.D. and Halvorson, A.D. 1978. Saline seep chemistry. In: Dryland Saline Seep Control. 11th Int. Congr. Soil Sci. Ed. Monion, 2: 7-29.

Overstreet, R., Martin, J.C. and King, H.M. 1951. Gypsum, sulphur and sulphuric acid for reclaiming an alkali soil of the Fresno series, Hilgardia, 21: 5-15.

Pandey, R.N., Narayana, V.V.D., Tyagi, N.K. and Gupta, S.K. 1977. Management of sloping borders for rice cultivation in alkali soils. Indian J. agric Sci., 47: 613-616.

Pandey, R.N. and Singh, O.P. 1977. Evaluation of irrigation method under salt affected soils. Ann. Rep., CSSRI, Karnal, p. 95.

Paikray, S.P. and Singh, K. 2012. Microbial reclamation of saline and sodic soil. WO 2012046098 http://www.google.com/patents/WO2012046098A1.

Ponnamperuma, F.N. 1977. Screening for Tolerance to Mineral Stresses. IRRI Res. Pap. Ser. No. 6. International Rice Research Institute. Los Banos, Philippines. 21 p.

Ponnamperuma, F.N. Martinez, E. and Loy, T. 1966. Influence of redox potential and partial pressure of carbon dioxide on pH values and the suspension effect of flooded soils. Soil Sci. 101: 421-431.

Poonia, S.R. and Bhumbla, D.R. 1973. Effect of gypsum and calcium carbonate on plant yield and chemical composition and calcium availability in a non saline sodic soil. Pl. Soil, 38: 71-80.

Pratt, P.F., Branson, R.L. and Chapman, H.D. 1960. Effect of crop, fertilizer and leaching on carbonate precipitation and sodium accumulation in soil irrigated with water containing bicarbonate. Trans. 7th Intern. Cong. Soil Sci. 2:185-192.

Prihar, S.S., Sandhu, B.S. and Singh, N.T. 1975. A critical appraisal of research on irrigation scheduling to crops. Proc. 2nd World Congr. Water Research, New Delhi, 1: 343-351.

Puntamskar, S.S., Mehta, P.C. and Seth, S.P. 1971. Note on the inducement of salt resistance in two wheat varieties by presoaking with different salts of varying concentration. Indian j. agric. Sci., 41: 717-718.

Qadar, A., Joshi, J.C., Bal, A.R. and Dwiwedi, R.S. 1980. Effect of calcium salts on induction of sodicity resistance in rice. Curr. Agric., 4: 83-86.

Qadar, A., Joshi, Y.C., Bal, A.R. and Rana, R.S. 1983. Effect of seedling treatment with calcium salts on sodicity induced cation accumulation in rice Indian. J. Plant Physiology, XXVI: 93-98.

Qadir, M., Oster, J.D., Schubert, S., Noble, A.D. and Sahrawat, K.L. 2010. Phytoremediation of sodic and saline-sodic soils. Advances in Agronomy. 96: 197–247.

Ram, B., Karuppaiyan, R. and Pandey, S.K. 2011. Sugarcane Cultivation in Sub-tropical India: Farmer's Guide. Sugarcane Breeding Institute, RRS, Karnal. 29 p.

Ramadoss, D., Lakkineni, V.K., Bose, P., Ali, S. and Annapurna, K. 2013. Mitigation of salt stress in wheat seedlings by halotolerant bacteria isolated from saline habitats. Springer Plus, 2: 1–7.

Ramamoorthy, B. 1968. The scope for living with soil salinity and alkali conditions in India. Proc. 7th NESA Irrig. Pract. Sem., Lahore, 345-360.

Ram Babu, Tejwani, K.G., Agarwal, M.C. and Bhusan, L.S. 1979. Rainfall intensity duration return period equations and monographs of India. Bull. 3, CSWCRTI, Dehradun.

Ranade, D.H. 1984. Leaching Phenomenon in Sandy Loam Soils. Unpublished M. Tech. Thesis, I.I.T. Kharagpur.

Rao, D.L.N. 1986. *Sesbania* for Green Manuring. Better Farming in Salt Affected Soils, No. 6, Central Soil Salinity Research Institute, Karnal. 7 p.

Rao, K.V.G.K. 1977. Study on fluctuations and salinity status of water table in the research farm, Ann. Rep. CSSRI, Karnal.

Redly, M., Darab, K. and Csillag, J. 1980. Solubility and ameliorative effect of gypsum in alkali soils. Int. Symp. Salt. Affected Soils, Karnal, 313 - 321.

Reeve, R.C. and Bower, C.A. 1960. Use of high salt waters as a flocculant and source of divalent cations for reclaiming sodic soils. Soil Sci., 90 : 139-144.

Reeve, R.C., Bower, C.A., Brooks, R.H. and Gschwemd, F.B. 1954. A comparison of the effects of exchangeable sodium and potassium upon the physical condition of soils. Proc. Soil Sci. Soc. Am., 18: 130-132.

Reeve, R.C. and Doering, E.J. 1966. The high-salt-water dilution method for reclaiming sodic soils. Soil Sci. Soc. Amer. Proc. 30:498-504.

Reeve, R.C., Pillsbury, A.F. and Wilcox, L.V. 1955. Reclamation of a saline and high boron soil in the Coachella Valley of California. Hilgardia. 24: 69-91.

Rengasamy, P. 2010. Soil processes affecting crop production in salt-affected soils. Functional Plant Biology, 37: 613–620

Rhoades, J.D. 1968. Leaching requirement for exchangeable sodium control. Proc. Soil Sci. Soc. Am., 32: 652-656.

Rhoades, J.D. 1977. Drainage for salinity control. In: Drainage for Agriculture. Agron. Mono. 17 (Ed. van Schilfgaarde). 433-461.

Rhoades, J.D. 1983. Using saline water for irrigation. Proc. Intern. Workshop Salt Affected Soils of Latin America, Maracay, Venezuela, 233-264.

Richards, L.A. (Ed.). 1954. Diagnosis and Improvement of Saline and Alkali Soils. USDA Handbook 60, 160 p.

Rohtagi, B.B., Saxena, S.K. and Singh, R. 1977. Availability of quality gypsum from indigenous sources as a soil amendment. Proc. Sem. Use of Gypsum in Reclamation of Alkali Soils. Lucknow, 37-55.

Saini, K.S. and Brar, N.S. 2018. Crop and water productivity of sugar beet (*Beta vulgaris*) under different planting methods and irrigation schedules. Agric. Res., 7: 93-97.

Sandhu, S.S., Acharya. C.L. and Abrol, I.P. 1980. The effect of exchangeable sodium on the water retention, transmission and modulus of rupture of a soil. Proc. International Symp. Salt Affected Soils, Karnal, India. CSSRI, Karnal. 185-194.

Satyanarayana, K.V.S. 1961. Studies on the soils of Rann of Kutch I, Description of some profiles and the distribution of salts and gypsum in the profiles. Bull. Indian Soc. Soil Sci., 6: 125-53.

Satyanarayana, M. 1953. Crop and land improvement under the Krishna East Bank Canal (Madras State). Madras agric. J., 40: 293-294.

Saxena, C.K., Gupta, S. K., Purohit, R.C., Bhaka, S.R. and Upadhyay, B. 2013. Performance of okra under drip irrigation with saline water. Journal of Agricultural Engineering, 50: 72-75.

Saxena, P.N. 1960. Algae flora of usar lands in Uttar Pradesh. Sci. Culture, 25: 542-543.

Schoonover, W.R. 1952. Examination of soils for alkali. Univ. of California. Extension Service, Berkley, California.

Sen, H.S. 2011. Management of coastal salt affected soils. A communication to CSSRI.

Shah, R.K. 1975. A note on quick and cheap reclamation method for saline alkali soils using dairy waste sulphric acid. Agril. Res. Newsletter, 3: 1-2.

Shanon, M.C. and Francois, L.E. 1977. Influence of seed pretreatments on salt tolerance of cotton during germination. Agron J., 69: 619-622.

Sharma, D.C., Puntamkar, S.S., Jain, S.V. and Seth, S.P. 1977. Effect of different frequencies of irrigation with saline water on the yield of wheat and salts accumulation in saline sodic soil. Indian J. agric. Sci., 4: 485-488.

Sharma, D.P. and Agarwal, R.P. 1979. Crust strength in relation to sodicity and dispersion of soils. Agrokem, Talajt., 28: 39-44.

Sharma, D.P. and Gupta, S.K. 2006. Subsurface drainage for reversing degradation of waterlogged saline lands. Land Degradation and Development, 17: 605-614.

Sharma, D.P., Mehta, K.K. and Yadav, J.S.P. 1981. Effect of reclamation practices on soil properties and crop growth on farmer's fields – a case study. J Indian Soc. Soil Sci., 29 : 356-360.

Sharma, D.P., Mehta, K.K., Yadav, J.S.P. and Singh, G. 1983. Effect of crop rotation on growth and soil properties in sodic soils on farmers' fields. Indian J. Agric. Sci., 53: 681-685.

Sharma, D.K., Ashok Kumar and Singh, K.N. 1990. Effect of irrigation scheduling on growth, yield and evapotranspiration of wheat in sodic soils. Agricultural Water Management, 18: 267-276

Sharma D.K., Singh, A. and Sharma, P.C. 2016. Role of ICAR-CSSRI in sustainable management of salt-affected soils-achievements, current trends and future perspectives. In: Souvenir 4th International Agronomy Congress, November 22-26, 2016 (Eds. R. Prasad, G. Singh, R.L. Yadav and I.P.S. Ahlawat), Indian Society of Agronomy, New Delhi. 91-103

Sharma, D.K. and Singh, K.N. 1993. Effect of scheduling of irrigation using pan evaporation on growth, yield and water use efficiency of mustard

(*Brassica juncea*) in reclaimed sodic soils Agricultural Water Management, 23: 225-232.

Sharma, P.C., Thimmappa, K., Kaledhonkar, M.J. and Chaudhari, S.K. 2016. Reclamation of Alkali Soils through Gypsum Technology. ICAR-CSSRI/Karnal/Technology Folder/2016/01, Central Soil Salinity Research Institute, Karnal. 4 p.

Shrotiya, G.C. and Mishra, U.N. 1976. Utilization of phospho-gypsum for agricultural purposes. Fertilizer News, 21: 37-38.

Singaraja, C., Chidambaram, S. and Jacob, N. 2018. A study on the influence of tides on the water table conditions of the shallow coastal aquifers. Applied Water Science, 8: 11-24.

Singh, A. 1980. Studies on fluorine retention movement and uptake by plants in sodic soils and its interaction with phosphorus. Ph.D. Thesis, Agra University, Agra.

Singh, A., Chhabra, R. and Abrol, I.P. 1980. Fluorine in phosphogypsum and reclamation of sodic soils. Fertil. News, 25: 18-23.

Singh, A.K. 2009. Nutrient management in problem soils. In: Enhancing Nutrient Use Efficiency in Problem Soils (Singh *et al.* Eds.). CSSRI, Karnal. 1-21.

Singh, A. K. and Verma, S.K. 2016. Hydro-physical properties and solute movement in black alkali soils. In: Innovative Saline Agriculture, Dagar *et al.* (Eds). Springer India Ltd. 119-144

Singh, I.S., Singh. R.K. and Khanna, S.S. 1994. Management of Sodic Soils through Plantations. Bulletin, N.D. University of Agriculture and Technology, Faizabad, UP. 51 p.

Singh, I. S. 1996. Agro-forestry for Salt Affected Lands. Bulletin, N.D. University of Agriculture and Technology, Faizabad, UP. 23p.

Singh, G. 1995. An agroforestry practice for the development of salt lands using *Prosopis juliflora* and *Leptochloa fusca*. Agroforestry Systems, 29: 61–75.

Singh, G., Singh, N.T. and Tomar, O.S. 1991. Agroforestry in salt affected soils. Technical bulletin 17, CSSRI, Karnal. 65 p.

Singh, G. and Dagar, J.C. 1998. Agroforestry in salt affected soils. In: Agricultural Salinity Management in India (Tyagi and Minhas Eds.). CSSRI, Karnal. 473-487.

Singh, G., Singh, N.T. and Tomar, O.S. 1993. Agroforestry in Salt-affected Soils. Technical bulletin No.17, CSSRI, Karnal. 65 p.

Singh, K. N. 1988. Cultural practices for crop production in alkali soils. Indian Farming, 38: 21-23, 43.

Singh, K.N. 1985. Cultural practices for growing rice in alkali soils. Better Farming: Salt- Affected Soils, CSSRI, Ext Series no. 9, 10 p.

Singh, K.N., Chillar, R.K. and Sharma, D.K. 1980. Studies on comparative performance of oil seed crops under different levels of gypsum. Ann. Rep. CSSRI, Karnal.

Singh, K.N. and Sharma, D.K. 1984. Effect of nitrogen on rice. IRRI News Letter, 12:16

Singh, K.N., Sharma, D.K. and Chillar, R.K. 1988a. Forage yield and chemical composition of pearl millet as influenced by exchangeable sodium. Journal of Agricultural Science, Cambridge, 111, 465–467.

Singh, K.N., Sharma, D.K. and Chillar, R.K. 1988b. Growth, yield and chemical composition of different oilseed crops as influenced by sodicity. Journal of Agricultural Science, Cambridge, 111, 459–463.

Singh, M. V. and Chhabra, R. 1978. Effect of levels of zinc on yield of crops. Ann. Rep., CSSRI, Karnal, p. 30.

Singh, M.V., Chhabra, R. and Abrol, I.P. 1987. Interaction between application of gypsum and zinc sulphate on the yield and chemical composition of rice grown in an alkali soil. Journal of Agricultural Science (Cambridge), 108: 275-279.

Singh, N. and Singh, M. 1980. Effect of sulphur and selenium on nitrogen, sulphur, selenium and phosphorus in raya (*Brassica juncea* Coss), grown in some soils. J. Indian Soc. Soil Sci., 28: 491-500.

Singh, N.T. 1974. Physico-chemical changes in sodic soils incubated at saturation. Pl. Soil, 40: 303-311.

Singh, N.T., Hira, G.S. and Singh R. 1978. Comparative effect of pyrites and gypsum on physico chemical properties of a sodic soil. Proc. Sem. Use of Sedimentary Pyrites in Reclamation of Alkali Soils. Lucknow, 69-80.

Singh, O.P. 1980. Analysis of the Hydrologic Changes taking place in Alkali Soil Watersheds during Reclamation. Ph.D. thesis, JNTU, Hyderabad.

Singh, O.P., Narayana, V.V.D. and Tiwari, A.K. 1979. Analysis of the hydrologic changes in alkali watersheds during reclamation. Canadian agric. Engg., 21: 147-150.

Singh, O.P. and Pandey, R.N. 1977. Collection of basic hydrologic data required for drainage design. Ann. Rep., CSSRI, Karnal, p. 80.

Singh, O.P. and Tyagi N.K. 1980. Estimating rainwater availability in reclaimed alkali soils. Int. Symp. Salt Affected Soils, Karnal, 549-554.

Singh, O.P., Tyagi, N.K. and Tiwari, A.K. 1978. A conceptual model for determining reclaimed alkali watershed system response. J. Instn. Engrs. (India), 59: 1-4.

Singh, P. 1978. Creditworthiness of reclamation of alkali soils. Financing Agriculture, 1-4.

Singh, P. 1979. Reclamation of alkali soils and its employment potential. Eastern Economist. 1-4.

Singh, P. 1980. An appraisal of the economics of reclamation of alkali soils. Int. Symp. Salt Affected Soils, Karnal, 555-561.

Singh, P. and Kumar, R. 1989. Comparative performance of trickle irrigation for tomato. J. Agril. Engg., 26:39-48.

Singh, P. and Kumar, R.1994. Effect of irrigation methods for cauliflower grown in heavy soils with shallow water table. J. Agril. Engg., 31: 36-43.

Singh, P., Kumar, R., Agarwal, M.C., and Mangal, J.L. 1990. Performance of drip and surface irrigation for tomato in heavy soils. Proc. XI International Convention on Use of Plastics in Agriculture, NCPA, New Delhi. 67-72.

Singh, R., Chaudhari, S.K., Tripathi, R.S., Joshi, P.K., Dey, P., Sharma, S. K. Sharma, D.K. and Singh, Gurbachan. 2014. Resource Conservation Technologies in rice–wheat system. ICAR-CSSRI, Karnal. Technical Bulletin: CSSRI/Karnal/2014/02, 24 p.

Singh, R., Kumar, A., Kharub, A.S., Kumar, V., Chhokar. R.S., Selvakumar, R., Chand. R. and Sharma, I. 2014. Barley Cultivation in India (Pocket Guide). Directorate of Wheat Research, Karnal. Extension Bulletine, 53: 18 p.

Singh, R., Singh, N.T. and Arora, Y. 1980. The use of spent wash for the reclamation of sodic soils. J. Indian Soc. Soil. Sci., 28: 38-41.

Singh, R.N. 1950. Reclamation of *usar* lands in India through blue green algae, Nature, London, 165: 325-326.

Singh, S. and Singh, B. 1966. Trace elements studies on some alkali and adjoining soils of Uttar Pradesh II - Profile distribution of cobalt. J. Indian Soc. Soil Sci., 14: 177-181.

Singh, Y.P., Sharma, D,K., Singh, Gurbachan, Nayak, A.K., Mishra, V.K. and Singh R. 2008. Alternate Land Use Management for Sodic Soils. Technical Bulletin 2/2008. CSSRI Regional Research Station, Lucknow. 16 p.

Singh, Y.P., Singh, G. and Sharma, D.K. 2014. Bio-amelioration of alkali soils through agroforestry systems in central Indo-Gangetic plains of India. Journal of Forestry Research, 25: 887–896.

Singhla, S.K., Bhargava, G.P. and Gupta, I.C. 1974. Relationships between certain diagnostic indices of salt affected soils of Indo-Gangetic alluvial region. Ann. Arid Zone, 13: 218-223.

Soil Science Society of America. 1974. Glossary of Soil Science Terms, Soil Sci. Soc. Am., 677, Madison, USA.

Stanhill, G. 1957. The effect of differences in soil moisture status on plant growth - a review of analysis of soil moisture regime experiments. Soil Sci. 84: 205-214.

Strogonov, B.P. 1964. Physiological basis of salt tolerance in plants. IPST, Jerusalem.

Suarez, D.L. 1981. Relationship between pH and SAR and an alternative method for estimating SAR of soil or drainage water. Soil Sci. Soc. Am. J. 45:469-475.

Swaroop, A. 1976. Transformation of iron and manganese in sodic soil under moist and submerged conditions. Ann. Rep., CSSRI, Karnal, p. 10.

Swarup, A. 1980. Effect of submergence and farmyard manure application on the yield and nutrition of rice and sodic soil reclamation. J. Indian Soc. Soil Sci. 28: 532-534.

Swaroop, A. 1981. Effect of iron and manganese application on the availability of micro nutrients to rice in sodic soil. Pl. Soil, 60: 481-485.

Swarup, A.1991. Effect of gypsum, green manure, farmyard manure and zinc fertilization on the zinc, iron and manganese nutrition of wetland rice on a sodic soil. Indian Soc. Soil Sci. 39: 530-536.

Swarup, A. 2000. Nutrient management for crop production in alkali lands. In: Salinity Management in Agriculture (Gupta *et al.* Eds.). CSSRI, Karnal. 99-108.

Swarup, A. and Yaduvanshi, N.P.S. 2000. Effect of integrated nutrient management on soil properties and yields of rice and wheat in alkali soils. J. Indian Soc. Soil Sci., 48: 279-282.

Swarup, A., and Yaduvanshi, N.P.S. 2004. Response of rice and wheat to organic and inorganic fertilizers and soil amendment under sodic water irrigated conditions. IRRI Newsletter. 29.1: 49-51.

Szabolcs, I. 1977. Salinity and alkalinity of soils, extension, classification and main properties of salt affected soils. In Proc. Indo-Hungarian Sem. Management of Salt Affected Soils, CSSRI, Karnal, 20-30.

Szabolcs, I. 1989. Salt-affected soils. Boca Raton, Fl: CRC Press.

Talsma, T. 1963. The Control of Saline Groundwater. Meded Landb Hoogesch, Wageningen. 63:1-68.

Thimmappa, K., Tripathi, R.S., Raju, R. and Singh, Y.P. 2013. Livelihood security of resource poor farmers through alkali land reclamation: an impact analysis. Agricultural Economics Research Review, 26 (Conference Number): 139-147.

Thokal, R.T. 1990. Strategies of leaching in highly saline soils. Unpublished M. Tech. Thesis, G.B.P.U.A.T., Pantnagar.

Tiwari. K.N., Sharma, D.N., Sharma, V.K. and Dingar, S.M. 1992. Evaluation of fly ash and pyrite for sodic soil rehabilitation in Uttar Pradesh, India. Arid Land Res. Manage., 6: 117-126.

Tiwari, K.N., Sharma, D.N. and Tripathi, S.K. 1989. Salt-Affected Soils of U.P.: Their Reclamation and Management. C.S. Azad University of Agriculture and Technology, Kanpur.

Todd, D.K. 1959. Ground Water Hydrology. John Wiley and Sons. Inc. New York, 336 p.

Tomar, O.S. and Gupta, R.K. 1987. Annual Report, CSSRI, Karnal. 101-103.

Tomar, O.S. and Yadav, J.S.P. 1980. Effect of saline irrigation water of high salinity with and without application of fertilizers on the growth of *Eucalyptus hybrid*. Indian J. Range Mgmt., 2: 121-127.

Tomar, O.S. and Yadav, J.S.P. 1985. Effect of saline water irrigation and fertilizer application on growth of tree seedlings. Annals of Arid Zone, 24: 94-100.

Tripathi, R.S. and Sharma, D.K. 2011. *Kshariya Bhumi Sudhar evam uske samajik-aarthik Labh.* Tech. Bull. CSSRI/Karnal/2009/02. Central Soil salinity Research Institute, Karnal. 12 p.

Tyagi, N.K. 1976. Studies on performance of closed level and graded borders in alkali soils. Ann. Rep. CSSRI, Karnal.

Tyagi, N.K. 1978. Some design aspects of ponded borders with cut back stream in alkali soils. Cur. Agric., 2: 13-18.

Tyagi, N.K. and Singh, O.P. 1979. Investigations on effect of soil moisture on performance of land levelling implements. J. agric. Engg., 15: 67-72.

Uppal, H. 1955. Green manuring with special reference to *Sesbania aculeata* for treatment of alkaline soils. Indian J. agric. Sci., 25: 211-235.

Van Alphen, J.G. and Rice Romero, F. de Los. 1971. Gypsiferous Soils. Bull. 12. Int. Inst. Ld Reclam Improve, Wageningen.

Vadjunina, A.F. El Saaby and El-Kouni. 1975. Chemical and electrical methods for reclaiming soda-saline soils. Int. Symp. New Develop. Field Salt Affected Soils. 505-512.

van Hooren, J.W. 1979. The effect of capillary flow on salinization and the concept of critical depth for determining drain depth. In: J. Wesseling (Ed.). Proc. of the Int. Drainage Workshop. ILRI Pub. 25, Wageningen. 686-700.

Venkat Krishnan, S. 1979. Effect of soil incorporation of pyrites on its oxidation and improvement of sodic soil. Curr. Agric., 3: 87-90.

Voelcker, J.A. 1897. Report on the improvement of Indian Agriculture.

Wahid, A., Perveen, M., Gelani, S. and Basra, S.M. 2007. Pretreatment of seed with H_2O_2 improves salt tolerance of wheat seedlings by alleviation of oxidative damage and expression of stress proteins. J Plant Physiol., 164: 283-294.

Whitcomb, E. 1971. Agrarian conditions in Northern Indian Vol. I (1880-1900), Thomson Press, New Delhi.

Wiegand, C., Anderson, G., Lingle, S. and Escobar, D. 1996. Soil salinity effects on crop growth and yield - Illustration of an analysis and mapping methodology for sugarcane. Journal of Plant Physiology, 148: 418-424.

Wilcox, L.V., Blair, G.Y. and Bower, C.A. 1954. Effect of bicarbonates on suitability of waters for irrigation. Soil Science. 77:259-266.

Wild A. 2003. Soils, Land and Food: Managing the Land during the Twenty-first Century. Cambridge, UK: Cambridge University Press.

Yadav, B.R. 2006. Status of management of saline and alkali water for crop production in arid and semiarid regions. In: Enhancing Water Use Efficiency in Arid and Semiarid Areas for Sustainable Agriculture (Yadav, B.S. Ed.). Agricultural Research Station, Sriganganagar. 236-241.

Yadav, D.V., Verma, R.P., Prasad, K., Sah, A. K., Gupta, R. and Singh, K.P. 2012. Ring-pit Method of Sugarcane Planting for Saving Irrigation Water. Ext. Brochure: EPARP No. 1. Indian Institute of Sugarcane Research, Lucknow. 5 p.

Yadav, J.S.P., Bandyopadhya, A.K., Rao, K.V.G.K., Singh, T.S. and Biswas, C.R. 1979. Coastal Saline Soils of India. Bull. 5, CSSRI, Karnal.

Yadav, J.S.P. and Girdhar, I.K. 1980. Effect of varying Mg/Ca ratios and electrolyte concentrations in the irrigation water on the soil properties and wheat crop. Pl. Soil, 56: 413-428.

Yadav, J.S.P. and Girdhar, I.K. 1981. The effects of different magnesium : calcium ratios and sodium adsorption ratio values of leaching water on the properties of calcareous versus non-calcareous soils. Soil Sci., 131: 194-198.

Yadav, J.S.P. and Sharma, B.M. 1981. An appraisal of alkali tolerance of gram (*Cicer arietinum*) in the field. Agrokem. Talajt, 29: 456-62.

Yaduvanshi, N.P.S. 2017. Nutrient management for sustained crop productivity in sodic soils. In: Salinity Management in Agriculture: Technological Advances and Applications. Apple Academic Press, Waretown, NJ, USA. 365-394.

Yaduvanshi, N. P. S. and Dey, P. 2009. Nutrient management in sodic soils. In: Enhancing Nutrient Use Efficiency in Problem Soils (Singh *et al.*, Eds.). CSSRI, Karnal. 84-108.

Yaduvanshi, N. P. S. and Yadav, D.V. 1990. Effect of sulphitation press mud and nitrogen fertilizer on biomass and nitrogen economy, plant composition in sugarcane and on soil chemical properties. J. Agri. Science. 114: 259- 263.

Zhawar V. Kaur and Kaur, K. 2018. Physiological and biochemical changes in plants under soil salinity stress: a review. In: Engineering (Gupta *et al.* Eds.), Apple Academic Press, USA (In Press).

■■■■

GLOSSARY

1:1 or 1:2 Soil Solution: 1 part water is added to 1 or 1/2 part soil (weight : weight) to measure pH or electrical conductivity. The measured EC/pH is designated as EC_1/pH_1 or EC_2/pH_2.

Adjusted R_{Na}: A parameter similar to SAR describing a sodic water to take care of precipitation of Ca due to presence of HCO_3.

Adjusted SAR: A water quality parameter similar to SAR. In this case, Ca concentration of the water is adjusted for the HCO_3 present in the water and the EC of the water. Also see SAR. The term is now obsolete

$$\text{Adj. SAR} = \text{SAR}\ (1 + 8.4 - pH_c)$$

Alkali Soil: A soil that contains sufficient exchangeable sodium to interfere with the growth of most crop plants, either with or without appreciable quantities of soluble salts. A soil having EC less than 4 dS/m, ESP more than 15, and pH_s more than 8.2. Also called sodic soil.

Alkali Water: A water having variable SAR and RSC more than 2.5. EC may or may not be high.

Alkalization: The process whereby the exchangeable sodium content of the soil is increased. Sometimes referred as sodication or sodification.

Amendment Soil: Substances that are added to the soil such as lime, gypsum, sawdust, etc. to make them more suitable for production of plants. Strictly speaking even fertilizers constitute a special group of soil amendments.

Anion: Anions are negatively charged particles (ions) such as chloride (Cl^-), sulphate (SO_4^{2-}), carbonate (CO_3^{2-}), or bicarbonate (HCO_3^-).

Anion Exchange Capacity: The total amount of exchangeable anions that a soil can absorb. It is expressed as milli-equivalents per 100 g of soil.

Artificial Drainage: Drainage constructed artificially by men in contrast to natural drainage which develops naturally in a region.

Available Water: The portion of water in a soil that can be readily absorbed by plant roots. Defined as the quantity of water retained in the soil that is smaller than field capacity (FC) and larger than the permanent wilting point (PWP). Available water is the difference between the water content at FC and PWP.

Bar: A unit of pressure. One bar equals 14.5 pounds per square inch (psi) or about 0.987 standard atmospheres.

Base: A compound resulting in a pH greater than 7 when in aqueous solution, a molecule that can accept a proton from an acid, donate an unshared pair of electrons to an acid or react with an acid to form a salt, a substance that has fewer free hydrogen ions, H^+, than hydroxyl ions, OH^-.

Black Alkali: Alkali soils also called black soils because black patches appear on these soils due to the deposition of organic matter that gets dissolved at high pH and accumulates at the soil surface. The term is now obsolete.

Benefit: Cost Ratio: The discounted values of benefits divided by the discounted values of costs adjusted to a common time basis.

Bio-drainage: Use of water from the shallow water table by quick growing plants to cause lowering of the water table; drainage with the help of biological material, the plants.

Calcareous Soil: Soil containing sufficient free calcium carbonate or calcium-magnesium carbonate to effervesce visibly when treated with cold 0.1 N hydrochloric acid.

Calcite: Calcium carbonate ($CaCO_3$). A trade name for finely ground limestone, very high in calcium carbonate, which is used to raise the pH of acidic water. The insoluble form of calcium precipitated by waters high in Ca, HCO_3^- and CO_3^{2-}. Sometimes naturally occurring in calcareous soils in arid regions. Insoluble until reacted with an acid

Capillary Fringe: Soil zone above the water table where water rises due to surface tension and remains almost saturated. It may consist solely of capillary water or it may be combined with gravity water in transit to the water table.

Capillary Rise: The upward movement of water from a free water table due to adhesion of water to the soil capillaries and the cohesion of water molecules.

Cation: Cations are positively charged particles of atoms or molecules. Calcium (Ca^{2+}), magnesium (Mg^{2+}), sodium (Na^+), potassium (K^+), or ammo-nium (NH^{4+}). The ratios of these cations as well as the amounts of the various cations present are important for plant nutrition and the structure of the soil.

Cation Exchange: The interchange between a cation in solution and another cation on the surface of any surface-active material such as clay colloid or organic colloid.

Cation Exchange Capacity (CEC): A measure of the net negative charge of a soil. Measured as the total quantity of cations that can be exchanged on a unit of soil material, expressed as milliequivalents per 100 grams of soil (me/100 g), millimoles of charge per kilogram of soil (mmol [+ charge]/ kg) or centimoles of charge per kilogram of soil (cmole [+ charge]/kg).

Collector Drain: For surface drainage system, it is an open drain into which a number of field drains discharge their water. For subsurface drainage system, it is usually a closed conduit, larger in size than the lateral drains into which a number of laterals discharge their water.

Conjunctive Use: Use of multi-quality, multi-sources water in conjunction to get better returns in comparison to when these waters are used separately.

Critical Depth: i) The depth of flow in a channel of specified dimensions at which specific energy is minimum for a given discharge. ii) The depth to water table at which rate of evaporation through capillary rise is reduced to almost zero. In practice, it is 1/10 of the rate which occurs when the water table is at or near the soil surface.

Crop Tolerance: Tolerance of crops to adverse abiotic stresses such as water logging, soil salinity, alkalinity, heat and cold waves etc.; it may be to the biotic stresses as well

Crop Tolerance Period: It is the duration up to which a specific crop can withstand adverse stresses such as water logging condition without significant damage to the quality and quantity of yield.

Crust: A surface layer of soil, from a few mm to 2.5 cm thick, caused by rain drops impact, clay dispersion and clay translocation. When dry, it is much more compact, hard, and brittle than the material just under it.

Cumulative Infiltration: Total sum of the depth of water absorbed by a soil in a specified elapsed time in reference to the time of initial water application.

Cyclic Use: A term used in conjunctive use of multi quality waters when the two or more qualities of waters are used to irrigate the crop in some cycle or sequence, helps to avoid stress at salt sensitive stages.

Deep Percolation: Water which percolates downward below the root zone and cannot be used by plants.

Degradation: The process of soil change from one type to another resulting in increased production constraints. Soils salinization or the process of alkalization or even the process of replacement of sodium by hydrogen ions during leaching constitutes degradation.

Desalinization: Removal of excessive soluble salts from the saline soil, usually by leaching.

Discounted Benefits: Actual benefits occurring in future adjusted through the use of discount rates to a common time base.

Discounted Costs: Actual costs occurring in future adjusted through the use of discount rates to a common time base.

Discounting: The use of interest factors to adjust cash flows to a common time basis.

Dispersed soil: Soil in which clay readily forms a colloidal solution, usually have low permeability, tend to shrink, crack and become hard on drying and become plastic on wetting.

Dispersion: Dispersion is the breakdown of soil clay minerals into single particles in solution. The opposite is flocculation. Dispersion occurs when the bonds in a soil are not strong enough to keep the particles together. It is usually because a soil has too great a proportion of exchangeable sodium in it, or is too low in organic matter. However, working a soil too wet or too dry can also increase its tendency to disperse on wetting. Soil aggregates form larger, more continuous soil pores than do individual soil particles. The larger pores provide better water and air movement.

Distillery Spent Wash: A by-product of the molasses based distilleries, nutrient rich wastewater with high biological oxygen and chemical oxygen demand (BOD and COD). Used as an amendment to reclaim alkali soils and water.

Double-main system: A system of drains used where seeps appear on both sides of a broad flat, also where a stream separates an open flat into two parts.

Drain: (a) A channel, such as open ditches or drain tile, to remove excess water by surface or by internal flow. (b) To lose water (from the soil) by percolation.

Drain Depth: For open surface drains, it is the depth from the ground surface to the drain bottom. For subsurface drains, it is the distance between the ground surface and the drain tube centre.

Drain Spacing: The horizontal distance between the centrelines of adjacent parallel drains.

Drain Tile: Concrete or ceramic pipe used to conduct water from the soil.

Drainage: Process of removing groundwater or surface water by artificial or natural means.

Drainage Coefficient: It is the amount of water to be removed in 24 hours to provide favourable conditions for crop growth. It is usually expressed in mm/d. It is also known as drainage modulus.

Drainage System: A system comprising a main drain, its branches and subsidiary drains, used to collect and dispose of excess surface or subsurface water.

Economic Analysis: assessment of an investment's profitability to a whole society or economy regardless of who in the society contributes to the costs and receives the benefits, conventionally measured in 'opportunity' rather than 'market' prices.

Electrical Conductivity: It is the property of a medium expressing its capacity to carry electrical charge from one point to another point. The reciprocal of the electrical resistivity. The resistivity is the resistance in ohms of a conductor which is 1cm long and has a cross-sectional area of $1cm^2$. This property, usually expressed in milli or micro ohm/cm, is used to express the salinity of the soil extracts, drainage water, irrigation water, etc. International units of EC are deci-Simens per meter or dS/m. One dS/m is equal to 1 mmhos/cm.

Elemental Sulphur (S): A yellow, inert crystalline mineral that is finely ground. In soil, elemental S is oxidized to sulphuric acid via microbial activity. The rate of elemental S oxidation in soil is most rapid in warm, moist soils. Complete oxidation of elemental S to sulphate often takes one to several years. Elemental sulphur as an amendment is applied whenever sodium is in excess and there is an excess of lime. Gypsum can be substituted for elemental sulphur. If available iron is low, elemental sulphur is recommended to help acidify the soil and make iron available to crops.

Equivalent Depth: It is fictitious reduced depth of effective flow zone due to increased length of streamlines of flow when the tile or open drains do not reach down to the impermeable layer.

Exchange Complex: The surface active constituents of soils (both inorganic and organic) that are capable of cation exchange.

Exchangeable Sodium Percentage (ESP): This term indicates the degree of saturation of the soil exchange complex with sodium and is defined as follows:

$$ESP = \frac{\text{Exchangeable sodium (me/100 g soil)}}{\text{Cation exchange capacity (me/100 g soil)}} \times 100$$

Exchangeable-cation Percentage (ECP): The extent to which the adsorption complex of a soil is occupied by a particular cation. See Exchangeable Sodium Percentage.

Field Capacity (Field Capillary Moisture Capacity and Specific Retention): The amount of water retained in a soil against the force of gravity at any specified time (about 24 to 36 h) after flooding. It is approximated by moisture equivalent and one-third atmosphere percentage.

Final Infiltration Rate: Rate at which water is absorbed by soil when the infiltration has become essentially constant. Also called basic intake rate (See infiltration).

Flocculation: The joining together of smaller individual particles of soil, especially clay, into larger units or flocs. In soils, the presence of calcium tends to flocculate clays, maintaining structure, while sodium tends to disperse clays in solution. Gypsum and lime, both calcium salts, are used to achieve flocculation in some alkali soils.

Flood Irrigation: Method of irrigation where water is applied to the soil surface without flow controls, such as furrows, borders or corrugations. Also known as wild flooding or field to field irrigation.

Flushing: A process to remove salts from the top few cm by impounding and draining the water through surface drainage. Used for heavy textured soils where leaching may pose problems. See leaching.

Furrow irrigation: Method of surface irrigation where the water is supplied to small ditches or furrows for guiding across the field.

Grid-iron System: A system in which parallel laterals enter into the branch or main drain only from one side.

Groundwater Balance: It is a concept *which* states that all inputs of water in a defined space and time are equal to the sum of all outputs of water, and the changes of water storage, in the same space and time.

Groundwater Quality: Quality of groundwater in relation to various purposes. For the purpose of this book purpose is irrigation.

Groundwater Table: The level of water storage (above mean sea level) in an aquifer; hence, the point at which the soil is fully saturated with water.

Gypsum: Gypsum (calcium sulphate, $CaSO_4.2H_2O$) occurs naturally as crystals and is mined. This gypsum is of variable quality and contains impurities. Gypsum is also available as a by-product of phosphate fertilizer production (phosphorgypsum), or of hydrofluoric acid production (calcium sulphate anhydrite). Calcium sulphate anhydrite has the highest purity of commercially available gypsum. Gypsum is used as an amendment to reclaim alkali soils.

Gypsum Requirement (GR): The approximate amount of gypsum needed per hectare to lower the ESP of the soil to a desired level. It is calculated as:

$$GR = (\text{present ESP} - \text{desired ESP}) \times CEC \times 0.052.$$

The factor of 0.052 assumes CEC is in me/100g or cmol (+ charge) kg^{-1} units. If CEC is in cmol (+ charge)/kg the factor is 0.0052. These factors assume 90% reclamation efficiency and a low SAR of the irrigation water.

Herringbone System: A drainage layout in which drain laterals enter the collector from both sides at an angle to ensure adequate slope.

Horizontal Drainage: A method of drainage to control water table through tile or pipes placed horizontally beneath the soil surface.

Humus: Organic matter, also called 'humus', forms from the decay of leaves, plants and other biological material.

Hydraulic Conductivity: The rate of flow of water through a porous material of a unit cross-sectional area under a unit hydraulic gradient, at a specified temperature. It is also called unit of permeability, transmission constant or coefficient of transmission. It is the coefficient, K in the Darcy's equation $v = -Ki$, such that v is the flow velocity of a specified liquid under saturated conditions, and i is the hydraulic gradient. Thus, unit of K is L/T.

Hydraulic Radius: It is ratio of cross sectional area of flow to the wetted perimeter.

Impeded Drainage: Condition in which the downward movement of gravitational water in soil is restricted.

Infiltration: The process of entry of water into the soil profile through the ground surface.

Infiltration Rate: The maximum rate at which a soil, in a given condition at a given time, can absorb water impounded on the surface when adequate precautions are taken regarding border or fringe effects. Defined as the volume of water passing into the soil per unit of area per unit of time, it has the dimensions of velocity (L/T).

Interceptor Drain: (i) Surface/subsurface drains or their combination, installed to collect subsurface flow before it resurfaces (ii) Surface/subsurface drains to intercept seepage from canals.

Internal Rate of Return: The rate of discounting at which the present value of benefits becomes equal to the present value of costs.

Ion: Ions are electrically charged particles made up of atoms or molecules. Cations are positively charged ions. Anions are negatively charged ions.

Ion Exchange: The interchange between an ion in solution and another ion on the surface of any surface-active material such as clay or humus.

Land Forming: Changing the micro-topography of the land to meet the pre-decided requirements say of drainage or irrigation.

Land Grading: A process of land levelling to reshape the field surface to a planned grade.

Land Reclamation: Changing general physico-chemical characteristics of the land to make it capable of more intensive use for agriculture.

Land Uniformity Co-efficient (LUC): It represents the magnitude as well as frequency of occurrence of successively larger undulations in the field. It shows the degree of land uniformity achieved following a land leveling operation.

Lateral Drain: It is a field drain to remove excess water from root zone. It can be unlined open channel, mole drain, tile drain, perforated pipe drain, etc.

Leaching: The process of removal of soluble material from soil by the passage of water through the soil. A primary step in the reclamation of saline soils.

Leaching Fraction: The fraction of water applied on the soil that actually passes through the root zone as against the leaching requirement which must pass through the root zone to prevent soil salinity from exceeding a specified value.

Leaching Requirement: The fraction of the water entering the soil that must pass through the root zone in order to prevent soil salinity from exceeding a specified value. Leaching requirement is used primarily under steady state or large-time average conditions.

Levelling: (1) The surveying process of determining the difference in elevation between two or more points by measuring the vertical distance between two points; the determination of elevation of points above a datum. (2) The tillage operation in which the soil is moved to establish a desired soil elevation slope.

Lime: Strictly, calcium oxide (CaO), but as commonly used in agricultural terminology, calcium carbonate ($CaCO_3$) and calcium hydroxide $Ca(OH)_2$ are included. Agricultural lime refers to any of these compounds, with or without magnesia, used as an amendment for acid soils to make them more alkaline. The carbonate in lime neutralizes the acidity in soils and raises the pH. However, lime may also be useful on some acid sodic soils, because of its calcium content.

Matric Potential: The energy by which soil water is held by soil particles. This represents the energy required to be expended by plants to take up water from the soil. Matric potential is always negative indicating that energy has to be applied to remove water from the soil.

Mhos: Reciprocal of Ohms. A unit of Electrical Conductivity to measure the quality of water, See Electrical Conductivity.

Milliequivalents per Liter (me/l): Equal to ppm / molecular weight.

Non-saline Alkali Soil: A soil that contains sufficient exchangeable sodium to interfere with the growth of most crop plants and contains variable quantities of soluble salts. The exchangeable sodium percentage is greater than 15, the conductivity of the saturation extract is less than 4 dS/m (at 25°C) and the pH of the saturated soil usually ranges between 8.2 to 10.0 or even more.

Osmotic Effect: Reduction in the ability of plants to extract water from the soil as the osmotic pressure of the soil water solution increases with increasing salinity.

Osmotic Potential (ψo): The osmotic potential (ψo) of the constituent water in soil at a specified temperature is the amount of useful work per unit mass of pure water, in J/kg, that must be done to transfer reversibly and isothermally an infinitesimal quantity of water from a pool of pure free water to a pool of free soil solution (identical in composition with the soil liquid phase at the point under consideration, and at the same height). The potential may be expressed in terms of the experimentally accessible osmotic pressure of the solution (π). Osmotic stress/tension/ suction is identical to osmotic pressure as defined above. It can be interpreted as the negative of the negative osmotic pressure, in pascal.

Parallel Field Ditch System: A system similar to bedding, except that the drains are spaced farther apart and have a greater capacity than the dead furrows. The minimum depth of the ditch is 22.5 cm and minimum cross-section area 0.4 sq m.

Parallel Lateral Open Ditch System: A system similar to field ditch system, except that the ditches are deeper and cannot be crossed with farm machinery.

Parts Per Million (ppm): Weight units of any given substance per one million equivalent weight units of oven-dry soil; or in the case of soil solution or other solutions, the weight units of solute per million weight units of solution.

Payback Period: It is the time period for an investment to generate sufficient incremental cash to recover its initial capital outlay in full.

Permeability: A soil property, which is a measure of the readiness with which the soil transmits water. The permeability is a velocity, and is expressed in cm/h. For most soils, both the chemical nature of the water used and the history and treatment of the sample greatly affect the permeability.

Permeability Index: The permeability Index (PI) given by Doneen is calculated by the following equation:

$$PI = 100 *(Na + \sqrt{HCO_3})/ (Na + Ca + Mg + K).$$

The plot of PI and total salt concentration describes the problems of permeability if any with the continuous use of waters.

pH: The common logarithm of the reciprocal of the hydrogen ion concentration of a system. Soil pH refers to the acidity or alkalinity of the soil with a pH of 7.0 being neutral. A pH below 7.0 being acid, a pH above 7.0 being alkaline and a pH_3 above 8.2 being alkali because of excess sodium. Most soils with a pH above 7.0 will usually have an excess of undissolved free lime. The optimum pH for most plants is between 6.3 and 7.5, and pH's below 6.0 require lime for best growth. pH_s refers to the pH of soil saturation paste while pH (1:1 or 1:2 soil solution) refers 1 part water added to 1 or 2 parts soil (weight : weight) in order to measure pH.

Descriptive terms commonly associated with certain ranges in soil pH measured in distilled water are:

Extremely acid	:	< 4.5
very strongly acid	:	4.5 – 5.0
strongly acid	:	5.1 – 5.5
medium acid	:	5.6 – 6.0
slightly acid	:	6.1 – 6.5
neutral	:	6.6 – 7.3
mildly alkaline	:	7.4 – 7.8
moderately alkaline	:	7.9 – 8.4
strongly alkaline	:	8.5 – 9.0
very strongly alkaline	:	> 9.0

Phosphogypsum: A kind of gypsum available as a by-product of phosphate fertilizer industry.

Precipitation: The total measurable supply of water of all forms of falling moisture, including dew, rain, mist, snow, hail and sleet; usually expressed as depth of liquid water on a horizontal surface in a day, month, or year, and designated so daily, monthly or annual precipitation.

Press Mud: A by-product of sugar industry, used in agriculture as a source of nutrient as well as an amendment to reclaim alkali soils if press mud is from industries that use sulphur to clarify sugarcane juice, called sulphitation press mud.

Primary Benefits or Primary Effects: Benefits from project including the value of any reduction in costs due to the project and of any increase of farm products marketed or consumed by the farm family.

Reconnaissance Survey: An exploratory study of the area to identify problems, degree and extent of the problem and even possible solutions although only qualitatively.

Rainfall: 1- precipitation in the form of liquid water. 2- The amount of rain, usually expressed in mm (inches) depth of water on an area, that reaches the surface of the earth. The term is sometimes expanded to include other forms of atmospheric precipitation, such as snow, dew, but technically the term precipitation would be used in this broader scope.

Residual Sodium Bicarbonate (RSBC): It is defined as the excess of HCO_3 ions. Over Ca ions present in the water. RSBC is calculated as below, ions expressed as me/l.

$$RSBC = HCO_3 - Ca$$

Residual Sodium Carbonate (RSC): It is defined as the excess of CO_3+HCO_3 ions over the Ca+Mg ions present in the water. The RSC is calculated with the following relation having units of me/l.

$$RSC = (CO_3 + HCO_3) - (Ca + Mg)$$

Resodification: Sodification of reclaimed lands; the process commonly encountered in areas where water table is shallow, irrigation water is alkali and where lands are poorly managed by the farmers.

Reuse: Use of same water several times often in other sectors such as use of domestic wastewater in agriculture. Often recycle and reuse is used synonymously.

Runoff: That portion of precipitation which is not absorbed by the deep strata but finds its way into the streams after meeting the demands of evapotranspiration including interception and other losses. It includes surface runoff received into the channels after rainfall, delayed runoff that enters the streams after passing through the earth and other delayed runoff that has been temporarily detained as a snow cover and stored in natural lakes or swamps.

Runoff Coefficient: (i) The ratio of the peak rate of runoff to the uniform rate of rainfall with a duration equalling or exceeding the time of concentration which produced this rate of runoff. (ii) Also used incorrectly for 'percentage runoff'.

Saline-Alkali Soil: A soil containing sufficient exchangeable sodium to interfere with the growth of most crop plants and containing appreciable quantities of soluble salts. The exchangeable sodium percentage is greater than 15 and the conductivity of the saturation extract is greater than 4 dS/m (at 25ºC). The pH of the saturated soil is usually more than 8.2.

Saline Soil: A non-alkali soil containing soluble salts in such quantities that they interfere with the growth of most crop plants. The conductivity of the saturation extract is greater than 4 dS/m (at 25ºC), the exchangeable sodium percentage is less than 15, and the pH of the saturated soil is less than 8.2.

Saline Water: A water having excessive amount of soluble salts such that if used for irrigation would interfere with plant growth. The EC is usually more than 4 dS/m. SAR or RSC may or may not be high.

Salinity: Concentration of salts in a given fluid or material.

Salinization: The process of accumulation of salts in the soil particularly in the root zone.

Salt Balance: The salt balance of an irrigated area is commonly defined as the difference between the total dissolved solids brought to the land in a pre-decided duration by the irrigation water and the total solids carried away by the drainage water during the same period.

Salvage Value: It is the price at which a machine/equipment can be sold at the time of its disposal.

Saturated Soil Paste: A mixture of soil and water; at saturation it reflects light, flows slightly when the container is tipped, and the paste slides freely and cleanly from a spatula for all soils at its saturation percentage.

Saturation Extract: The solution obtained by pressure or vacuum filtration of a soil paste that has been made up to a saturated condition by adding water while stirring.

Saturation Percentage: The moisture content of a sample of soil that has been brought to saturation by adding water while stirring. Expressed as grams of water per 100 g of dry soil.

Selectivity Coefficient: A value describing equilibrium conditions for an exchange reaction. It may be compared with the thermodynamic equilibrium constant, but differs from it in that assumptions are made regarding the activities of the ions participating in the exchange reaction.

Sensitive Crops or Plants: A qualitative term used to designate the plants which are sensitive to water, salts or some toxic element. Also see tolerant.

Silica: The chemically resistant dioxide of silicon: SiO_2. It occurs naturally in five crystalline polymorphs (the minerals quartz, tridymite, cristobalite, coesite and stishovite); in cryptocrystalline form (as chalcedony); in amorphous and hydrated forms (as opal); in less pure forms (such as sand, diatomite, tripoli, chert and flint) and combined in silicates as an essential constituent of many minerals.

Slick Spots: Barren areas having puddled or crusted, very smooth, nearly impervious surfaces, usually because of high salinity or sodicity.

Smoothening: An operation required nearly every cropping season, particularly where cultivation and harvesting processes significantly disrupts the field surface. It is also a final process in the land levelling to smooth the topographic variations that remains following the operation.

Sodic Soil: (a) A soil containing sufficient exchangeable sodium to interfere with the growth of most crop plants. Used synonymously with alkali soils. See Alkali soils

Sodic water: A water with residual sodium carbonate more than 2.5. EC/SAR may or may not be high.

Sodication or Sodification: See Alkalization

Sodium Adsorption Ratio (SAR): A ratio for soil extracts and irrigation waters used to express the relative activity of sodium to calcium and magnesium. SAR is calculated as follows:

$SAR= Na/((Ca+Mg)/2)^{1/2}$ (all ions expressed in me/l)

$SAR= Na/(Ca+Mg)^{1/2}$ (All ions expressed in mmol/l)

Expressed in units of $(mmol/l)^{1/2}$.

Sodium to Calcium Activity Ratio (SCAR): A ratio for soil extracts and irrigation waters used to express the relative activity of sodium to calcium. SCAR is calculated as below:

$SCAR = Na/Ca^{1/2}$ (all ions expressed as me/l)

Soil Aggregation: The cementing or building together of several soil particles into a secondary unit or ped, aggregate or granule. Water-stable aggregates, which will not disintegrate easily, are of special importance to soil structure.

Soil Auger: A tool for boring into the soil and withdrawing a small sample for field or laboratory determinations. Soil augers are classified into several types.

Soil Extract: The solution separated from a soil suspension or from a soil by filtration, centrifugation, suction or pressure (may or may not be heated prior to separation).

Soil Profile: A vertical section of the soil through all its horizons and extending to the underlying material.

Soil Survey: The systematic examination, description classification, and mapping of soils in an area. Soil surveys are classified according to the kind and intensity of field examination.

Soil Texture: The texture is the term indicating the coarseness or fineness of the soil. It is decided on the basis of the amount and quantity of each of the grain size group of particle that constitute the soil.

Soil Water Balance (Field Water Balance): The sum of all gains and losses of water over a given period of time; mm/per unit time.

Soluble Sodium Percentage: The proportion of sodium ions in solution in relation to the total cation concentration, defined as follows :

$$SSP = \frac{\text{Soluble sodium concentration (me/l)}}{\text{Total cation concentration (me/l)}}$$

Specific Ion Effect: Any effect of a salt constituent in the substrate on plant growth which is not due to the osmatic properties of the substrate.

Sub Soil: That part of the soil beneath the top soil, usually not having an appreciable organic matter content.

Subsurface Drainage: The removal of surplus or excess groundwater accomplished by means of natural drainage (for gravelly subsoil) or by artificial means, such as drains placed under a surface.

Subsurface Drains: Subsurface channels used primarily to remove subsurface water from soil. Also known as tile drains or pipe drains.

Surface Drainage: The removal of surplus or excess surface water collected on land accomplished by natural or artificial means, such as open ditches and terracing.

Surface Irrigation: Type of irrigation where water is distributed to the plant material by a surface distribution network possibly including rows or dikes or broad class of irrigation methods in which water is distributed over the soil surface by gravity flow.

Surface Runoff: Overland runoff when it has reached a definite stream channel.

S-value: The sum of exchangeable (as opposed to soluble) Ca, Mg, Na and K ions expressed in Cmolc/kg soil.

Swelling: The increase in bulk volume of a substance, usually upon addition of water. A swelling soil is one that contains expanding-lattice clay minerals which cause the soil to swell or heave upon wetting and shrink (with the formation of cracks) upon drying.

TDS, Total Dissolved Solids: A quantitative measure of the total dissolved organic and inorganic solids concentration in water, an indicator test used for water analysis and also a measure of the mineral content of bottled water and groundwater, one can monitor electrical conductivity quickly in the field and estimate total dissolved solids or TDS. Also sum of all dissolved materials such as salts, which are non-filterable and remain following evaporation of the water.

Tile Drain: Concrete or ceramic pipe placed at suitable depths and spacing in soil or subsoil to provide water outlets from the soil.

Tile Drainage System: Drainage system consisting of a drainage outlet, tile main, sub-main and laterals. The laterals remove free water from the soil and the sub-mains and main convey water to the outlet.

Tolerant: Capable of withstanding a condition usually un-favourable. Some plants are tolerant to soil acidity and others are tolerant of soil salinity or alkalinity. **See** crop Tolerance

Toxic: Poisonous; tending to reduce the yield of crops below the normal.

Toxic water: Water containing one or more of toxic elements. Yield reduction is attributed to toxic effects of the ions rather than the osmotic potential of the soil solution.

Vertical Drainage: (1) Disposal of drainage water through wells into a porous layer of earth or an open-rock formation. (2) The term vertical drainage is also used for a drainage system in which tube wells are used to lower the water table.

Wasteland: Land not suitable for producing material or services of value. A miscellaneous land type.

Water Application Efficiency: Water application efficiency is the measure of efficiency with which water delivered to the field is stored in the root zone.

Water Distribution Efficiency: The extent to which water is uniformly distributed in an irrigation system. Or, it is defined as the percent of difference from unity of the ratio between the average numerical deviations from the average depth stored during the irrigation.

Water Logging: It refers to the condition of land in which it cannot be put to its normal use as a result of either or both of shallow water table and surface ponding. In arable lands depending upon the extent and duration of water logging, the crop yield may reduce substantially or even become zero.

Water Stable Aggregates: Soil aggregates which are stable to the action of water such as falling drops, or agitation as in wet sieving analysis.

Water Stress: The stress of water caused to the plant due to inadequate supply of water through the soil. Also known as matric stress. Usually expressed in bars. One bar is equal to 1023 cm of water.

Water Table: The upper surface of groundwater, locus of points in soil water at which the hydraulic pressure is equal to atmospheric pressure.

Water Use Efficiency: The amount of dry matter that can be produced from a given quantity of water. Ratio of the crop yield per unit area per unit of applied irrigation water. Expressed in units of kg/ha- mm.

Well: An artificial excavation (pit, hole, tunnel) often walled sunk (drilled, dug, driven, bored or jetted) into the ground to such a depth so as to penetrate water yielding rock or soil and allow the water to flow or to be pumped to the surface.

Wetted Perimeter: Length of the wetted contact of a liquid, measured along a plane at right angles to the direction of flow.

■■■■

IMPORTANT CONVERSIONS AND APPROXIMATIONS

Salt Concentration

To convert from	To	Multiply by
TDS (mg/l)	TDS (ppm)	1.0
Salinity (dS/m)	salinity (mmho/cm)	1.0
Salinity (mmho/cm)	salinity (μmho /cm)	1000
Salinity (dS/m) (3-30)	osmotic pressure	0.36
Salinity (dS/m) (0.1-5.0)	ppm	0.64
Parts of salt per million parts of irrigation water	t/ha-cm	0.001323
Salinity (dS/m) (<5 dS/m)	TDS (mg/l)	640
Salinity (dS/m) (>5 dS/m)	TDS (mg/l)	800
Salinity (dS/m) (< 5 dS/m)	meq/l	10
Water, NO_3 N, SO_4 S B (ppm)	kg/ha-cm of water applied	0.1015

TDS = Total dissolved solids

Elemental Conversions

Percent or kg	Multiply by	Percent or kg
P_2O_5	0.437	P
K_2O	0.826	K
CaO	0.710	Ca
MgO	0.600	Mg
$CaCO_3$	0.400	Ca
NO_3-N	0.226	N
NH_4-N	0.777	N

Useful Approximations

Value	Approximation
1 in	2.5 cm
1 ft	30 cm
33 ft	10 m
1 ha	2.5 ac
1 ac	0.4 ha
1 in^2	6.5 cm^2
1 m^2	11 ft^2
1m^3	35 ft^3
1 in^3	16.4 cm^3
1 m^3 soil	1200 kg to 1650 kg (Bulk density 1.20g/cm^3 to 1.65 g/cm^3)
1 m^3 water	1000 kg
Average particle density (soil)	2.65 g/cm^3
1 ha-cm soil weighs	2000 t/ha to 2250 t/ha (Bulk density 1.33g/cm^3 to 1.5 g/cm^3)

■■■■

SUBJECT INDEX